EDITED BY
CATHERINE OWEN AND JENNIFER MONK

WHAT IS GEOGRAPHY TEACHING, NOW?

A PRACTICAL HANDBOOK FOR ALL GEOGRAPHY TEACHERS AND EDUCATORS

Together we unlock every learner's unique potential

At Hachette Learning (formerly Hodder Education), there's one thing we're certain about. No two students learn the same way. That's why our approach to teaching begins by recognising the needs of individuals first.

Our mission is to allow every learner to fulfil their unique potential by empowering those who teach them. From our expert teaching and learning resources to our digital educational tools that make learning easier and more accessible for all, we provide solutions designed to maximise the impact of learning for every teacher, parent and student.

Aligned to our parent company, Hachette Livre, founded in 1826, we pride ourselves on being a learning solutions provider with a global footprint.

www.hachettelearning.com

Although every effort has been made to ensure that website addresses are correct at time of going to press, Hodder Education cannot be held responsible for the content of any website mentioned in this book. It is sometimes possible to find a relocated web page by typing in the address of the home page for a website in the URL window of your browser.

Hachette UK's policy is to use papers that are natural, renewable and recyclable products and made from wood grown in well-managed forests and other controlled sources. The logging and manufacturing processes are expected to conform to the environmental regulations of the country of origin.

To order, please visit www.johncatt.com or contact Customer Service at education@hachette.co.uk / +44 (0)1235 827827.

ISBN: 978 1 036004 85 9

© Catherine Owen and Jennifer Monk 2025

First published in 2025 by
Hachette Learning,
An Hachette UK Company
Carmelite House
50 Victoria Embankment
London EC4Y 0DZ
www.HachetteLearning.com

The authorised representative in the EEA is Hachette Ireland, 8 Castlecourt Centre, Castleknock Road, Castleknock, Dublin 15, D15 YF6A, Ireland

All rights reserved. Apart from any use permitted under UK copyright law, no part of this publication may be reproduced or transmitted in any form or by any means, electronic or mechanical, including photocopying and recording, or held within any information storage and retrieval system, without permission in writing from the publisher or under licence from the Copyright Licensing Agency Limited. Further details of such licences (for reprographic reproduction) may be obtained from the Copyright Licensing Agency Limited, www.cla.co.uk

Typeset in the UK.
Illustrations by DC Graphic Design Limited, Hextable, Kent.
Printed in the UK.
A catalogue record for this title is available from the British Library.

*To all current and future geography teachers,
we hope this book gives you support and inspiration.*

Catherine Owen

During her 30 year career, Catherine has been privileged to remain a classroom practitioner while also being involved in writing and presenting, supporting a group of schools in Uganda in developing their practice and working with both geography subject associations on a range of projects. Long may this continue!

Jennifer Monk

Jen is currently head of geography at a school in the north-west of England but has previously been a lead practitioner for humanities. In 2019 she won the Royal Geographical Society award for Excellence in Geography Teaching. She is passionate about supporting others, which led to the co-creation of @GeogChat.

REVIEWS

This is a 'state of the art' volume – a wonderful collection of short and readable essays written by some of the UK's most gifted, creative and critically-aware geography teachers. Each of the contributions contains an invigorating mix of theory and practical advice – things that are tried and tested that you can introduce into your classroom with immediate effect. This book will become like a best colleague – after half an hour or more in its wise company you will come away enthused and full of new ideas

Prof Alastair Owens
Queen Mary University of London and Chair of Trustees and Former President of The Geographical Association

A great read for any teacher in any geography department! This book covers so many teaching ideas and concepts in a way that can support teaching in the classroom. It is written in an engaging style that makes it difficult to put down, I have little time to read but found the time with this. The chapter on climate change has generated many questions which I can't wait to discuss further within my department. As an experienced teacher there was still so much to take from this book. I will certainly use lots of the ideas in my teaching going forward.

Karen Robertson
Head of Geography, The King Alfred School and Academy

Jen and Catherine have done an incredible job of pulling together the collected wisdom of dozens of geography teachers and others working in geography education. Between them they have managed to not only capture what geography teaching is, but also make a compelling case for what it could be. The book is filled with practical and actionable ideas and concrete examples of brilliant practice that will make geography come to life. I have no doubt that this book will make me a better teacher and a better head of department.

Mark Enser
Head of geography, author and former geography subject lead for Ofsted

ACKNOWLEDGEMENTS

To Gary, thank you for being my rock, my companion in adventures and the best husband I could ask for. Here's to many more adventures!

To Jack and Megan, I couldn't be more proud of the adults you've become. Keep caring and striving for a better world. You mean everything to me.

To all the wonderful geography teachers I've worked with and encountered over the years, thank you for the ideas and support. What an amazing community we are! With particular thanks to Karen for being generally amazing.

Catherine Owen

For my parents, my intention in life has always been to try my best and that stems from your constant and unfailing love and support. I hope I make you both proud and I am forever grateful for all you do for me.

For Lewis, you have always championed and supported me, I know that wherever I go you will be by my side and for that I am endlessly grateful.

For Harry and Joseph, I love you with all my heart and I hope that you know that anything is possible if you work hard. I hope I inspire you to be the best version of yourselves and to know that being kind is the best thing you can be.

For Nic and the staff at Stretford High School, you have inspired me more than you will ever realise, you reminded me why I love teaching. SHS is a special place and one I feel lucky to be a part of.

Jennifer Monk

CONTENTS

Introduction – why this, why now?..1

1. What is geography? – Alan Parkinson...5

2. Where has geographical enquiry taken me? – Catherine Owen 13

3. Writing a key stage 3 geography curriculum from scratch – Joseph Milton..........25

4. Lesson planning in geography – Ben Newborn 41

5. Assessment at key stage 3 – Nicola Dowling 49

6. Purposeful feedback – Helen Pipe..67

7. Questioning in geography – Michael Chiles...75

8. Where does fieldwork fit in the 11-14 geography curriculum? – Fiona Sheriff........81

9. Using homework to add breadth and depth to the curriculum – Jennifer Monk..95

10. Incorporating geographical skills within a curriculum
    – Louis Vis and Mason Davies...103

11. Making GIS 'part of the furniture' for the teaching of geography – Brendan Conway and Alistair Hamill 117

12. Using models in the geography classroom - Hannah Steel 141

13. Oracy strategies to improve geographical writing - Louise Holyoak 149

14. Literacy in geography - Bethany Aldridge 161

15. Writing like a geographer - Catherine Owen 171

16. Choosing and using case studies - Ramya Rajkumar 183

17. Supporting students with SEND in geography - Amy Cushing 197

18. Teaching 'disadvantaged' students - Catherine Owen 207

19. Retrieval practice in geography - Jennifer Monk 219

20. Using Instagram and an iPad to engage students - Meg Picken 229

21. Leading a geography department - Jo Payne 243

22. Supporting geography early career teachers - Hannah Gowling 255

23. Supporting non-specialists to teach geography - Jennifer Monk 265

24. One approach to key stage 2 geography - Moat Farm Junior School - Jon McCormick 277

25. Thinking synoptically in the A-level geography classroom - Paul Logue 287

26. Teaching careers in geography - Rouna Ali 301

27. The wonder of physical geography - Catherine Owen 319

28. The diagram, the visualiser and the vignette: How to teach about the multiplier effect - Abdurrahman Perez-McMillan 333

29. Exploring gender in the secondary geography classroom - Briley Habib 341

30. Geography ... is it queer? - Jacob James Profitt..353

31. Why voices matter now - Chantal Mayo-Holloway 365

32. The power of storytelling in geography: a decolonial perspective
- Iram Sammar..379

33. What is geography teaching today as a Black British geography teacher? -
Shanique Harris..389

34. How can we make our UK geography content less England-centric?
- Alice McCaughern...399

35. Lantinx visibility - Izzy Wood 405

36. Environmental Education: An integrated approach - Sandra Zoe Patterson 419

37. Climate change, super wicked problems and antifragility
- Catherine Owen and Sebastian Witts..429

38. At and beyond the chalkface: the role of the geography teacher in the
age of climate crisis - Kit Marie Rackley 439

39. Who owns the Earth? Property rights and environmental conflict
- Andrew Taylor and Catherine Owen...453

40. Hopeful geography - David Alcock...463

INTRODUCTION - WHY THIS, WHY NOW?

As geography teachers we face challenges in teaching students about a rapidly changing world, but are also privileged to have the opportunity to work with young people as they develop their understanding, appreciation and confidence in navigating their way in the world.

We also have fantastic opportunities to collaborate, whether online or face to face, allowing us to share and discuss ideas, resources and issues. This can happen through subject associations such as The Geographical Association and The Royal Geographical Association, multi-academy trusts, local networks and social media.

This book is a collaboration between 36 geography teachers, the vast majority of whom are still in the classroom, who are keen to share ideas, start conversations and support others. We feel immense pride to be part of a geography community so committed to working together to provide the best geographical education we can for our students and agree with The Geographical Association (2020) that:

- 'everyone is entitled to a geographical education, to value and be responsive to the world in which we live'
- 'a vibrant, diverse and knowledgeable subject community secures high-quality geography teaching'.

Geography as a subject has changed dramatically over the years, from knowing facts about places, through the quantitative years when everything seemed to be about models, past the skills and themes-based curriculum, when there was danger of losing sight of what geography actually was, to today's geography (as explored so well by Alan Parkinson

in Chapter 1). We continue to respond to external factors which influence our subject and teaching, including:
- Developments in technology, including teaching resources and GIS.
- Increased awareness of the need to decolonise geography and to embed environmental issues in our curricula.
- Inspection frameworks and DfE policies.
- Inspiration from the work of authors, speakers and activists, an example being how 'Disaster by Choice' by Ilan Kelman has prompted so many of us to change the ways we perceive and teach about 'natural' hazards.
- Changes in society which mean that we are now often pushing at an open door when teaching about environmental ideas such as recycling, but also have to deal with online misinformation, deflection and denial.
- A world which has changed since the COVID pandemic, which particularly affected geography fieldwork.

The 'Getting our Bearings: Geography Subject Report' (2023) notes improvements to the geography curriculum in many schools, but also highlights issues including topics being taught in isolation, limited formative assessment, place being poorly planned in the curriculum, fieldwork being underdeveloped, limited use of GIS, ineffective support for non-specialist teachers and a lack of subject specific continual professional development (CPD). This book includes chapters focussed on these themes, as well as questions to consider and suggestions for resources to 'delve deeper', aiming to provide subject specific CPD for geography educators.

This is a book written by teachers for teachers, trainees and educators, putting the classroom at the centre of everything we do. We want the reader to be able to find practical advice to support them in the classroom, but also ideas which challenge their thinking and enable them to develop as geographers and educators. We hope that this book will stimulate conversations in geography departments, staff rooms and the wider geography community. There will inevitably be gaps in terms of chapters not written and encourage people to fill these gaps with blogs, articles in journals, conference sessions, etc - this is a conversation which has been going on for a long time and will never be finished!

We are extremely grateful to the contributors to this book and those who have supported them in writing their chapters. We are delighted to provide a platform for so many new writers, teachers from non-selective state schools and other teachers from diverse backgrounds. It is vital that different voices are part of the conversation.

Thank you for choosing to read this book - it means the world to us.

REFERENCES

OFSTED (2023) 'Getting our bearings: Geography subject report'. https://www.gov.uk/government/publications/subject-report-series-geography/getting-our-bearings-geography-subject-report

The GA (2020) Strategic Plan Overview. https://geography.org.uk/wp-content/uploads/2023/02/GA_Strategy_2020-2025_A4_portrait.pdf

1. WHAT IS GEOGRAPHY?
ALAN PARKINSON
@GEOBLOGS

'Geography speaks directly to young people's curiosity, wonder and concern for the world around them. It is a subject that can provide them with the knowledge and competencies they need to understand and contribute to the world they live in.' (Rawling et al., 2022)

GEOGRAPHY: THE WORLD DISCIPLINE

One challenge for geography teachers when we are asked to define our subject is its sheer scale. We have the whole Earth as our object of study: posing quite a challenge for both teachers and students. Where *should* we start with such a breadth of possible subject matter, and what (if anything) is to be excluded? For Bonnett (2008) the subject's ambition as the world discipline is 'absurdly vast'. Its global scale allows for a huge variety of different 'geographies' to emerge, further complicating the task of defining what it is. As geography teachers, we each need to have our own conception of the subject but be mindful of its changeable nature: which is both a strength and a weakness. If we struggle to define geography, how can we expect our students to understand what it is without our support? (Parkinson, 2020) Is history a much easier subject to define? Does that matter?

The late Professor Ron Johnston's *Encyclopaedia Britannica* definition of geography described it as 'the study of the diverse environments, places, and spaces of Earth's surface and their interactions', and that it is therefore the study of 'the Earth as the home of mankind'.[1]

Geography's development has been traced back to the Ancient Greeks who mapped the world and studied the processes they observed. Eratosthenes used the word 'geography' over 2000 years ago it seems. With 'geo' meaning the world, and 'graphy' broadly meaning to write about, geography can therefore be seen as the science (or art) of 'earth description' or 'earth writing'. *Geography is therefore concerned with the creation of an evolving narrative for the planet.*

Historically, the question of *who* wrote the prevailing narratives gained particular significance. Many geographers are currently in the process of reconsidering the diversity and accuracy of the stories they have been telling about groups of people and places which are partial – and at worst derogatory and inaccurate. Globally, landmarks are being returned to their indigenous naming, as maps are literally being redrawn, and traces of previous colonial outrages erased. This decolonising process also involves revisiting the stories told by those who might be considered the first 'geographers' – early explorers or travellers. Each of them, whether it be Ibn Battuta, Abu Rayhan Al-Biruni, Egeria, Isabella Bird or Jeanne Baret, explored the world from a particular perspective, and varying sensitivities. The stories we tell in geography hopefully include the grand Polar narratives of Shackleton (plus Felicity Aston and Matthew Henson[2]) for example, but also 'smaller' individual stories. If students are only exposed to a narrow range of stories their worldview will be similarly prejudiced and partial. They will, in the words of Chimamanda Ngozi Adichie's much-quoted 2009 TED talk, only have a 'single story'[3] to work with. Geography should therefore involve critical thinking and a careful selection of texts, just as historians are critical of the sources they use. We must also involve the students' own experiences here and

---

1. https://www.britannica.com/science/geography – he would perhaps have used the phrase humankind today (Accessed: 28/05/2025)
2. https://www.nationalgeographic.com/adventure/article/the-legacy-of-arctic-explorer-matthew-henson (Accessed: 28/05/2025)
3. Can be viewed here: https://www.youtube.com/watch?v=D9Ihs241zeg (Accessed: 28/05/2025)

allow them to feel that the details of their everyday lives are worth closer attention. They are all consumers and will have an impact on the planet through their past, present and future actions. They deserve a geography education which presents a rigorous, coherent and creative version of the subject.

Geography is also about (re)presenting places of courses. The concept of 'place' is central to geography: broadly meaning 'spaces which people have made meaningful' (Cresswell, 2013, p.2). Maps of places are an important part of geography's identity to non-geographers – and part of its particular contribution to the development of knowledge of the world. Maps remain important, and the growing ease of use of geographical information systems (GIS) has democratised the production of maps. Should we still include Ordnance Survey map skills in our teaching? (Parkinson, 2021) In the 1970s, Professors WVG Balchin and Alice Coleman gave the name 'graphicacy' to this key ability of geographers to understand and present information in the form of sketches, photographs, diagrams, maps and graphs. This remains part of the subject's identity today, when visual information is increasing in quantity, and the appearance of artificial intelligence (AI) means we must also be critical of the origin of images or texts we use in the classroom (Hamilton, Wiliam and Hattie, 2023). Geography is often defined by its approach as much as by the objects of its study, and the use of our critical lens on the world is a key aspect of geography which needs to be built into our pedagogy.

THE NATIONAL CURRICULUM AND SCHOOL GEOGRAPHY

For some teachers, hopefully not you, geography might be limited to what is included in the English national curriculum, which is now more than ten years old. The Purpose of Study statement it contains suggests that:

A high-quality geography education should inspire in students a curiosity and fascination about the world and its people that will remain with them for the rest of their lives. Teaching should equip students with knowledge about diverse places, people, resources and environments, together with a deep understanding of the Earth's key physical and human processes.

Department for Education (DfE), 2013

The words 'curiosity' and 'fascination' are important, but overall this is a fairly unambitious and bland vision for the subject. Do such statements 'written by committee' fail the profession? (Rawling, 2020) Young people should feel empowered by our teaching to change the world for the better, be able to consider knowledge critically and feel they have agency. For many students, the question of 'what is geography?' is perhaps overtaken by their personal experiences of it and becomes 'why bother with geography?'. A growing number of students may also be taught by a teacher whose first specialism is not geography. Departments therefore need a particular vision for geography suitable for their own school context, and external inspectorates will now ask for this to be articulated in a statement of curriculum intent and will proceed to assess its implementation and impact.

As mentioned earlier, there are many geographies and different national curricula define it differently. It may be worth looking at the work of other countries. In some, geography is placed within the social sciences rather than retaining its own identity. The German Geographical Society (DGfG, 2023) recently published an information brochure called *Geography: Key to the Future*, illustrating the relevance of geography as a school subject and setting out some educational policy demands for strengthening geography in the German system.[4]

Teachers in England also have a choice of different specifications for GCSE and A-level (or equivalent) teaching. All geography teachers have their own particular favourite areas of the curriculum, which may lead to inconsistencies between and within schools prior to the start of key stage 4. Over the decades, choices have been made as to which particular knowledge is important for students to learn, and powerful to have when it comes to making decisions. Geographers have always explored the relationship between people and the environment of course, which is one of its four 'traditions' (Murphy, 2018).

---

4. https://geographiedidaktik.org/geography-keytothefuture/ (Accessed: 28/05/2025)– the booklet is available in English.

Finding a place for geography

Geography is everywhere (Matthews and Herbert, 2008) and geographical study 'offers a critically important window into the diverse nature and character of the planet that serves as humanity's home' (Murphy, 2018). Geography has evolved as the world has become increasingly complex, and new subdivisions have appeared to reflect new areas of study. This places an additional pressure on those who teach the subject to keep broadly abreast of new approaches and knowledges. If our curriculum is to have epistemic quality, we need to remain as cognisant of new approaches to cognitive science and teaching and learning as we are about events happening in the news: whether that be the geopolitical complexities of the war in Ukraine or the most recent weather extreme to visit itself upon an unprepared region (see chapter 37).

Geography is often described as an enabling subject. Sitting at the junction of both physical and social sciences, and within the humanities, a key aim of modern geographical study is to help students to 'think geographically'.

'As geographers our work is always interdisciplinary. It interacts with, borrows from, and informs other disciplines by default,' says Peter Kraftl from the University of Birmingham.[5]

In *The Handbook of Secondary Geography* published by the Geographical Association (GA), Alastair Bonnett suggests that geography helps us to make sense of the world we live in. 'Geography surrounds us. We walk, drive and fly over and through it. At the same time, geography is distant: geography is the landscape beyond the horizon and the intriguing distance between here and there.' (Jones, 2017)

A key component of geography is that it enables students to explore our 'messy' world first hand and 'ground truth' the models in the textbooks. For many adults, geography fieldtrips remain an abiding memory – often for reasons unrelated to learning. They offer a chance to develop specific geographical skills in measurement, recording and data presentation as well as more general social skills. Ofsted (2023) is clear that the provision of fieldwork is an expectation of any geography curriculum, although rising costs and some haziness over the expected extent of this fieldwork means that the experiences vary widely between schools (see chapter 8).

---

5. https://www.birmingham.ac.uk/research/idr/case-studies/children-and-childhood.aspx

Where is geography going next?

Geography's relevance is becoming increasingly apparent through the challenges that are emerging in this period which some call the Anthropocene: an era where humanity has altered natural physical systems. If we are going to ask our students to *do* geography (earth writing), we need to continue to provide them with stories which are hopeful (see Chapters 32 and 40). The National Trust published a research report in 2023[6] exploring the importance of 'nature connectedness'. Of the thousands of people interviewed, 85% agreed that 'humans are severely abusing the planet'. This concern can be converted into informed action through geographical study, and it is geography which young people feel is the subject which best covers topics such as climate change. The GA's *A framework for the school geography curriculum* (Rawling, 2022) lays out a vision for what school geography could be, and in the process includes a working definition for the subject, where 'the educational value of geography can be enhanced by developing young people's capabilities as human beings, to enable them to use their geographical understanding to live in harmony with others and to share responsibility for the well-being of the planet' (Kinder and Rawling, 2023). The document suggests some key concepts that need to be part of a course of study in geography: space, place, earth systems and environment. In a world where climate breakdown is becoming increasingly apparent, the subject of geography is coming to the foreground, just as it did during the pandemic, when everyone suddenly became aware of their own personal circumstances and the importance of space. The world needs more geographers, and their training begins in your classroom. Make the most of this opportunity and responsibility.

---

6. Available from: https://www.nationaltrust.org.uk/our-cause/nature-climate/nature-conservation/everyone-needs-nature (Accessed: 28/05/2025)

Five reflection questions

1. How could you best communicate the meaning of geography to the students that you currently teach?
2. How would you set up a discussion to elicit the students' own conception of geography as a subject?
3. Think back to your own experiences of learning geography. How has it changed since you were at school? Have those changes been for the better or the worse?
4. What stories do you think you have told in the past which were perhaps problematic, and which you should now avoid repeating with students? What steps are you taking to revisit your curriculum to uncover and address these problematic areas?
5. What is the best definition of geography that you've found, and how does your department's intent statement define the subject?

DIGGING DEEPER – THREE RESOURCES TO DELVE FURTHER

- Dorling, D. and Lee, C. (2016) *Geography: Ideas in profile*. London: Profile Books – a very useful book which summarises the development of the subject and its potential. Dorling's wider work is also worthy of investigation.
- The Geographical Association National Curriculum Framework and supporting documents (2022) – available at https://geography.org.uk/ga-curriculum-framework/ – provides a structure for exploring the subject and considering teachers' own curriculum-making process.
- Owens, J. (2023) *Dust: The modern world in a trillion particles*. London: Hodder and Stoughton – an investigation by a geographer into the significance of dust of various kinds. A reminder of the hidden geographies in everyday situations and our global interconnections.

REFERENCES

Bonnett, A. (2008) *What is Geography?* London: Sage Publications Ltd

Bonnett, A. (2012) 'Geography: What's the big idea?' *Geography*, 97, pp.39–41

Cresswell, T. (2013) *Geographic Thought: A critical introduction*. Hoboken, NJ: Wiley-Blackwell

DfE (2013) National curriculum for England: geography programmes of study, 11 September 2013. Available at: https://www.gov.uk/government/publications/national-curriculum-in-england-geography-programmes-of-study/national-curriculum-in-england-geography-programmes-of-study (Accessed: 28/05/2025)

German Geographical Society (DGfG) (June 2023). Geography: Key to the Future. https://geographiedidaktik.org/geography-keytothefuture/ (Accessed: 28/05/2025)

Hamilton, A., Wiliam, D., and Hattie, J. (8 August 2023) The Future of AI in Education: 13 Things We Can Do to Minimize the Damage. Available at: https://doi.org/10.35542/osf.io/372vr

Jones, M. (ed.) (2017) *The Handbook of Secondary Geography Teaching*. Sheffield: Geographical Association

Kinder, A. and Rawling, E. (2023) The GA's framework for the school geography curriculum, *Geography*, 108(2), pp.101–106

Matthews, J. A. and Herbert, D. T. (2008) *Geography: a very short introduction*. Oxford: Oxford University Press

Murphy, A.B. (2018) *Geography: why it matters*. Cambridge, UK: Polity Press

Ofsted (2023) 'Getting our bearings' in *Geography Subject Report*. Available at: https://www.gov.uk/government/publications/subject-report-series-geography/getting-our-bearings-geography-subject-report#pedagogy

Parkinson, A. (2020) *Why Study Geography?* London: London Publishing Partnership

Parkinson, A. (2021) 'I know where I'm going' – teaching map and GIS skills. *Teaching Geography*, 46(1), pp.7–10

Rawling, E. (2020) How and why National Curriculum frameworks are failing geography. *Geography*, 105(2), pp.69–77

Rawling, E. et al. (2022) *A framework for the school geography curriculum*. Geographical Association. Available at: https://geography.org.uk/wp-content/uploads/2023/07/GA-Curriculum-Framework-2022-WEB-final.pdf (Accessed: 28/05/2025)

2. WHERE HAS GEOGRAPHICAL ENQUIRY TAKEN ME?

CATHERINE OWEN
@GEOGMUM

'In a minority of schools, leaders had thought carefully about how to teach geography in a distinctive way. At its best, this involved students being taught the processes of geographical enquiry. They were introduced to geographical questions, and presented with data (in its widest sense, including images, maps and people's testimonies as well as graphs and charts). They then used this data to reach conclusions.' (Ofsted, 2023)

Is it wrong to start a chapter in a book by telling the reader to read another book? I hope this chapter will be useful to those looking for an overview of one teacher's experience of the enquiry approach, but for those wanting to delve deeply into this fascinating aspect of geography I strongly recommend reading *Geography Through Enquiry*, 2nd Edition (2023) by the wonderful Margaret Roberts, the queen of geographical enquiry.

Roberts (2023) states that there are four key elements of geographical enquiry in the classroom:

- Enquiry should be question driven.

- Enquiry should be supported by evidence.
- Enquiry should give students the opportunity to make sense of geographical information for themselves.
- Students need to reflect on what they have learned from the enquiry.

ENQUIRY AS A STUDENT

I don't remember a great deal about being at school in the 1980s, but I do remember the geography enquiries I took part in. I still have my old geography books, which show that, while most lessons relied on teacher talk and a textbook, the lessons I remember gave us a question to investigate, draw conclusions about and reflect upon. An example is shown in Figure 2.1 – we were taught about motorways, used maps to investigate the stretch of the M5 motorway from Bridgwater to Weston-super-Mare, then set the question 'What is the best route for a new motorway?' and given a map to scrutinise before adding our chosen route. We evaluated our route, discussing potential conflicts and how these could be minimised.

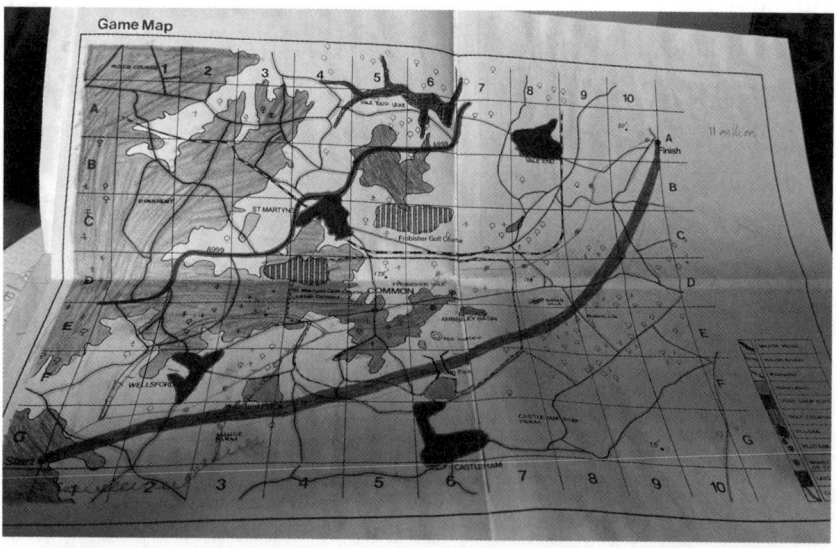

Figure 2.1 'What is the best route for a new motorway?' project by Catherine in around 1985

Mine was the first year-group to sit GCSEs, with coursework crucial across all subjects. In geography, I completed several mini-enquiries (some involving fieldwork) and my own project in which I investigated the impact of the rapid growth of my home village in the previous 20 years. I loved these enquiries and the opportunity they gave me to make sense of the world around me. I have wanted to be a teacher since I was small, but had set my heart on history teaching. It was a combination of these geographical enquiries and my excellent teacher, Mr George, who changed my mind to geography. However, sometimes things go wrong. Unfortunately, my first major geography enquiry disaster was my A-level individual study. I focused on the Mendip Hills as a local landscape resource, investigating use of the limestone for leisure, settlement and industry, but my work became a project about the Mendips, rather than a decisive investigation. It was also only tenuously linked to the specification. Despite being devastated by this failure at the time, I now see it as an important learning experience – it is a great non-example of how to complete an enquiry!

The enquiry approach was very significant as I studied geographical sciences at Polytechnic South West/University of Plymouth (the status changed while I was there). I delighted in my access to so much information from the library and carried out regular primary data collection on Dartmoor for my hydrogeomorphology course.

ENQUIRY AS A TRAINEE TEACHER

My Postgraduate Certificate of Education (PGCE) at Marjon in 1993/94 was at the tail end of the first national curriculum (NC), with all the folders of attainment targets that involved. With five attainment targets, including 183 statements of attainment, teachers were faced with the challenge of covering the prescribed content while also tracking the attainment of students. Rawling (2001) describes how 'Essentially, in those schools with curriculum development experience and with confident and creative leadership, it was possible to impose a more appropriate curriculum structure and to develop good quality teaching and learning, despite the Order.' I was fortunate to find myself in schools (in Plymstock and then Bodmin) with excellent curriculum practice – they were able to balance the demands of the NC with their desire to keep enquiry at the heart of the geography curriculum. Having started by teaching enquiry lessons designed by other teachers, I began to design and teach my own enquiry lessons. I found it easy to come up with enquiry questions, but much harder to source material for students to use for research. I scoured national and local newspapers and picked up leaflets, cutting out articles and storing them in folders for future use. These could then be combined and photocopied to produce well focused and informative resources for students. This meant that students had the opportunity to make sense of geographical information for themselves and support their points with evidence, as recommended by Roberts (2023).

Most of my enquiries went well; students seemed to develop their learning and enjoy answering the questions. An exception was an enquiry disaster when Marjon took us to East London to work in a contrasting place for a week. Many schools had bought the *Key Geography* (Waugh and Bushell, 1991) textbook series, using this to help them cover the vast content of the NC. This series took an enquiry approach in that each spread had a question as its title, but activities mainly involved comprehension questions. I spent two days at a school in Rotherhithe and was tasked to work with a partner to plan and teach a lesson based on a spread in the 'Foundations' textbook with an enquiry question about where to build a bypass. I had only been teaching a few months and was in an unfamiliar environment, so should have stuck to the book, as the teacher advised. Instead, I decided to create an activity similar to that in Figure 2.1, providing students with a map showing a fictional town with residents demanding a bypass and asking them to work in groups to plan a route before feeding back to the class. This went horribly wrong! The students had no idea what a bypass was and found it difficult to get their heads around the idea – they lived in the depths of urban London and had no experience of bypasses. This taught me an important lesson – enquiries need to take the prior learning of students into account and ensure that they have a firm foundation in the knowledge needed to complete the enquiry successfully.

ENQUIRY AS A TEACHER

I started teaching at Holyrood Community School in September 1994, with the revised NC of 1995 giving us the opportunity to review our overloaded geography curriculum. Geography had previously been part of a humanities curriculum at key stage 3, so it was exciting to work with the new head of geography to develop a standalone geography curriculum. We used *Key Geography* (Waugh and Bushell, 1991) as our core text, but developed enquiries to make our lessons more relevant in our context and to challenge our students' thinking about the world. Again, I faced challenges – for example, when my Year 9 class carried out an enquiry looking at ways different countries were managing their populations, I received a complaint from a parent as their son had drawn conclusions going against the teachings of their Catholic faith.

On reflection, my enquiry lessons in the 1990s were limited in terms of Roberts' (2023) fourth key element – students reflecting on what they had learned. Roberts (2023, p.10) suggests that:

> *If students are to make sense not only of the evidence provided but of a unit of work as a whole, then they need to reflect on what they have learned. They need to reflect on the extent to which they have answered or explored the questions posed at the outset, the extent to which evidence was sufficient, on whether the techniques they have used to analyse or interpret source materials were appropriate and whether the conclusions and judgements they reached were sufficiently supported by the evidence.*

I participated in a course for new teachers in Somerset being hosted at Millfield School in my second year of teaching. This led to me studying an MEd in School Effectiveness and School Improvement in my third and fourth years of teaching, with many opportunities to study curriculum and the place of enquiry within this. Taking an academic approach to studying the curriculum forced me to become more evaluative – having churned out reflections on lessons with little thought during my PGCE, I started to truly engage in reflection and to support my students in doing the same. It was fortunate that my evaluation epiphany happened at the same time as *Thinking through Geography* (Leat, 1998) was published. This book was developed by teachers from the north-east of England who started out with the aim of making geography lessons more interesting and challenging, but soon realised that the most important aspect was the debrief following the 'thinking' activity. The book therefore included thinking activities alongside detailed instructions on how to debrief students, considering what they had learned, how they learned it, how they could apply this in different situations and building bridges to other geographical concepts. Not only did this book improve my use of evaluation in lessons, it also helped me take my first steps towards developing metacognition skills with my students.

ENQUIRY AS A HEAD OF GEOGRAPHY

In January 1999 I took on the role of head of geography at The King Alfred School, gaining even more freedom to shape the geography curriculum. Again, the school was using *Key Geography* heavily in lessons (partly to support non-specialists), but each year group completed a fieldwork enquiry and there was enthusiasm to further develop the enquiry approach. We tweaked schemes of learning and started to develop new approaches, but by 2007, with a new NC being introduced for 2008, we felt ready to carry out a full restructure of our KS3 curriculum. We worked on this as a team, starting by coming up with possible enquiry questions, putting them on sticky notes and moving them around to find the best sequence, creating a document we called 'the global plan'. We showed the plan to a group of students, who received it with enthusiasm and offered to help develop it. Having just joined the Geographical Association's International Special Interest Group, I was particularly interested in developing the international dimension and wrote about this for the journal *Teaching Geography* (Owen, 2008).

Working with the Geographical Association, reading *Teaching Geography* journal and attending the annual GA conferences supports me in being a life-long learner, helping me to develop my understanding of geographical concepts and pedagogy, including enquiry. I cascade my learning to my department so that we can regularly evaluate our curriculum and see if aspects need to change in light of our new understanding. With a new NC introduced in 2014 and an increasing focus on knowledge-rich curricula in schools, we have been keen to maintain enquiry in our lessons. In *Powerful Geography*, Mark Enser discusses how a focus on powerful knowledge doesn't exclude enquiry, saying enquiry 'is an important part of the discipline and should infuse the way in which we put our curriculum into practice' (2021, p.135). Enser suggests six steps to take to embed a culture of enquiry in the curriculum, as shown in Table 2.1.

Table 2.1 Six steps to embed a culture of enquiry in the curriculum (adapted from Enser, 2021)

Step 1: Frame through a question	Create a fertile question to sit at the heart of the topic. This question should have more than one possible answer and involve students working hard to answer it, but also needs to be within their ability to answer.
Step 2: Source the information	This information could take many forms, including articles, graphs, photos, statistics and more, but it is vital that students are able to make sense of these and that they are up to date. Enser also reminds us to think about whose story we are telling and whose voices we are hearing.
Step 3: Create the need to know	Creating a fertile question means the question should intrigue students. Using images with the question to illustrate it may also give students intrinsic motivation to find out more.
Step 4: Using data	Students should build upon their prior learning by finding out new information about the topic. It may be better for students to encounter sources of information one by one to build their knowledge.
Step 5: Reaching a conclusion	Students need to answer the question geographically – they may need scaffolding or to see models of answers to do this.
Step 6: Evaluate the process	This could involve students comparing their answers and seeing if everyone reached the same conclusion (if not, why not?), questioning reliability and validity of information used or exploring the implications of their conclusions.

ENQUIRY AS AN AUTHOR

Having had my first article published in *Teaching Geography* in 2000, with others following, the Geographical Association asked me to write for their *Geography Toolkit* series in 2009. I highly recommend becoming an author; you get to work with editors who challenge your thinking, meaning you learn a great deal. In this series each book included ten enquiry lessons and the series editors were Justin Wooliscroft and Ruth Totterdell. Justin and Ruth were the perfect editors for a novice writer, taking time to discuss ideas and refine them with me. Working on a book in this series with John Widdowson led to me co-authoring a GCSE textbook with John and Andy Crampton. I have since been part of the

Progress in Geography team with David Gardiner, written enquiries for the Geographical Association's Online Teaching resources website and more. The power of collaboration should never be underestimated; each time I work on a project with others my understanding of geography content and pedagogy, including the enquiry approach, grows. *What is Geography Teaching, Now?* is my first foray into editing as well as writing!

ENQUIRY NOW AND IN THE FUTURE

Enquiry learning in my department took a big hit during COVID-19 lockdowns and while we were teaching in 'bubbles' in early 2020. With students unable to share resources between groups or go on field visits, we stripped back our curriculum. Following the pandemic, we recognise that the world has changed, with schools across England facing issues related to poor attendance, families living in poverty, students in inappropriate housing, the impact of social media and a teaching recruitment crisis (Weale, 2024). We are currently working to improve our KS3 curriculum, bringing back the creativity and enquiry we lost and implementing new ideas. This gives us the opportunity to build in aspects such as supporting students' wellbeing (including ecoanxiety) and decolonising the curriculum.

I was inspired by Daryl Sinclair's presentation entitled *'Power: a new and essential lens for geography education'* at the 2024 Geographical Association Conference and am planning to include power as a concept in future enquiries. Castree et al. (2023) suggest three modes of power:

> **Power to** *is simply the ability to exercise power of some kind – as when a local government gives planning permission to build new apartments on a brownfield site in central Manchester.*
>
> **Power over** *occurs when a person, organisation or element of the non-human world (e.g. wild animals) is subjected to power. This may cause resistance, opposition or protest, as with Black Lives Matter demonstrations about police brutality in cities like Los Angeles. There are entire discourses that contest 'power over', employing words like oppression, domination, anti-colonialism and injustice. But subjection to power is not always negative – for instance, land managers exert power over invading species for sometimes good reasons.*

Power through *refers to the mechanisms or the media that propagate or transmit power. For instance, the unintended power of humans to change the world's atmosphere has been exercised incrementally by emitting greenhouse gases from power stations, steel factories and so on. Climate change represents a marriage of human and physical power on a very large scale.*

As I sit down tomorrow to review my Year 8 scheme of work related to our local city, Bristol, I will be keeping both enquiry and power in mind. My blog 'Bristol: The Danger of the Single Story' (Owen, 2022) was an attempt to get teachers thinking about how they present Bristol to students, but how can I explore power to, over and through using these locations? Filwood, in Knowle West, has recently been blighted by knife crime and residents are campaigning for the Eagle House Youth Centre, which closed ten years ago, to be reopened (Cork, 2024) – how can I turn this into a powerful enquiry, challenging my students to think about who had the '*power to*' close the youth club and how the local residents are responding to '*power over*' them through resistance? Tomorrow will tell!

Five reflection questions

1. How effectively do you use Roberts' four key elements of geographical enquiry?
2. How do you keep up to date with developments in geographical content and pedagogy?
3. To what extent has the COVID-19 pandemic and resulting changes in education affected your use of enquiry in lessons?
4. How could you bring consideration of power into your enquiries?
5. Where has geographical enquiry taken you?

> **DIGGING DEEPER – THREE RESOURCES TO DELVE FURTHER**
> - Roberts, M. (2023) *Geography Through Enquiry*, 2nd edition. Geographical Association.
> - The Geographical Association's 'Planning for geographical enquiry', available at: https://geography.org.uk/ite/initial-teacher-education/geography-support-for-trainees-and-ects/learning-to-teach-secondary-geography/geography-subject-teaching-and-curriculum/geography-knowledge-concepts-and-skills/geographical-practice/geographical-enquiry-2/planning-for-geographical-enquiry/ (Accessed: 28/05/2025)
> - The Geographical Association's Online Teaching resources from the Geographical Association https://geography.org.uk/online-teaching-resources/ (Accessed: 28/05/2025)

REFERENCES

Castree, N., Oakes, S. & Sinclair, D. (2023) Exploring "power" as a concept in geographical education. *Teaching Geography*, Autumn 2023.

Cork, T. (2024) Renewed calls for Knowle West youth club to reopen after tragic deaths of teens. https://www.bristolpost.co.uk/news/bristol-news/renewed-calls-knowle-west-youth-9081309.

Enser, M. (2021) *Powerful Geography*. Carmarthen: Crown House Publishing.

Leat, D. (1998) *Thinking Through Geography*. Cambridge: Chris Kington Publishing

Ofsted (2023) 'Getting our bearings' in *Geography Subject Report*. https://www.gov.uk/government/publications/subject-report-series-geography/getting-our-bearings-geography-subject-report#pedagogy.

Owen, C. (2008) Developing the international dimension at KS3. *Teaching Geography*, Autumn 2008.

Owen, C. (2022) Bristol: The danger of the single story. https://geogmum.wordpress.com/2022/01/09/bristol-the-danger-of-the-single-story/ (Accessed: 28/05/2025)

Rawling, E. (2001) *Changing the Subject*. The Geographical Association.

Roberts, M. (2023) *Geography Through Enquiry*, 2nd edn. Sheffield: Geographical Association

Waugh, D. & Bushell, T. (1991) *Key Geography*. Cheltenham: Stanley Thornes (Publishers) Ltd.

Wheale, S. (2024) Hunger, homelessness and gang grooming: just a normal week at one London academy. *Guardian,* 3 April 2024. https://www.theguardian.com/education/2024/apr/03/hunger-homelessness-and-gang-grooming-just-a-normal-week-at-one-london-academy (Accessed: 28/05/2025)

3. WRITING A KEY STAGE 3 GEOGRAPHY CURRICULUM FROM SCRATCH

JOSEPH MILTON
@GEO_DOUGIE

'What matters for a curriculum is that teachers come together and agree on a set of aims that underpin their vision for student development.' (Standish, 2021)

This chapter is about ripping up a curriculum and starting from scratch. I started my current role as head of geography in 2019, just before the COVID-19 pandemic, with a whole new team of geography staff. Through teaching the curriculum that was there before, I, and my team, quickly identified the key stage 3 (KS3) curriculum as something that needed to change for the department to have the kind of success that we all wanted and could be proud of. This is a story of that journey.

VISION: THE SOLID FOUNDATION TO BUILD A CURRICULUM ON

Curriculum planning is like an ouroboros – an eternal cycle of life, death and rebirth. Before any curriculum that has any real meaning behind it is given life, you need to decide what you want it to be. What you want it to look like. You need to have an aim, or to use Ofsted parlance 'intent', in mind. Something to focus the efforts of not only you, but every colleague within your department. A foundation upon which everything is built and the funnel that channels all the differing strengths of your department into creating a coherent end goal. This is your vision and it is incredibly important that you get it right. So, how do you do that?

MAKE IT COLLABORATIVE

If you lead a geography department then you have ultimate responsibility for what is written down as your vision and everything that follows from it. However, geography is a broad church and encompasses different specialisms and ideas of what school geography should be. Should it be about raising awareness of global events? Should it be about producing change makers? Should it be about producing students who can see the world about them in a synoptic way? Should it be about creating people fit for the next stage of their education? Ultimately, there is not a right answer, but there is a right way to get a vision that everyone can get behind: collaboration (Standish, 2021). Working together to create a vision that everyone can get behind gives momentum to any changes you wish to make and drives a collective effort to make the curriculum the best it can be.

MAKE IT SUCCINCT

There's no point in being verbose. Your vision needs to be at the forefront of you and your colleagues' minds not only for planning, resourcing and evaluating your curriculum, but also for when you are asked about it by anyone with an interest, be it parents, senior leadership team (SLT), potential new staff on interview or Ofsted. Verbosity allows for mission creep in your curriculum design. Succinctness does not.

MAKE IT SUBJECT SPECIFIC

Your vision shouldn't veer towards the generic but should have a monomaniacal focus on being subject specific and laden with what student development in your subject means. Your curriculum will hang off your statement, it should therefore guide those who experience it into the finer detail of your subject rather than leave them perplexed by a series of genericisms.

> **Our vision**
> So, having made it collaborative, succinct and subject specific, the foundation vision from which our curriculum spawned was:
> To equip students with the knowledge, skills and attributes to be able to make sense of the world around them, both past, present and future.

THE TRICKY WORLD OF GEOGRAPHICAL CONCEPTS

'A programme of study articulated without concepts, runs the risk of focusing entirely on knowledge or skill acquisition'. (Brooks, 2017)

WHAT IS A CONCEPT?

Upon settling on a vision for your curriculum, you then need to start the process of fleshing this out and here-in lies the first consideration when creating what you hope will be a coherent curriculum – what concepts are you going to build your curriculum around and what organisational structure are you going to use to allow your chosen concepts to match your vision? For the sake of clarity, a concept, which is sometimes referred to as the grammar of the subject, is something that helps coherently organise the knowledge that you hope to convey to the students in your subject.

CONCEPTS TO ORGANISE CONTENT

While Ofsted shouldn't be the sole purpose for organising a good KS3 curriculum it was noted that effective curriculum design and delivery considered concepts, not only as a means of considering how to teach disciplinary knowledge, but also as a means of effectively organising the content that you wish to teach (Ofsted, 2023). This brings us to a difficult juncture. Geography needs some underpinning concepts to hang your curriculum on and weave your geographical story together. What concepts are you going to build your curriculum upon? Quite simply, there is no simple answer to this question and indeed, the role of concepts and what concepts to select is contested (Brooks, 2017).

As shown below in Table 3.1, there are many different concepts that geographers have used to devise their curricula around.

Leat (1998)	Geography Advisors' and Inspectors' Network (2002)	Rowley and Lewis (2003)
• Cause and effect • Classification • Decision making • Inequality • Location • Planning • Systems	• Bias • Causation • Change • Conflict • Development • Distribution • Futures • Inequality • Interdependence • Landscape • Scale • Location • Perception • Region • Environment • Uncertainty	• Describing and classifying • Diversity and wilderness • Patterns and boundaries • Maps and communication • Sacredness and beauty
Holloway et al. (2003)	Jackson (2006)	UK 2008 KS3 Curriculum (QCA, 2007)
• Space • Time • Place • Scale • Social formations • Physical systems • Landscape and environment	• Space and place • Scale and connection • Proximity and distance • Relational thinking	• Place • Space • Scale • Interdependence • Physical and human processes • Environmental interaction and sustainable development • Cultural understanding and diversity

Table 3.1 Concepts

Source: Fogele, (2016). Article printed under Creative Commons License CC BY 4.0

While there is no hard and fast rule about the concepts you use in devising your curriculum, the maxim 'less is more' applies well here. Too many concepts and you risk your curriculum becoming a slave to a loose set of concepts that lack coherence. However, the risk of taking a reductionist approach is that your concepts become too broad and, ultimately, meaningless and again make your curriculum lack coherence. The concepts we decided upon that appear later we felt walked that fine balance between being reductionist or too numerous. However, in deciding your concepts you must also consider how organising these concepts will help structure your curriculum.

Organising your concepts

Geography is a content-rich subject and therefore requires this content to be sorted, or organised, into the various silos that allow it to not only be coherent to the teachers who are teaching it, but also to the students it is being delivered to as well. How you choose to organise the concepts within your curriculum will dictate how your curriculum is planned and, ultimately, delivered.

Hierarchical concepts

Arranging your concepts as a hierarchy centres upon selecting a small number of concepts, which are abstract and/or technical, that are gradually fleshed out to become concrete and/or vernacular. An example that forms in my mind to exemplify this would be thinking about the concept of pressure, which would be a technical/abstract concept that would be fleshed out in a series of lessons to the concrete/vernacular concepts of rainfall, biome, etc. A hierarchy of concepts would work well, particularly where you have a large number of discrete topics to work through, such as physical geography at A-level, and would work well if you would like to share the concepts underpinning the curriculum, topics and lessons with your students, as the reigning in of the number of concepts certainly works for a geography teacher who is skilled at seeing the linkages between concepts, but not necessarily a student. I however, decided not to use hierarchical concepts in the curriculum, but a different approach. The rationale was that the topics I thought would form at least part of our KS3 offer would revolve around an interwoven curriculum of themes and regions as outlined by Enser (2021).

Organisational concepts

The organisational model view takes a fundamentally different approach to organising your concepts with the aim being to develop geographic thought through having wide-ranging concepts that link everyday experiences to those higher-level geographical ideas. This moves in the opposite direction to hierarchical concepts, which widen as you move from the abstract to the concrete, having a smaller range of overarching abstract concepts that are allowed to sit without being fleshed out further with concrete concepts. To my mind this would mean that an overarching concept would be something along the lines of 'processes and systems' which facilitates for a wide range of processes and higher-level geographical ideas (or threshold concepts), such as global atmospheric circulation and air pressure, to be taught in a unit on weather and climate and then returned to several times in later units on regions to explain the climates of these regions. This method of organising concepts certainly on the face of it seemed to suit the initial idea for our themes and regions curriculum. It was felt that this structure for organising our concepts allowed for greater time to be spent on planning learning and not getting bogged down in thinking about the concrete concepts that could be fleshed out for every abstract concept. With this in mind, we decided to go for the six concepts below, which are by no means something that everyone will agree upon but worked for how we planned to organise our curriculum.

Our six concepts

These are:
- Place
- Scale
- Sustainability
- Interdependence
- Risk
- Processes and systems.

TOPICS: SELECTION AND SEQUENCING

Once you've outlined your vision and the concepts that will act as the foundation and linkages for your curriculum, the next step is deciding topics and the sequencing of them. Your choice of topics at KS3 needs to be mindful of, though certainly not straight-jacketed or limited by, the national curriculum document for KS3 (though this depends on what type of school you work at). It should also ideally play to the specialisms and interests within your department, the place and locale your school sits in and, importantly, your Key Stage 4 and 5 curricula, as you should avoid repeating topics through every key stage where possible. As exciting as the formation of a wave cut platform is (though trying to prove a point, I don't want to leave you with any doubt that I look forward to days when I can explain with diagrams the formation of a wave cut platform) no student will want to hear about it in three different key stages. What sort of reward is that for opting to continue with your subject?

How you choose to meet the requirements of the national curriculum is your (and your schools) choice but the ultimate aim should be to show the relevance and importance of your subject and allow every student to increase their knowledge of the world and its interactions as they progress through it. So, how do you go about doing this?

TOPICS

How you decide upon the topics you wish to teach outside the national curriculum is a choice for you and your department and will rest upon not only where you are within the UK, but also upon the specialisms within your department. It may also be tempered by the topic coverage of your KS4 and KS5 courses. This, of course, is all underpinned by a continual requirement to keep on top of your subject knowledge through a variety of means, something which I think is often under-represented as a planning and workload issue within the subject, but also from those outside the subject too. Our initial planning was a piece of A3 paper that dumped all the topics that we would like to teach, but, importantly, also matched with our mission statement and aims. It was at this juncture that, subconsciously, the decision was taken to move away from topics that might appeal to students and be 'current', such as fast fashion or the geography of sport or crime. While there is nothing inherently *wrong* with these topics and approach, we felt that these topics have a fleeting nature to them and have a limited shelf life, and also wouldn't allow us to have coherency within our curriculum as we couldn't easily identify a thread that could weave these disparate topics together. After getting down all the topics we would like to teach that matched with our vision, and importantly before we slimmed the topic list down, we added the 'meat to the bones' of the topics with everything we would like to see taught within every discrete topic. This proved to be a valuable part of the planning process as it allowed us to triage topics and select those that had depth and therefore a basic, but coherent structure that we could develop to answer a big geographical question. It should also be said at this juncture that the length of a topic is something to consider too. While allowing them to fit into a half term is neat, it doesn't necessarily follow that every topic will fit into this. Some will be longer, some will be shorter. You really need to think about the coherency and depth of the topic rather than the length. However, topics within themselves are not a curriculum and there needs to be some careful consideration of other factors before you start planning lesson sequences and resourcing lessons themselves. Firstly, whether you are having a discreet or interwoven curriculum.

BIG GEOGRAPHICAL QUESTIONS

A big geographical question is, as it suggests, an overarching question that you can build your topic around and enable every lesson to slowly add pieces of knowledge that will facilitate your students to arrive at an answer at the end. Some of these are easy to come up with, some less so. They should be something that allows for extended thought and long-prose answers, but cannot ultimately be answerable until the whole topic has been completed and, even when the students are in a position to answer them, they may arrive at different answers even though they have sat in the same lessons and digested the same materials. It's all a matter of the importance you place on differing aspects of what is studied. It's even better if a question can be worded in a way that means that the students need to wrestle with their views on answering it with every lesson. Some of the best examples we have of this are from our Year 7 topic on population, which asks students to consider 'Was Malthus right?' and from our Year 9 topic on Haiti, which asks 'Is Haiti a victim of its geography?'.

A DISCRETE OR INTERWOVEN CURRICULUM

A discrete curriculum has topics that are independent of any other, that will have all the necessary requirements of good topics, such as skills, knowledge and application, and can be taught in an order of your choosing and ultimately takes less planning to create. However, the downside of doing this is that your curriculum might have an awesome set of topics that don't necessarily weave together to develop the students in front of you as geographers. An interwoven curriculum requires more thought so that the knowledge, content and skills gradually weave together and build upon each other over time. Our curriculum (see Figure 3.1) operates a hybrid model of the two and moves from a discrete set of topics in Year 7 that can operate as topics in their own right, to an interwoven set of topics that build up through Years 8 and 9 that can function without the previous topics being delivered but works a lot better with them having been taught. This marries with the idea of developing mastery within our curriculum as well through the revisiting of content and applying it in different contexts, and avoids the pitfalls identified by Enser (2018) in opting for a thematic curriculum – coherency.

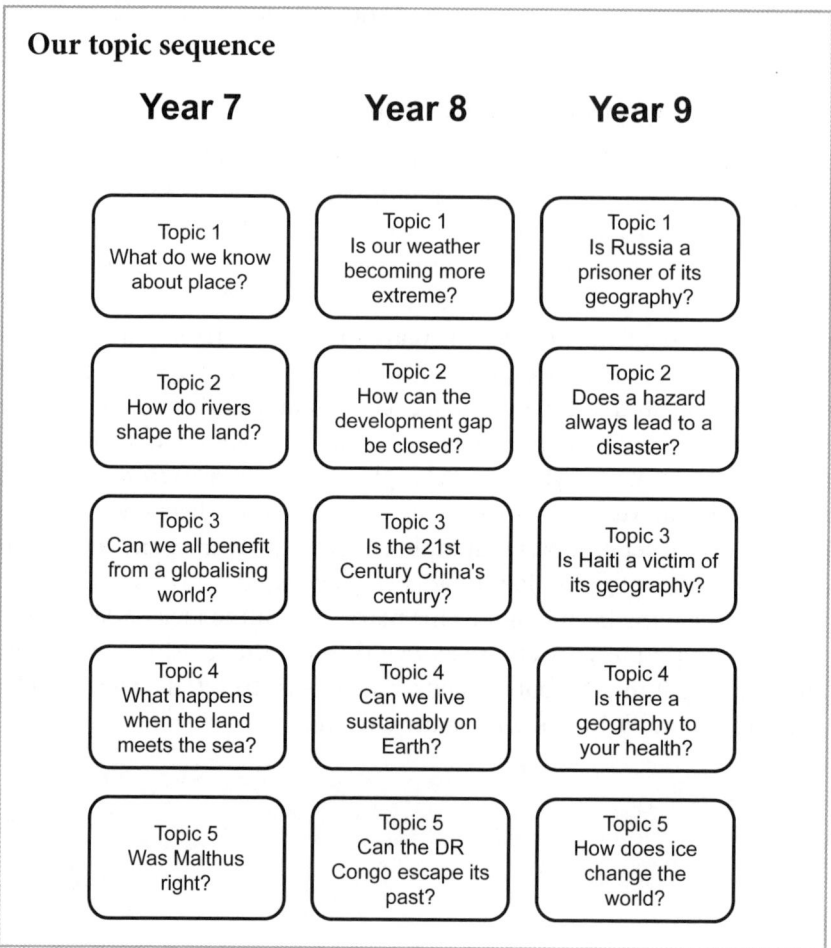

Figure 3.1 Our topic sequence for KS3

THE PHYSICAL: HUMAN TIGHTROPE

In my experience, the majority of students have a preference for one side of geography or the other. For our curriculum we made the decision early on that we would have a fair balance between human and physical in all three year groups where possible to fairly reflect geography as a whole. However, our discussions in the early phases of planning our new curriculum had also made clear that we wanted to show, again where possible, how the two sides of geography can be bridged by having a regional/country focus interwoven through the curriculum as well. This was where the decision to have an interwoven curriculum of discrete topics that weave together in Years 8 and 9 with a regional focus allowed for human and physical geography to be explored together to investigate place.

SEQUENCING

Once we had decided upon the topics and the countries/regions in our curriculum we then needed to look at how we sequenced it, so it made sense and built as complete a picture of the study of geography as possible. What we were looking for was for a foundational core of discreet knowledge and skills to be built up through time, but for it to be interspersed with regional or country topics that allowed for the application of this discreet knowledge to a place. Easy in theory, but trickier in reality as there are a number of factors to consider in an appropriate sequence. This was our list of factors and though not exhaustive, I'd imagine the majority of these (and some we maybe didn't consider) are those that are discussed at any curriculum meeting. I'm not going to give any answers to these questions, as this will depend on a range of factors, but you should be able to see from the curriculum map that appears at the end what decisions we made.

- **What are the foundation topics that are required before others, e.g. should globalisation come before development or vice versa?**
- **Where should the more conceptually challenging topics, e.g. hazards, weather, climate change, go?** Should they be taught in Year 9 or do you decide to teach them in Year 7 but scaffold appropriately because of other topics you'd like to have later?
- **Are there topics that require a greater level of maturity to deliver effectively, e.g. population?** Will talking about contraceptive availability with Year 7 mean you spend most of the lesson quelling giggles?
- **When do GCSE options take place?** We all need bums on seats at GCSE to ensure we teach our specialism instead of picking up the

slack elsewhere. Do you have topics you'd like to deliver around this that still fit with your ideal sequence?

- **Where do you slot the region and/or country studies in?** What content needs to be covered before you teach these to ensure you allow for more than a single story of a place?

THE CULMINATION OF A THREE-YEAR JOURNEY

After three years of hard work we now have a curriculum at KS3 that we are proud of. We are not complacent though. Every single curriculum meeting we have as geographers has a section of time devoted to discussing the planned curriculum and where it could be tweaked with new information, made better with different lessons or even topics that we just feel don't hit the mark as we intended. The curriculum map will no doubt be different when you read this now than when I wrote this sentence. For example, I'm practising what I've preached and for next year we are amalgamating rivers and coasts into one physical landscapes topic to avoid repetition and, in doing so, are creating space for a topic on the Middle East, which we've finally been able to plan into a coherent topic fitting with our aims. In the last three years we have seen a number of metrics move in the right direction too. The number of students opting for geography at KS4 is now the highest it has ever been, with over 160 students now studying geography in Year 10. We are also seeing an increase in numbers studying geography at A-level owing to the fact that not only are there simply more students in the pipeline, but also enjoying geography as a result of the experience we have given them. Enjoyment isn't everything though. They need to achieve as well. Building the foundations of an excellent geography curriculum in KS3 has seen results improve year on year to the point where my department and I are now being congratulated on our results each September. This is all because of a shared vision filtering through to excellently planned and sequenced lessons that focus on building knowledge and skills progressively over time, but with a focus on the geography, not on the exams. Student voice backs this up too. As if numbers opting at KS4 isn't enough to say students are enjoying the experience, when students are asked about their experience in geography they speak highly of their learning and the progress they make. However, it's not all been plain sailing. Changing a curriculum is tough on those who are in the years of transition. Though numbers are high now, the first year of the new curriculum saw numbers opting for GCSE drop to 65, which was tough to take. In hindsight, the changes made were too much too quickly for year groups that had started out their geography journey on the previous curriculum. The majority couldn't see success in the new curriculum yet and therefore didn't enjoy the learning and couldn't, at that point, see themselves making progress in geography. The SLT at the school were supportive of the changes we were making and could see the 'green shoots' emerging from these changes, but I imagine they must have been concerned by this initial drop in numbers.

THE FRUITS OF AN INTERWOVEN CURRICULUM

The best example I can give of an interwoven curriculum having the desired impact is our Year 9 topic on Haiti (see Figure 3.2). This topic requires knowledge of topics studied in Year 7 (globalisation and population), Year 8 (development and DR Congo) and Year 9 (hazards) in order to answer the big question 'Is Haiti a victim of its geography?'. The final lesson in the topic looks at HIV/AIDS in Haiti. The lesson starts by looking at the distribution of HIV/AIDS globally, a skill our Year 9 geographers will be well versed in. After this, they read a short, guided reading piece that links the decolonisation of the DR Congo with that of Haiti, globalisation and development to suggest why Haiti is an anomalous country in the Western Hemisphere. The drawing together of discrete pieces of knowledge about processes and places previously studied not only to answer the question about HIV/AIDS being an anomalously high prevalence disease in Haiti, but also considering whether Haiti is a victim of its geography, be it human or physical or both, is the product of a deeply considered curriculum.

Year: 9
Topic: Is Haiti a victim of its geography?

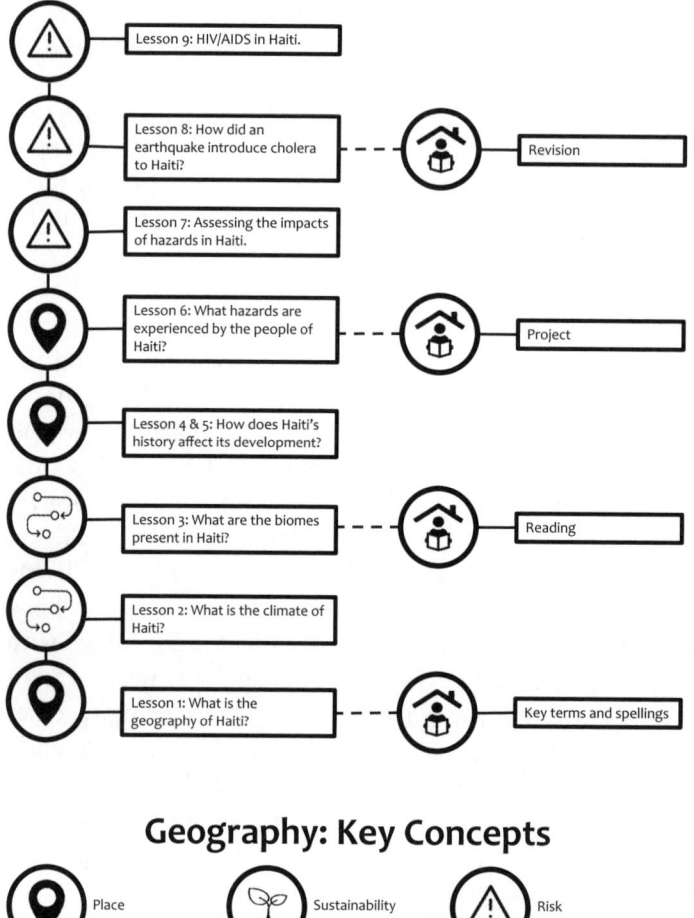

Figure 3.2 Year 9, Topic 3 'Is Haiti a victim of its geography?'

> **Five reflection questions**
> 1. Does your vision align with the curriculum you have in place?
> 2. Do your concepts allow for powerful geographical knowledge to be developed?
> 3. Do the topics you've selected for your curriculum allow for knowledge to be built upon as you progress through it?
> 4. What do your staff think of your curriculum? Do they feel like they have ownership of it?
> 5. What do your students think of your curriculum? Do they feel like they make progress through it?

> **DIGGING DEEPER – THREE RESOURCES TO DELVE FURTHER**
> - Sehgal Cuthbert, A. & Standish, A. (2021) *What Should Schools Teach? Disciplines, Subjects and the Pursuit of Truth (Knowledge and the Curriculum)*. UCL Press.
> - Bustin, R. (2019) *Geography Education's Potential and the Capability Approach: GeoCapabilities and schools.* Palgrave Macmillan.
> - Enser, M. (2021) *Powerful Geography: A curriculum with purpose in practice.* Crown House Publishing.

REFERENCES

Brooks, C. (2017) 'Understanding conceptual development in school geography' in Jones, M. & Lambert, D. (2017) (eds) *Debates in Geography Education (Debates in Subject Teaching)*. Abingdon: Routledge.

Enser, M. (2018) How to take a thematic approach to the curriculum. *TES*, 30 November 2018. https://www.tes.com/magazine/archive/how-take-thematic-approach-curriculum (Accessed: 28/05/2025)

Enser, M. (2021) *Powerful Geography: A curriculum with purpose in practice.* Carmarthen: Crown House Publishing.

Fogele, J. (2016) From content to concept. Teaching global issues with geographical principles. *European Journal of Geography*, 7(1), pp.6–16.

Ofsted (2023) 'Getting our bearings' in *Geography Subject Report*. https://www.gov.uk/government/publications/subject-report-series-geography/getting-our-bearings-geography-subject-report#pedagogy (Accessed: 28/05/2025)

Standish, A. (2021) 'Geography' in Sehgal Cuthbert, A. & Standish, A. (2021) *What Should Schools Teach? Disciplines, Subjects and the Pursuit of Truth (Knowledge and the Curriculum)*. UCL Press. https://discovery.ucl.ac.uk/id/eprint/10118307/ (Accessed: 01/07/2025)

4. LESSON PLANNING IN GEOGRAPHY

BEN NEWBORN
@MRBNEWBORN

'Lesson planning is not the same as lesson preparation ... . Lesson preparation is critical whether you have written your own lesson or are using one developed for you.' (Lemov, 2019)

SOME CAVEATS BEFORE WE START EXPLORING GEOGRAPHY LESSON PLANNING

There is no such thing as a 'perfect' lesson. The planning and preparation for a single lesson could take any amount of time you have to sink into it. There will always be more potential tweaks to be made, adaptations for certain individuals or subgroups to add challenge to or scaffold activities that are likely to make the learning better. It is a cliché, but when it comes to lesson planning, particularly in a subject like geography, the mantra 'good is good enough' has to be adhered to less you want to run the risk of burning out early on in your career.

A second important point behind lesson planning is there is never a good enough argument that you *must* plan your lessons from scratch. I often think about the number of teachers across the country at any given point googling the same diagram of longshore drift or perfect photograph that illustrates interlocking spurs (note: it doesn't exist!) and it frustrates me to think how much of that planning time could be better spent on preparing an amended worksheet or tweaking an activity to better suit the class who needs more scaffolding or challenge to work at the best of their ability. The majority of departments these days have shared materials on their local system. Where they don't, there are many places online that can be a starting point for lesson planning, paid or free, available to us these days. If all else fails, a plea for a specific lesson or unit on any of the social media sites will often yield a menu of kind offers. So, when it comes to planning, you shouldn't feel you need to start from scratch every time.

WHAT DOES THE AVERAGE GEOGRAPHY LESSON LOOK LIKE?

I spend a lot of time visiting schools across the country, dropping into many different geography lessons; and the nature of our subject in particular seems to lead to some consistent challenges:

- The complexity of the content we teach can lead to wide learning gaps between high and low prior-attaining students. This is exacerbated as most geography classes are mixed ability.
- The point above can lead to middle portions of lessons where pace is at high risk of dropping off as teachers try to maintain positive progress with difficult content across a huge range of abilities in the room.
- Timing can get away from us in lessons. This can lead to the last task(s) in the lesson being narrowed, pushed into subsequent lessons, or even scrapped completely. If this happens, students never get to apply the knowledge and understanding they have gained in the lesson, and this runs the risk of the knowledge never becoming embedded in their long-term memory.

Trying to pin down the 'average' geography lesson is an impossible task. An individual lesson can range from 35 minutes to 2 hours. I have seen classes ranging from 6 up to 60 students. It is very common to find students on track for a GCSE level 7 or above in the same class as, and often sitting next to, those aiming to secure a grade 1–3.

I am a firm advocate of Doug Lemov's idea of planning versus preparation (Lemov, 2021). The wide array of sources of baseline lesson planning materials we have available to us now means we can prioritise time to amending and tweaking these to suit the needs of individual groups and students rather than what is often the busy work of finding images or formatting slide decks for lessons. Teacher autonomy is vital in tailoring materials to suit the needs of individual groups because the lesson for 8E on a Tuesday morning can look very different from that of 8B on a Tuesday afternoon. This chapter cannot substitute for experience, nor can it tell you the best amendments to get the most out of your actual classes with all their nuances, differences and needs – that preparation is on you.

However, there are some principles that emerge for me following years of teaching and leading in departments across England and Wales, and in my current role observing geography lessons across the country and having discussions with professionals ranging from PGCE interns through to seasoned veteran geography specialists. These principles have evolved over time into a sequential checklist for thinking about planning geography lessons. I find this helps thinking about where to start and where to go, to ensure the pitfalls outlined above are avoided and you are free to deliver a quality lesson that suits the needs of the learners in each particular class.

PRINCIPLES OF GEOGRAPHY LESSON PLANNING

STEP 1: WHAT'S THE POINT?

Before considering activities or searching out the perfect image or designing any slides or pages in a booklet, the first thing to do is figure out what you are teaching and why. You need to be crystal clear what you want all students to know and/or be able to do when they leave your classroom. With this focus, you can now judge your lessons as a success or a failure in concrete, unequivocal terms. While doing this, you need to look ahead and look backwards to see where your lesson fits in your students' geography experience (see chapter 3). A lesson evaluating the impact of groynes in a case study requires prior knowledge and understanding about different sea defences, longshore drift, swash and backwash, fetch, etc. Any misconceptions in this prior knowledge have the potential to derail your lesson, or at least lead to a drop in pace where you have to (necessarily and quite rightly) take a step back to secure students' understanding before moving on. In the same vein, looking forward is also key. You are setting the foundations for the next step in their geography understanding, so you must be clear what you are building towards next lesson/week/unit etc. in order to set a solid knowledge and skills base to build on.

STEP 2: HOW ARE THEY SHOWING THEY GET IT?

Once the why and the what are pinned down, the next step is to secure the final part of your lesson. More often than not, this should take the form of an independent writing task at the end of the lesson. Generally speaking, geography lessons should contain at least one protected 10–15 minutes where students are consolidating their understanding of what they have learned in the lesson and producing independent work that is a product of their own thinking. For exam classes, this is most likely going to be an exam question because ultimately the best way to explain the formation of a waterfall is to complete the task: 'Explain the formation of a waterfall.'

In lessons lacking pace, it is very common that the first thing to go is the final activity. This is problematic because if students do not have the opportunity to consolidate their understanding, they are less able to embed the new learning effectively in their long-term memory (Riches, 2022). This means it is important that we protect this time and prioritise this step as the second part of lesson planning.

STEP 3: HOW DO THEY LEARN ALL THE NEW INFORMATION? (HINT – IT'S YOU!)

So now we know what we want our students to leave the room knowing and being able to do, and we have set up the opportunity to show they can do it in a way that we can test to evaluate the effectiveness of the lesson, we just need to get them there – the teaching bit!

Although there are other structures and systems to build your planning around, I personally advocate for Rosenshine's Principles of Instruction (Rosenshine, 2012). At a fundamental level, you can view geography as a collection of concepts that need delivering precisely and discretely before students can embed them into their long-term memory. The most efficient and practical way of achieving learning therefore is for the teacher to communicate the information through direct instruction, before checking students' understanding in order to gauge whether the class and individuals within it are ready to move on to an independent task. Lessons therefore can become series of 'I ..., We ..., You ...' phases or, perhaps better, 'Teach, Check, Practice'. The best learning resource that students have is their teacher. You have the knowledge they need to understand, and you are best placed to communicate this.

However, direct instruction must be efficient and effective. An often overlooked but worthwhile planning task is to rehearse explanations and models to refine the delivery of particular concepts and ideas during direct instruction. The more convoluted and disjointed an explanation, the greater the risk of cognitive overload for students because they are struggling to take on the knowledge you want them to. Taking time to observe experienced teachers, discussing best practice, and rehearsing the delivery and drawing of fiddly diagrams on the board step-by-step can be a valuable element of the lesson planning process.

For this same reason, questioning should be used sparingly in the I-phase of your lesson. Students learning about the stages of the Demographic Transition Model need to hear you talk about each stage in turn. Hearing another student guessing what they think is happening next and why in this part of the lesson at best slows the pace of the lesson, but often can strain their working memory, leading to cognitive overload and/or misconceptions bedding in (Shibli and West, 2018). To repeat, you as the teacher are their best source for learning new content and if you want them to understand the often complex and varied ideas we tackle in geography, they need to hear it from you clearly and efficiently.

STEP 4: DO THEY GET IT? ARE THEY READY TO MOVE ON?

You have a clear sight of the finish line. You have planned and rehearsed your direct instruction. Now you need to plan how you are going to help students transition into their final task. This means checking what they know and don't know so that you can make the big call: do you go forward towards the You-phase or do you go back and re-teach the concepts they haven't learned yet?

There are many strategies to gather data about your class's understanding available to you. Cold call questioning, thumbs, purposeful circulation, use of mini-whiteboards and others. I would advise you find a relatively small collection that works for you and stick to them for consistency and impact as students will learn your systems. One of my personal favourite strategies for assessing understanding is using mini-whiteboards for students to write a word, letter, number or draw a simple picture on that they can hold up as a group together for me to check. This allows quick and easy assessment of their understanding to make the judgement on the spot whether I can move on to the next part of my lesson or whether there are elements of understanding that are still lacking. Should the latter be the case, it is straight back to the I-phase to come at things another way and repeat my explanation of the content with clarity and efficiency as described previously.

The best assessment for learning (AfL) in a lesson is semi-structured. You must be responsive to and reflective of the group in front of you as a priority. If you get a hunch that there is a problem from peering over some shoulders as you circulate around the room, check it with an AfL strategy! The likelihood is that your gut feeling is right and there is a problem somewhere; and you cannot let students go forward with a misconception. However, this phase of the lesson is semi-structured in that there are clear opportunities in your planning to design hinge questions you can ask that best check the understanding. At this point in your planning, you can create slides or activities that will help to assess more precisely what students need to successfully complete the task that is coming up. There is lots of guidance available (William, 2015) on what these questions should look like and when in lessons they are most impactful.

SUMMARY

At the start of this chapter, I made clear that pinning down the 'average' geography lesson is an impossible task. The steps above outline a sequence you can follow to make sure you are thinking hard about what your students should learn and how best they will secure that information into their long-term memory:

Step 1	What's the point?	What do you want them all to leave your classroom knowing and/or being able to do?
Step 2	How are they showing they get it?	What independent task will allow students to illustrate Step 1? Protect adequate time in your lesson for this.
Step 3	How do they learn all the new information?	How will you ensure you deliver the new information efficiently and effectively?
Step 4	Do they get it? Are they ready to move on?	What strategies will you use to assess their learning? What questions can you ask?

Five reflection questions

Think back over a lesson you taught recently or flip these questions to consider an upcoming lesson.

1. Did all students leave the room knowing what you wanted them to know?
2. Did you preserve enough time for quality independent practice?
3. If 'no', where in your lesson could you have been more efficient and effective in your direct instruction or checking for understanding?
4. Was your understanding of the content secure?
5. Are you confident that your explanation and modelling is efficient and effective?

DIGGING DEEPER – THREE RESOURCES TO DELVE FURTHER

- Lemov, D. (2021) *Teach Like a Champion Field Guide 3.0: A practical resource to make the 63 techniques your own*. Hoboken, NJ: Jossey-Bass. – See Chapter 2: Lesson preparation and Chapter 10: Procedures and routines.
- The Geographical Association's 'Planning for geographical enquiry', available at: https://geography.org.uk/ite/initial-teacher-education/geography-support-for-trainees-and-ects/learning-to-teach-secondary-geography/geography-subject-teaching-and-curriculum/geography-knowledge-concepts-and-skills/geographical-practice/geographical-enquiry-2/planning-for-geographical-enquiry/– really good grounding in lesson structures and potential different approaches.

> - Continuity Oak website, available at: https://continuityoak.org.uk/Lessons (Accessed: 28/05/2025) – a potential source of resources to support Step 3 in the planning process. This site contains free-to-access recorded lessons, and you can also download the presentation slides for an entire KS3 and KS4 curriculum. This gives you the ability to see an experienced geography specialist breaking down complex ideas and concepts for KS3 and KS4 geography lessons, and shows how they chose to deliver content efficiently and effectively to their classes.

REFERENCES

Lemov, D. (2021) *Teach Like a Champion Field Guide 3.0: A practical resource to make the 63 techniques your own.* Hoboken, NJ: Jossey-Bass.

Riches, A. (2022) Closing time: how to end your lessons. *SecEd*, 16 November 2022. https://www.sec-ed.co.uk/content/best-practice/closing-time-how-to-end-your-lessons/.

Rosenshine, B. (2012) Principles of instruction: research-based strategies that all teachers should know. *American Educator*, Spring 2012, pp.12–39.

Shibli, D., & West, R. (2018) Cognitive load theory and its application in the classroom. *Impact*, 22 February 2018. https://my.chartered.college/impact_article/cognitive-load-theory-and-its-application-in-the-classroom/ (Accessed: 28/05/2025)

William, D. (2015) Designing great hinge questions. *Educational Leadership: Journal of the Department of Supervision and Curriculum Development*, 73(1), pp.40–44.

5. ASSESSMENT AT KEY STAGE 3
NICOLA DOWLING
@NICDOWLING16

'Teachers need to reclaim assessment and put it to work as the servant of the curriculum and of pedagogy so that assessment works for learning.' (Evidence Based Education, 2018)

Assessment is a thorny issue but an integral part of curriculum planning, nonetheless. Changing external influences over time, along with an accountability culture, have led to lack of clarity around assessment practices in the years preceding external exams. Coupled with the differing expectations from senior leaders, teachers, students and parents of what 'data' should be collected and reported, assessment has been a contentious issue.

For ease and consistency, schools have historically adopted a homogenous 'whole-school assessment policy' approach, with fixed assessment windows and minimal consideration of individual subject nuances. Following the removal of the national curriculum levels in 2014 and the associated pseudo-science of sub-levels, schools sought to continue to 'prove' that progress was evident and many resorted to a flawed approach of using GCSE grades and flight paths. The introduction of the content-heavy 9–1 GCSEs only added to this confusion and pressure, and many students were faced with learning GCSE content in the KS3 years and answering exam questions as their assessment.

Ofsted's Education Inspection Framework in 2019 provided clarity for leaders; 'assessment is a means of evaluating whether learners are learning/have learned the intended curriculum. In doing so, the curriculum becomes the progression model' (Ofsted, 2019). There is now a widely shared understanding that learning is an alteration in long-term memory and thus progress means 'knowing more (including knowing how to do more) and remembering more'. The framework explicitly references that progress is not about hitting the next data point. Put simply, if learners experience a well-sequenced, well-constructed curriculum, and attain well, then they are making progress. Our aim is not to 'do things for Ofsted', as this just makes sense, echoed here by Jeannie Fulbright; 'If the purpose of learning is to score well on a test then we've lost sight of the real reason for learning'.

Despite this clarity, Ofsted (2023) report that some schools are still using GCSE grades at KS3. David Didau (2022) argues that 'making KS3 assessments similar to GCSE assessments is not only unnecessary, it's actively harmful'. Daniel Koretz (2018) echoes this and urges us to 'teach to the domain, not the test'. The test measures the domain, rather than being the point of the learning. Mark schemes were never created to indicate progression and should not be used for anything other than GCSE practice assessments at the appropriate age group. Applying arbitrary scores or grades to assessments therefore, is not a valid or reliable record of students' knowledge.

THE CURRICULUM AS THE PROGRESSION MODEL

Knowledge matters at KS3; it provides a breadth of understanding and serves as crucial building blocks to enable students to further develop their subject knowledge in the latter years of their secondary education. If we are to assess effectively, then we must first set out the specifics of what we want students to know. This enables us to determine whether these specifics have been mastered, and mastering these specifics means getting better at geography (progress). Assessment enables us to 'know what they know' (Didau, 2019) and determine that progress.

Curriculum	Assessment	Instruction
Where the student is going	Where the student is now	How to get them from here to there
What to expect students to know	How we know what they know	What we do when they aren't there yet

Table 5.1 Assessment enables us to 'know what they know'

Source: Didau (2019)

Many geography departments are engaged with ongoing curriculum development at KS3 (see chapter 3). The dynamic nature of our subjects requires us to do so. In our trust, we opted for a co-constructed approach to our KS3 curriculum, involving all geography teachers. As such, there is a common and secure understanding of curriculum and assessment. Each unit has an enquiry question, underpinned by specific disciplinary concepts to focus the geographical lens, refining the subject content from a potentially infinite and ever-changing subject discipline. Examples include, 'Can the World cope with 9 billion people?' (underpinned by the concepts of patterns and sustainability) and 'Where is the hardest place to live on Earth?' (systems and interactions). All of our units include our core concepts of place, space and scale. Due to the lack of consensus among the geography community over disciplinary concepts, we spent significant but valuable time debating our choices. We did reflect on the Geographical Association's (GA)(2022) take on organising concepts, however, we found that our choice of concepts was appropriate for our curriculum and 'doing the disciplinary job'. Using concepts in this way enables our students to explore content and concepts in different contexts over time, securing their knowledge of, for example, the complexity and richness of place, and the intricacies of processes and systems.

Christine Counsell (2018) urges us also to carefully consider curriculum sequencing, so that the content has 'a proximal function, enabling students to grasp X now so they can understand Y later, and an ultimate function, enabling the knowledge to be secured and lasting'. While our curriculum will always be undergoing refinement and development, spending valuable time on designing a well-constructed curriculum has been essential. It is through this curriculum that our students have the best chances of completing their KS3 education as knowledgeable and successful geographers, no matter whether they choose to continue the subject at GCSE or not.

ASPECTS OF ACHIEVEMENT IN GEOGRAPHY

The GA's recently updated Progression and Assessment Framework (2023) is an invaluable tool in the development of assessment. The framework exemplifies the three types of geographical knowledge to be assessed, referred to as 'aspects of achievement': contextual world knowledge, geographical understanding, and geographical enquiry and skills (Figure 5.1). This helps to bring a sense of purpose to learning geography and clarity over what will be assessed. When we applied this framework to our KS3 curriculum, it was evident that the third component (geographical enquiry and skills) had not been as carefully considered. This was echoed in the recent review series report by Ofsted (2023), which stated it is 'the area of weakness most often seen in curriculum design'. In recognition of this, we created a procedural knowledge tracker to establish when we explicitly teach and retrieve different elements from the 'enquiry and skills' component across KS3, enabling us then to assess the security of this knowledge at appropriate times.

The framework also outlines dimensions of progress to help consider what progress looks like. It is our task to embed these into our curriculum to ensure students make progress. Contextualising the framework's age-related expectations into 'curriculum related expectations' for each individual unit ensures appropriate pitch and provides the clarity required for assessment purposes. Each 'aspect of achievement' in the framework provides a clear thread through the statutory curriculum and onto GCSE and A-level qualifications and their associated assessment objectives (Figure 5.1).

> The GA has identified three aspects of achievement or 'big objectives' of teaching geography that thread through all stages of the school curriculum.
> The three aspects of achievement are:
> - Contextual world knowledge of locations, places and geographical features.
> - Understanding of the conditions, processes and interactions that explain features and distributions, patterns and changes over time and space.
> - Competence in geographical enquiry, the application of skills in observing, collecting, analysing, mapping and communicating geographical information.
>
> The five dimensions of progress are:
> - Demonstrating greater fluency with world knowledge by drawing on increasing breadth and depth of content and contexts.
> - Extending from the familiar and concrete to unfamiliar and abstract ideas.
> - Making greater sense of the world by organising and connecting information and ideas about people, places, processes and environments.
> - Working with more complex information about the world, including the relevance of people's attitudes, values and beliefs.
> - Increasing the range and accuracy of investigative skills, and advancing their ability to select and apply these with increasing independence to geographical enquiry.

Figure 5.1 The three aspects of achievement and the five dimensions of progress in geography (© Geographical Association, 2023, reproduced with permission of the Licensor through PLSclear)

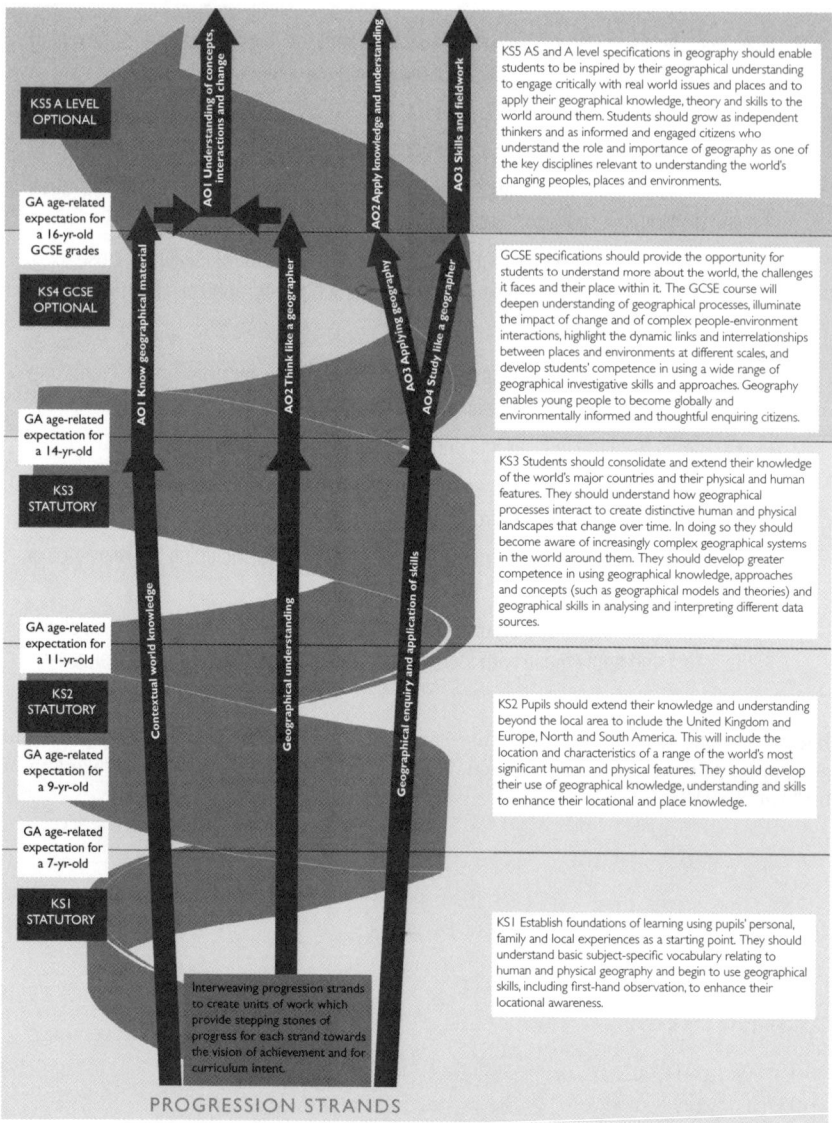

Figure 5.2 GA's progress strands (© Geographical Association, 2023, reproduced with permission of the Licensor through PLSclear)

A MIXED CONSTITUTION OF ASSESSMENT

Teachers and leaders typically speak of 'formative' and 'summative' assessment. The former enables us to check component knowledge, leading to responsive teaching, and the latter enables us to check composite knowledge and informs longer-term curriculum implications. Christodoulou (2017) reminds us that becoming secure in the building blocks does not necessarily mirror becoming secure in the final performance, and thus formative assessment should not merely be a practice of the summative assessment. Hence, using GCSE questions at KS3 is an invalid approach to KS3 assessment. Assessing building blocks as we go along requires multiple forms of assessment and enables us to test knowledge of the three diverse 'aspects of achievement' in geography. Using a 'mixed constitution' of assessment (both formatively and summatively) encourages us to make more careful inferences about students' progress through the curriculum and allows us to triangulate these inferences with information gleaned from different methods. Healy (2020) helpfully suggests a range of formative assessment approaches for geography in her article for *Teaching Geography*.

There are numerous approaches to formative and summative assessment at KS3 which can be powerful tools. The *Four Pillars of Assessment* report from Evidence Based Education (2018) suggests this will only be the case when assessments are consciously planned with purpose, reliability, validity and value in mind. In our medium-term plans, we refer to 'assessment opportunities' as suggestions of the types of assessment which could be used, taking the four pillars into account. Here, we explore some of the approaches we use, some of which are in their infancy and being refined, however, evidence of positive impact is emerging.

FORMATIVE ASSESSMENT

Many schools adopt a low-stakes retrieval practice approach to the start of lessons, which can be both responsive and strategic and should be carefully planned. Christine Counsell (2023) suggests that 'if the curriculum content is coherent then opportunities for retrieval practice are extensive and natural' (see chapter 19). Using a selection of questions requiring knowledge from short- and long-term memory to be retrieved is considered good practice. An often-forgotten element of retrieval practice is procedural knowledge. Previously, we recognised that there were not enough opportunities to practise this knowledge at KS3 when students were less secure at interpreting figures by the time they got to GCSE mocks. Including opportunities to interpret photos/graphs/diagrams and 'geographical enquiry' have become common features of our retrieval and are as crucial as the substantive 'contextual world knowledge'.

Increasingly, multiple-choice quizzes (MCQs) are being realised as a powerful tool in assessment. Writing a good MCQ, however, takes time. It requires inclusion of plausible distractors, multiple correct answers, non-examples and common misconceptions in order to assess students' residual knowledge. Done well, MCQs highlight precise gaps in students' knowledge to inform responsive teaching. We find that using technology to set quizzes, giving students automatic feedback and providing teachers with the information to help plan subsequent lessons, increases the efficiency and value of these.

Map tests have become a firm favourite of our students. Challenges from the beginning of Year 7 to learn specific countries, cities, rivers, oceans, etc. linked to the curriculum has significantly developed students' locational knowledge. Many students talked with confidence of their knowledge of flags (thanks FIFA) but now they know where these places are! Varying scales and projections of maps have also developed our students' ability to see beyond the Eurocentric Mercator World map and appreciate the complexity of the concept of space.

Schema maps are used to assess concepts in a formative or summative way in both our primary and secondary geography classrooms. Students have opportunities to explain their emerging knowledge of concepts in relation to a specific context through building their schema map for a specific unit. Over time, as they encounter familiar concepts in new contexts, their knowledge builds. At the end of the year or key stage, the schema maps reflect students' knowledge of the key concepts across a range of different contexts and can be assessed summatively. Answers can be written as annotations or provided verbally and added to the map by a teaching assistant or teacher.

SUMMATIVE ASSESSMENT

Recently, we have given more attention to our approach to KS3 summative assessments. It had been commonplace to assess at the end of a unit and once feedback had been provided, it was rarely returned to. When planning summative assessment, Chiles (2020) urges us to adopt a 'cumulative' approach, sampling from an ever-increasing knowledge domain. The spiral nature of a geography curriculum requires students to have opportunities to develop and revisit knowledge, understanding and skills, in different contexts over time. Assessment must then enable students to recall this wide-ranging knowledge over time, meaning cumulative assessment is key.

Adopting a cumulative approach, we assess the aspects of geographical knowledge we have given weight to in our curriculum, and reinforce the value assigned to them. We use a mixed-constitution approach including oracy, MCQs, short- and longer answer questions, and deliberately include questions to assess procedural knowledge (Figure 5.3). In our experience thus far, this approach has increased participation ratio and attainment in summative assessments. Ofsted's (2023) recent geography subject review found that where practice was strongest, summative assessments 'had a mix of shorter questions to check that component knowledge had been learned and then longer tasks in which students needed to apply this knowledge, often to novel situations'.

Our starting point in summative assessment provides an opportunity for oracy (see chapter 13). It is undeniable that oracy has a significant influence on writing, so in order for students to successfully 'write like a geographer', they require opportunities to 'think, read and speak like a geographer' first. Students begin by discussing specific key terms related to the content, in a low-stakes environment, thus 'unlocking' their knowledge. The Oracy All-Party Parliamentary Group 'Speak for Change' report (2021) reports that oracy often leads to writing with greater depth than would have been achieved without this preparatory work and is particularly important for students from low-income and disadvantaged backgrounds. Initial evidence from our assessments confirms this, alongside an increased participation ratio from boys and students with SEND. It sets students up for success with a low-stakes opportunity to retrieve knowledge, and in our experience, also reframes students' perspectives on assessments.

Next, we include multiple-choice questions which have been carefully constructed with all of the aforementioned considerations, and short-answer questions. Crucially, we also include opportunities to use and apply procedural knowledge. Sometimes this will take the form of fieldwork questions and other times it may be map/graph/diagram interpretation or mathematical calculations; all essential requirements of knowing how to 'work like a geographer'.

Throughout our curriculum, but also in summative assessments, we provide opportunities for students to develop their ability to 'write like a geographer' with an open question (see chapter 15). This is usually from the content in the most recent unit but students are also able to draw upon all knowledge secured to date. Using a 'how far do you agree' or 'to what extent' formula is recommended by Fordham (2014) as students find it easier to argue if they have something to argue for or against. Similarly, a decision-making exercise provides the opportunity to explore issues, without one correct answer. We do not use a scoring system for longer answers but instead use our 'curriculum related expectations' to inform assessment criteria which can be shared with students. Students are provided with formative feedback. We have found that this approach removes the ceiling of the knowledge that is drawn upon and applied, and encourages students to think more synoptically.

At the end of each year at KS3, we also have a synoptic, place-based unit; drawing upon a range of themes, applied to a new context. Our end of year assessment opportunities enable our students to demonstrate their holistic understanding of the subject through all three components of geographical knowledge, or 'aspects of achievement', in a format similar to the cumulative assessment in Figure 5.3.

The assessment example included (Figure 5.3) includes information from both Year 7 and Year 8 curricula. While marks are not included here, it is possible of course to do so for sections 2 and 3, and even the short-answer questions in section 4, and we do so in some cases to obtain quantitative assessment data. It is essential however, that these marks are explicitly driven by the curriculum-related expectations and not a replication of a GCSE mark scheme. In our experience, the removal of marks or scores has improved the response from students to assessments in both attempting the assessment, and afterwards when acting on feedback. As with most things, our cumulative assessments are a work in progress but evidence to date suggests this approach to summative assessment is providing us with the knowledge required to move students' learning and attainment forwards in all three components ('aspects of achievement').

Year 8 assessment
1 hour
Section 1: Discuss with your partner and write your answer.

1. What do we mean by 'resource'?	
2. When we refer to the 'Middle East', what do we mean?	
3. What is climate change?	

Section 2: Multiple-choice questions: circle the correct answer(s).

4. Which continent has the largest population? a. Europe b. Asia c. Africa d. North America	8. When water moves downwards through the soil, it is called: a. surface run off b. percolation c. through flow d. groundwater
5. What does the term 'distribution' mean when referring to population? a. The process of delivering goods from the manufacturer to the consumer. b. How many people there are per square kilometre. c. The total number of people in a given area. d. The way people are spread out.	9. Glaciers erode the land in two ways. Select the correct <u>two</u>. a. Abrasion b. Plucking c. Attrition d. Corrie
6. Is a high ecological footprint positive or negative? a. Positive b. Negative c. Neutral d. Depends on the country	10. What structure do geographers use to describe spatial and temporal patterns? a. CLOCC b. PEE c. TEA
7. Why does Dartmoor act as a carbon sink? a. Because it has a large number of trees that absorb carbon dioxide. b. Because it has a high altitude which allows it to absorb more carbon dioxide. c. Because it has lots of water. d. Because it has peat bogs that absorb carbon dioxide from the atmosphere.	11. What is the concept that refers to meeting the needs of today, without compromising the needs of future generations to meet their needs? a. Systems b. Patterns c. Sustainability d. Interactions e. Perspectives and values f. Processes

Section 3: Procedural knowledge

12. The diagrams below show the sequence of a formation of a waterfall. Complete the boxes with the time-sequencing words 'firstly, next, then, finally' to form the correct written sequence below.

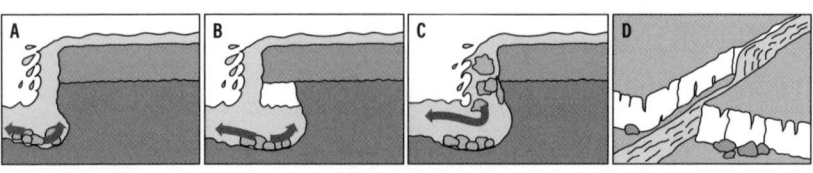

	The softer rock undercuts the harder rock above, due to continued erosion from the flow of the water, and leaves an overhang.
	Water flows over hard and soft rock. The softer rock erodes (abrasion and hydraulic action) more easily, leading to the beginning of an undercut.
	Erosion continues and the waterfall slowly retreats its way upstream, eventually leaving a gorge behind.
	The hard rock overhang collapses into the plunge pool and the load is eroded by attrition. It is then washed away through transportation processes such as traction and saltation.

13. Complete the climate graph below for Syria (November and December are missing). Data is provided.

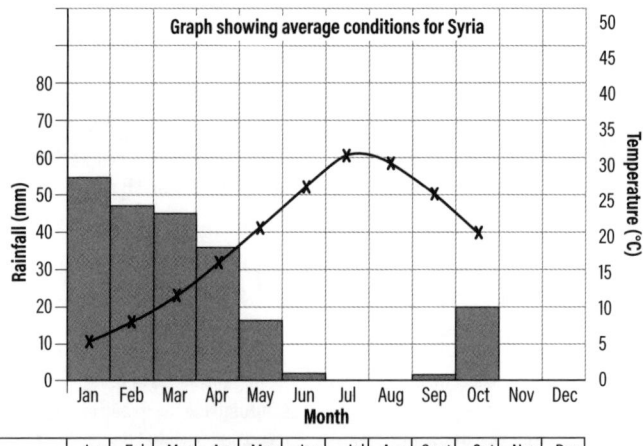

	Jan	Feb	Mar	Apr	May	Jun	Jul	Aug	Sept	Oct	Nov	Dec
Mean temp (°C)	6.1	7.9	11.5	16.4	21.6	26.2	30.2	28.7	25.4	19.7	12.8	7.7
Precipitation (mm)	55	47	46	36	16	2	0	0	1	20	32	50

14. Calculate the temperature range for Syria. ______________

15. Calculate the mean/average rainfall for Syria. ______________

16. Using the figure below, apply your knowledge to describe the distribution of resources within the Middle East. Think about how geographers describe locations and patterns.

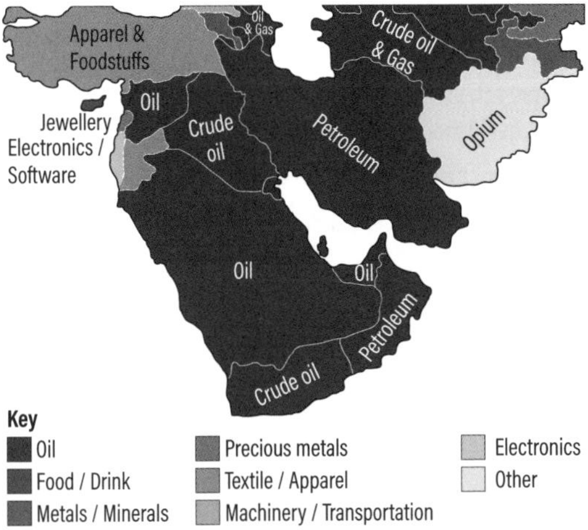

Apparel & Foodstuffs
Oil & Gas
Crude oil & Gas
Oil
Jewellery
Electronics /
Software
Crude oil
Petroleum
Opium
Oil
Oil
Petroleum
Crude oil
Key
Oil
Food / Drink
Metals / Minerals
Precious metals
Textile / Apparel
Machinery / Transportation
Electronics
Other

____
____
____
____
____
____
____
____
____
____
____
____

Section 4: Write like a geographer

17. What does 'water scarcity' mean?

____

18. Explain the difference between climate change and global warming.

____
____
____
____
____

19. Using your knowledge, suggest why parts of the Middle East have a hot, dry desert climate.

____
____
____

> 20. Name one global biome and discuss how it may be under threat from climate change.
> _____
> _____
> _____
> _____
>
> 21. Sir David Attenborough is calling for urgent action to tackle human caused climate change, which he describes as the 'biggest threat to this planet in thousands of years'.
> To what extent do you agree that climate change is causing a climate emergency?
> Discuss this statement using a range of evidence to include different stakeholders' viewpoints. End with a conclusion to summarise your opinion.
> Write your plan and answer on lined paper.

Figure 5.3 KS3 Cumulative assessment – Year 8 example

THE FORMATIVE USE OF SUMMATIVE ASSESSMENT

Summative assessments are more than something to mark the end of a period of study. Assessment information should be used to inform not only where students' knowledge is less secure but, more importantly, how teachers can respond to deepen their understanding. Reassuringly, the recent Ofsted (2023) report confirms that 'increasingly, schools are moving away from attempting to use assessment data to provide a grade or to say whether students were on track to meet a certain grade'. We do, however, teach in a system where external pressures around 'data' remain and in many cases, leaders in schools require data to be submitted at specific 'assessment points'. It is of course our responsibility to know whether students' knowledge is secure and understand how they can improve but often we are asked to quantity this, which can be challenging if we are to avoid arbitrary scores and mark schemes.

Our approach of a mixed-constitution of assessment enables us to build a holistic understanding of the security of students' knowledge over time, and not run the risk of relying too much on summative assessment data. We use a mix of questions, some of which can be scored, to provide the numerical data required by leaders but, more importantly, we also apply curriculum-related expectations to use assessment information in a more formative way. This can also work for short-answer questions where our curriculum-related expectation is that students can use a 'trend, example, anomaly' (TEA) approach to describe a graph or distribution. If necessary, marks can be allocated to this but, more importantly, we have decided what the crucial procedural knowledge is first, and our assessment allows us to check how secure students are in doing this. Ultimately then, it is by using a mixed constitution of assessment that we can gather a range of information and potential data to ascertain students' progress and inform our next steps. What happens with this assessment information should take precedence over any numerical scores.

Tomsett (2021) writes that 'teaching which is responsive to what the assessments tell you has or hasn't been learned is an essential aspect of the enacted curriculum'. As such, following assessment taking place, we identify short- and long-term curriculum and pedagogical implications/adaptations. If students do not demonstrate secure knowledge in an assessment, it is of utmost importance to spend time looking back. In the short term, we expect teachers to use assessment information to inform curriculum adaptations for the next lesson, next week, and the next unit of work. Longer term, we consider which adaptations we need to make to the curriculum for the next time we teach this and for future cohorts of students. Ultimately, we use assessment information to ensure that the curriculum is enabling knowledge to be secured. One of the ways in which we do this is through a teacher assessment log. This enables us to identify common misconceptions, aspects of less secure knowledge and also identify where any of our students from key groups, including those who are disadvantaged and with SEND, require specific and potentially additional attention. These logs are discussed across the department to identify any whole-cohort trends and subsequent curriculum adaptations required. As with our assessment model, this format is a work in progress, but it encourages us to make appropriate formative inferences from assessments.

Geography Assessment Log Sheet			
Teacher:	Assessment:		
Area of strength:			
DIRT/IACT activities:	So that ... (short-term curriculum adaptation)		
Disadvantaged:	SEND:	HPA:	Curriculum adaptation (long-term improvements):

Figure 5.4 Example assessment log sheet

What is clear is that assessment is complex. However, if we can keep in mind the crucial subject nuances, ensure assessment considerations are an integral part of curriculum design, and use assessment information in a formative way, then we serve our students well.

> ## Five reflection questions
>
> 1. Do you have absolute clarity over what you expect the students to know, and where they are going (the content choice and sequencing of your curriculum), so that you know what you are assessing, why and when?
> 2. How can you adopt a 'mixed-constitution' of assessment approach so that you obtain the best diagnostic information about the three prerequisites of knowledge ('aspects of achievement') we want our students to develop?
> 3. How can curriculum-related expectations help you to avoid the pitfalls of mini GCSE-style exams with arbitrary scores and associated mark schemes at KS3?
> 4. Are assessments used to diagnose issues at both an individual student level and at a cohort or class level?
> 5. Are assessments being used to inform judicious adaptations to the curriculum?

> **DIGGING DEEPER - THREE RESOURCES TO DELVE FURTHER**
> - Geographical Association (2023) *An Assessment and Progression Framework for Geography.* Geographical Association.
> - Gardner, D. (2021) *Planning your coherent 11-16 Geography Curriculum: a design toolkit.* Geographical Association.
> - Hayes, R. (2023) How do you solve a problem like a cross-trust, cumulative 50 minute KS3 History summative assessment? Sometimesiteachgood, 13 February 2023. Available at: https://sometimesiteachgood.wordpress.com (Accessed: 28/05/2025)

REFERENCES

Chiles, M. (2020) *The Craft of Assessment.* Woodbridge: John Catt Educational Ltd.

Christodoulou, D. (2017) *Making Good Progress: The future of assessment for learning.* Oxford: Oxford University Press.

Counsell, C. (2018). Blog. Senior Curriculum Leadership 1: The indirect manifestation of knowledge: (A) curriculum as narrative, The Dignity of the Thing. https://thedignityofthethingblog.wordpress.com/author/christinecounsell (Accessed: 28/05/2025)

Counsell, C. (2023) Implementing Opening Worlds – synoptic tasks and the assessment big picture. Delivered to Westcountry Schools Trust, 5 December 2023

Didau, D. (2019) How do we know students are making progress? Part 3: Assessment. https://learningspy.co.uk/assessment (Accessed: 28/05/2025)

Didau, D. (2022) Assessing English at KS3. https://learningspy.co.uk/assessment (Accessed: 28/05/2025)

Evidence Based Education (2018) The Four pillars of assessment: A resource guide. https://evidencebased.education (Accessed: 28/05/2025)

Fordham, M. (2014) as cited in Hayes, R. (2023*)* How do you solve a problem like a cross-trust, cumulative 50 minute KS3 History summative assessment?. Sometimesiteachgood, 13 February 2023. https://sometimesiteachgood.wordpress.com (Accessed: 28/05/2025)

Fulbright, J. (n.d.) https://www.jeanniefulbright.com/blog/incorporate-narration-homeschool

Geographical Association (2022) *A Framework for the School Geography Curriculum*. Geographical Association.

Geographical Association (2023) *Guidance on Assessment and Progression for Geography*. Geographical Association.

Healy, G. (2020) Placing the geography curriculum at the heart of assessment practice. *Teaching Geography*, 45(1), pp.30–33.

Koretz, D. (2018) *Measuring Up: What educational testing really tells us.* Cambridge, Mass.: Harvard University Press.

Ofsted (2019) *The Education Inspection Framework*. London: Ofsted.

Ofsted (2023) *Subject Report Series: Geography*. London: Ofsted.

Oracy All-Party Parliamentary Group (2021) Speak for Change. https://www.education-uk.org/documents/pdfs/2021-appg-oracy.pdf (Accessed: 28/05/2025)

Tomsett, J. (2021) This much I know about … Seven starting points for curriculum conversations between subject and senior leaders, 19 January 2021. https://johntomsett.com (Accessed: 28/05/2025)

6. PURPOSEFUL FEEDBACK
HELEN PIPE
@HELENPIPE_1

'Take a moment to consider the feedback that you have provided recently to students. How much of that feedback was heard? Did it make the 'sound' that you intended it to make?' (Chiles, 2021)

WHAT IS FEEDBACK?
In recent years there has been a shift from marking to giving feedback. Feedback is providing guidance to support students in improving their knowledge, understanding and application of threshold concepts. This encourages metacognition, as students are required to think about how they progress with their work, as opposed to just reading comments or marks. The Education Endowment Foundation (EEF) states that feedback has a very high impact at a low cost for improving student outcomes. Previously, teachers have spent hours marking each piece of work, writing the same comments for students to read, but not act upon. Providing feedback rather than marking ensures that students make improvements to their work; this could be adding to, making changes to or redrafting a section of a piece of work. By students acting on the feedback, the learning process continues, helping move knowledge from the student's working memory to their long-term memory where a lasting change has been made.

This chapter will discuss a variety of ways that feedback can be given effectively, at the same time as reducing the workload of teachers.

DIFFERENT TYPES OF FEEDBACK

KNOWLEDGE RECALL

Feedback is seen to be most impactful when given immediately. This can easily be achieved at the start of lessons with a knowledge recall task in the form of a 'Do Now'. The teacher can strategically ask questions which address misconceptions, tricky concepts or interleave prior knowledge from previous topics. In geography, retrieval tasks are inclined to revolve around specialist (sometimes known as tier 3) vocabulary, facts and figures or stating examples. While the students are answering the questions, the teacher can circulate to check for common errors. If most students are answering a particular question incorrectly, the teacher can then decide whether to reteach this item before moving on or to make a note for future planning. Students are then given feedback in which they check and correct their answers.

WHOLE-CLASS FEEDBACK

An example of whole-class feedback could follow a lesson where students have completed a piece of work evaluating the effects of an earthquake in a particular country. Throughout the lesson, the teacher would circulate to address misconceptions and adherence to the command word. At the end of the lesson, the teacher could then select a stratified sample of the exercise books and complete a whole-class feedback sheet, recording the following elements to share with the class in their next lesson:

1. Spellings – focusing on specialist vocabulary, e.g. epicentre, and common spelling errors, e.g. businesses.
2. Misconceptions.
3. A series of next steps.

MESSY MARKING

'Messy marking' is the concept of identifying misconceptions and next steps which drive student improvement, in a similar way to whole class feedback but live in the lesson, so immediate feedback is given. This strategy of assessing student misconceptions was introduced to my school's senior leadership team by Dixons City Academy, Bradford.

When planning the lesson, teachers need to decide if they will be fishing or hunting for students' next steps in learning. Fishing is unplanned and happens ad hoc, whereas hunting is carefully planned and targeted against core knowledge outcomes (Figure 6.1).

While students are completing an independent practice task, such as a written piece of work, the teacher circulates making a note of common misconceptions, SPAG errors and next steps. The teacher then decides whether to stop the class to address the misconceptions or select a piece of work and give live feedback. The live feedback consists of using a series of success criteria to annotate a piece of student work, facilitated by using a visualiser. Students then check their own work against these success criteria, make any necessary changes and then continue to add to their work in response to the live model.

LIVE FEEDBACK

This involves students receiving feedback on a piece of work during the lesson. This could be achieved through the use of 'Show Call', a 'Teach Like A Champion' technique (Lemov, 2021), whereby the teacher selects a student response to share with the class. A recommendation would be to have a consistent set of marking codes (as shown in Table 6.1) shared across the geography department. This helps with students moving classes or having a new teacher the following year. When the work is being projected using a visualiser, the teacher can narrate and annotate the strengths of the answer and suggest improvements.

Table 6.1 Marking codes

CS	Case Study
F	Facts
P	Point
Adv+	Advantages
C	Conclusion
J	Justification

With GCSE students, another strong recommendation is to encourage the use of the Cornell method of note taking which involves students presenting work in a structured way:
1. A double margin is created.
2. Five lines are ruled at the bottom for a summary of the lesson.
3. The remaining area is where the student makes their notes, completes independent practice, answers exam questions.

The double margin is the most important section for the purpose of feedback. During the lesson, students can record verbal feedback, e.g. use connectives, or advice, e.g. arrows show direction and movement. During live feedback, students can underline/highlight in their writing where they have met specific success criteria and then note down in the double margin which criteria have been achieved where. This is useful as students do not need to squeeze the annotations into their original writing and can visibly show where they have met the criteria and why.

EXAM FEEDBACK

When students have completed an assessment, instead of delivering a lesson in which the teacher goes through each question and gives the correct answer, try including 'twist it' questions which give students the opportunity to reapply their knowledge. For example, if most students successfully answered the question 'Explain two effects of a named earthquake in a developing country', an example could be shared and students then reapply their knowledge to the question 'Explain two responses to a named earthquake in a developing country'. Students who didn't complete the initial question well enough to attain full marks still have the opportunity to improve their response, but also to reapply their knowledge, while those who achieved full marks also have another chance to apply their learning.

KNOW - HOW - SHOW

A feedback strategy which has been introduced to me this academic year is Know – How – Show. Students complete a piece of extended writing or answer an exam question. Then, the teacher uses a marking grid such as the one in Figure 6.2 to:

- highlight what students 'know'
- highlight 'how' students can improve their answer
- give students an opportunity to 'show' they have understood the feedback by redrafting their answers using the advice identified in the 'how' section.

As a result, students are able to articulate what they need to do to improve their knowledge, understanding and application of key geographical concepts and are able to act on the feedback. The teacher can also identify any misconceptions and make a decision as to whether these can be addressed at a later date or if there is an element of reteaching necessary in the current or following lesson.

Discuss the arguments for and against the expanding of UK airports. (6 marks)	
KNOW:	**HOW:**
Mentioned an example of a London airport	One example is ... Heathrow Airport, Gatwick Airport, Stanstead Airport
Given reasons **for** London airports expanding	For example: • London airports are operating at full capacity. • Limited opportunities for more economic growth. • Creating vital global links. • Boost to economic growth
Begun to explain the reasons **for** expansion	Develop your explanation. Use phrases such as: consequently, as a result, this means that
Given detailed explanation of how expansion will **benefit** locally, regionally, nationally, globally	Link back to the question using the words you underlined when BUGing the question, e.g. this will benefit nationally because ...
Given reasons **against** London airports expanding	For example: • High cost of potential projects • Local people living nearby concerned about noise • Pollution and increased emissions from airplanes
Begun to explain the reasons **against** expansion	Develop your explanation. Use phrases such as: consequently, as a result, this means that
Given detailed explanation of how expansion will be **a disadvantage** locally, regionally, nationally, globally	Link back to the question using the words you underlined when BUGing the question, e.g. this will be a problem locally because ...
Given an evidence-based conclusion to answer the question overall	Conclude your argument to answer the question: Overall, the arguments for/against expanding London's airports outweigh the arguments for/agains because ...

Figure 6.1 Know – How – Show sheet for question: Discuss the arguments for and against the expanding of UK airports. (6 marks)

Note: Thank you to Barr's Hill School, Coventry and The Futures Trust for introducing me to Know–How–Show and to Dixons City Academy, Bradford for introducing me to messy marking.

EXAM WRAPPERS

Exam wrappers are used following an assessment, encouraging students to reflect on their preparation for the assessment. These ask students to record their score, identifying areas of focus and evaluate their preparation technique in order to self-regulate and improve on their technique for subsequent assessments. It is useful to provide success criteria for the students:

- On a five-point scale – self-evaluate revision.
- What are your strengths? For example, an element of the course, geographical skills, case study knowledge.
- What do you need to improve? Identify questions where you did not achieve full marks – what can you do to close these knowledge gaps?
- Identify areas of the course which you need to revisit. How can these be recapped – homework follow-up, self-quizzing?

CONCLUSION

Purposeful feedback aims to support students' learning while also contributing to teachers having a manageable workload. It is low cost and can make a significant contribution to closing the gap between different groups of students. Try these ideas in your context to see which work for you, sharing your experiences with your colleagues to widen the benefit. Make sure your feedback is being heard and that it makes the right 'sound'.

Five reflection questions

1. Are you fishing or hunting?
2. What will your marking codes look like?
3. How can students be encouraged to reflect on their assessments?
4. How can you collaborate with others to produce model answers?
5. How can you develop consistency across subject/faculty teachers which will support with giving feedback to students?

> **DIGGING DEEPER – THREE RESOURCES TO DELVE FURTHER**
> - Chiles, M. (2021) *The Feedback Pendulum, A Manifesto for Enhancing Feedback in Education.* Woodbridge: John Catt Educational Ltd.
> - Enser, M. (2019) *Making Every Geography Lesson Count: six principles to support great geography teaching.* Carmarthen: Crown House Publishing.
> - Education Endowment Foundation. Feedback. Available at: https://educationendowmentfoundation.org.uk/education-evidence/teaching-learning-toolkit/feedback (Accessed: 28/05/2025)

REFERENCES

Chiles, M. (2021) The key principles to effective feedback. *Impact*, 12 May 2021. https://my.chartered.college/impact_article/the-key-principles-to-effective-feedback/ (Accessed: 28/05/2025)

Education Endowment Foundation, Teaching and Learning Toolkit: 'Feedback', EEF. https://educationendowmentfoundation.org.uk/education-evidence/teaching-learning-toolkit/feedback (Accessed: 28/05/2025)

Jones, K. (2019) *Retrieval Practice: Research and resources for every classroom.* Woodbridge: John Catt Educational Ltd.

Lemov, D. (2021) *Teach Like a Champion Field Guide 3.0: A Practical Resource to Make the 63 Techniques Your Own.* Hoboken, NJ: Jossey-Bass.

7. QUESTIONING IN GEOGRAPHY
MICHAEL CHILES
@MCHILES

'I cannot teach anybody anything, I can only make them think.'
(Socrates)

QUESTIONING FOR DEEPER UNDERSTANDING
Teachers are often referred to as 'professional question-askers'. The act of asking questions and its role in the process of teaching and learning have deep historical origins, dating back to the Ancient Greek philosopher Socrates. Socrates believed that by employing the art of questioning, we stimulate thinking and promote ongoing reflection. In his perspective, questions were not merely a means of obtaining answers but rather tools for guiding learners on a journey from surface-level comprehension to profound insights.

Socrates was a leading advocate of employing questions as a conduit for learning. He contended that through thoughtful and deliberate questioning, educators could develop the right foundations for critical thinking and a deeper grasp of the subject matter. Socratic dialogues were centred around an ongoing exchange of ideas between the questioner (the teacher) and the recipient (the learner). The objective was to guide learners from superficial knowledge to a deeper understanding of the subject. However, even though Socrates believed that this should be the primary function of questioning, frequently the questions teachers pose in the classroom fail to go beyond surface-level understanding. Consequently, the skill of posing questions is frequently overshadowed by the intent behind the questions, creating a gap between the asking of a question and its actual impact on student achievement.

STRUCTURED QUESTIONING

In the classroom, teachers may pose in excess of a million questions over the course of their career, spanning from basic to advanced levels of thinking. These questions should be designed to encourage both surface-level and deeper-level thinking, depending on their nature and purpose. Striking the right balance between these question types can be complex, as research has not definitively determined which type is more effective. Nonetheless, a strategic approach that incorporates various question types in a structured sequence can serve as a foundation for nurturing and deepening students' understanding.

Think of the process of crafting questions as constructing a tower. To establish a robust and secure tower of knowledge, you need a solid foundation upon which to erect the pillars that support its height and structure. In the context of posing questions, this means commencing with questions that elicit initial recall and then progressing to questions that stimulate deeper reflection. This sequence prompts students to retrieve stored information from their long-term memory and apply it to express their understanding in new or familiar contexts.

Before looking at how we can sequence our questions in the geography classroom, we need to spend time crafting our planned questions. There should be time spent during department meetings to hold curriculum conversations around the content being taught and how best to approach the teaching of the content to ensure students build on prior knowledge effectively. For this to be effective, there needs to be time dedicated to discussing and generating the questions that are linked to the knowledge being taught in the curriculum.

QUESTION DIMENSIONS

When designing powerful questions, Vogt et al. (2003) suggested there are three dimensions to these questions: architecture, scope and assumptions.

- Architecture – is the question aimed to generate low or higher order thinking?
- Scope – to what extent does the question provide freedom for the respondent?
- Assumptions – what can be inferred from the context/meaning of the question being asked?

Let's consider a question that a teacher might ask in a geography classroom when introducing the term sustainability: ***What is sustainability?***

The architecture of this question is that it will generate low-order thinking because the aim is to gather information on whether students know the definition of the term. However, if you asked this question at the beginning of a unit it is likely that this question will be challenging and be high stakes. This is because we don't know if students have encountered this word during previous studies across other subjects and/or if it has been encountered in their home environment. Therefore, asking this type of question would be most beneficial after teaching the term to check knowledge recall. If this was the first time the teacher was teaching the concept, the question could be asked slightly differently to provide greater scope to check prior knowledge, reduce the stakes and increase the likelihood of participation: ***What do we know about sustainability?***

Changing the question slightly to add in 'do we know about' reduces the stakes and opens up the scope of the question when checking for prior knowledge. It allows the student to say what they know and gives the teacher an idea of the extent of student's prior knowledge.

But, if the teacher was looking to reflect on why we need to be more sustainable, the scope of this question is limited. To open up the scope and generate high-order thinking, the teacher could plan to ask a series of questions after teaching core knowledge to check for understanding. An example of a higher-order question with greater scope would be: ***to what extent will sustainable approaches reduce environment degradation?***

A minor change to the question asked allows for the teacher to prompt a deeper level of thinking for students. This is what Barrell (2003) referred to as 'moving beyond the immediate data or experience'. When you plan questions for lessons, create a sequence where the architecture and scope changes to layer knowledge over time so that deeper-level thinking can take place, which encourages students to use their knowledge and understanding.

Now that we have considered how the architecture and scope of a question can change it is important to plan out how to sequence questions over a series of lessons to build from surface- to deep-level thinking. As the professional question askers, we should use questions to check that students have a clear understanding of core concepts and processes in our subjects *before* we ask them to demonstrate higher-level skills.

In order to do this, we should look at what assumptions can be made about the context/meaning of the question being asked. Let's go back to the high-level question related to sustainability: ***To what extent will sustainable approaches reduce environment degradation?***

In order for a student to be able to answer this question successfully, they will need to have knowledge of the following:

- What 'sustainability' means.
- What 'environmental degradation' is.
- Examples of environmental degradation.
- The concept of global warming and climate change.
- What approaches can be used to make places more sustainable.

Once the student has this knowledge, they would then need to use this knowledge to understand:

- The negative impact environmental degradation is having on people and the natural environment.
- How people are contributing towards global warming and climate change along with its wider implications on the planet.
- Why sustainability is important to improve environmental degradation now and into the future.
- The barriers to implementing sustainable approaches to improving environmental degradation.

- The different successes of sustainable approaches already used to reduce environmental degradation.

CONCLUSION

In exploring this example of a potential higher-level question we might ask in our geography classrooms, it illustrates how the role of knowledge and understanding are the foundation to enable students to apply to more complex skills higher up in Bloom's taxonomy. The questions that lead up to students attempting this higher-order skill would be key to ensure knowledge and understanding is secure. When we plan our questions in a sequence it allows us to check for understanding and students to review and recall knowledge before applying to more complex thinking tasks. This example highlights that while asking higher-order questions will activate deeper thinking, the lower-order questions are fundamental to provide the right foundations for students to engage and be successful in the higher-order questions.

> *Questions that elicit responses in the knowledge, comprehension, and application domains are frequently considered lower-order questions, while questions in the analysis, synthesis, and evaluation domains are considered higher-order questions. Higher-order questions elicit deeper and critical thinking; therefore, teachers are encouraged to ask questions in these domains. This does not mean that lower-order questions should not be asked.*
>
> (Tofade, Elsner and Haines, 2013)

The practice of posing meaningful questions in the classroom is more than just a teaching technique; it is an art form. When we invest the time to carefully plan our questions, their impact goes well beyond obtaining correct responses; they actively involve students, stimulate thinking and promote a culture of learning.

To become a proficient questioner, our goal should be not only to ask questions but to do so with a strategic approach. We should ensure that our questions match the powerful knowledge within our curriculum and are adaptable to cater to our students' needs. Mastering the art of asking powerful questions can go a long way to aiding our students in becoming knowledge experts in geography.

Five reflection questions

1. How could you collaborate with other teachers to ensure questions asked are effective?
2. Do you ensure you use a variety of question types within a lesson?
3. When planning questions for a lesson, do you consider the sequencing of questions asked?
4. Do you design questioning to encourage both surface-level and deeper-level thinking from your students?
5. How could you consider the architecture, scope and assumptions when designing effective questions?

DIGGING DEEPER – THREE RESOURCES TO FURTHER

- Questioning in the Geography Classroom. "https://geography.org.uk/ite/initial-teacher-education/geography-support-for-trainees-and-ects/learning-to-teach-secondary-geography/ite-trainees-classroom-practice/questions-in-the-geography-classroom/.
- Enser, M. (2019) "Questioning: when and how to use it in the classroom, TES, 11 March. https://url.de.m.mimecastprotect.com/s/UqArClRJkqFz8wYpu9hDhzHJaP?.
- Doherty, J. (2018) "'Skilful questioning: the beating heart of good pedagogy', *Impact*. Chartered College of Teaching, June. https://url.de.m.mimecastprotect.com/s/de89CmqK0rCR3oQDcBiQhRmj3R?domain=my.chartered.college,

REFERENCES

Barell, J. (2003) *Developing More Curious Minds*. Alexandria, VA: Association for Supervision and Curriculum Development.

Tofade, T., Elsner, J. & Haines, S. T. (2013) Best practice strategies for effective use of questions as a teaching tool. *American Journal of Pharmaceutical Education*, 77(7), p.155.

Vogt, E., Brown, J. & Isaacs, D. (2003) *The Art of Powerful Questions: Catalyzing insight, innovation, and action*. Mill Valley, CA: Whole Systems Associates.

8. WHERE DOES FIELDWORK FIT IN THE 11-14 GEOGRAPHY CURRICULUM?

FIONA SHERIFF
@FIONA_616

'Geography wants to take children outside the school and into the streets and fields; it wants to take keyboard tappers out of their gloomy offices and into the rain or the sunshine.' (Bonnett, 2008, p. 80)

Fieldwork is the lifeblood of geographers. Having the opportunity to get outside and test the theories and knowledge learned in the classroom in the outdoors is a hugely important aspect of becoming a geographer. However, it is also one area of the curriculum which has become harder to put into practice over the last few years due to COVID, red tape and the decline in the number of specialist geography teachers.

While at Key Stages 4 and 5 schools give geography departments the time to complete the minimum days required for fieldwork, it is becoming less common to complete fieldwork at Key Stage 3. But does it need to be this way?

Key Stage 3 provides a plethora of opportunities for fieldwork which can be carried out both on-site and in the local area. Completing fieldwork at an earlier stage will not only add more depth to your curriculum, it will also encourage a deeper understanding of the outdoors and help to make the abstract, concrete. 'It is only through fieldwork that students can become engaged in the whole process of constructing geographical knowledge.' (Roberts, 2022)

WHAT DOES THE NATIONAL CURRICULUM SAY ABOUT FIELDWORK?

The geography Programme of Study for Key Stage 3 in the national curriculum in England simply states that 'Students should be taught to use fieldwork in contrasting locations to collect, analyse and draw conclusions from geographical skills, using multiple sources of increasingly complex information'. Similarly in the Welsh, Scottish and Northern Irish curriculums there are brief mentions with some degree of flexibility as to how fieldwork could be included in the wider geography curriculum. The curriculum statements from all devolved nations give very little away in terms of how frequently fieldwork should be carried out or how it should be embedded into the curriculum. This often leads to fieldwork being an 'add-on' rather than something that is part and parcel of our curriculum planning.

THE FIELDWORK CYCLE

The fieldwork cycle underpins the process of fieldwork, and we can use it to help us to decide how we embed the various parts of a fieldwork investigation into our KS3 curriculums. A whole investigation does not need to be completed at once; instead, a variety of elements could be included within each year, building up to the full investigation at the end of Year 9. Taking smaller parts of an investigation and building the skills up over time can help to make fieldwork feel more manageable.

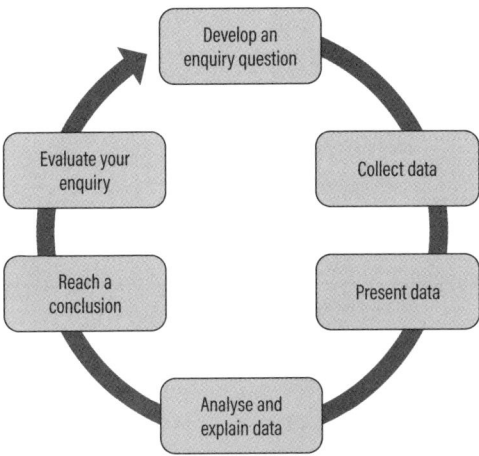

Figure 8.1 The fieldwork cycle

EMBEDDING FIELDWORK INTO THE CURRICULUM

If we start by considering where we can embed fieldwork opportunities into our current curriculums, it gives us a starting point. If we take the content from the English national curriculum, we can look for opportunities to embed fieldwork into the different areas, rather than having standalone skills or fieldwork topics which can serve as a bolt-on. While it isn't a problem to have stand-alone units of fieldwork, building in the opportunity to practise it over time enables students to see fieldwork as part of what geographers 'do' and how it helps geographers to further their knowledge and understanding in their field of expertise.

Table 8.1 Ideas for the inclusion of fieldwork in common KS3 curriculum topics

Example geography topics	Fieldwork ideas
Rivers	Rivers fieldwork off-site
	Infiltration survey
	Practising techniques on-site
Coasts	Coasts fieldwork off-site
	Practising techniques on-site
Weather and climate	Microclimate survey
	Using a weather station
Ecosystems	Biodiversity study
	Soil sampling survey
	Quadrat sampling
Local area study	Land use survey
	Traffic counts
	Environmental quality surveys
Tourism	Virtual fieldwork, using online mapping to visit tourist locations
Climate change	Sustainability survey
	Traffic counts
	Travel surveys

If we take the topics we currently teach and consider how we might embed fieldwork into them, we can start to prioritise fieldwork and showcase its importance. It is very easy to build a KS3 without fieldwork. With time constraints and more non-specialists teaching geography it is easy to put fieldwork to one side and to consider it only if you are allowed to plan an off-site trip. This was an issue we were facing at my school, and I hear it many times over when talking to other geography teachers. This then often results with students being ill-prepared at KS4 as they haven't experienced fieldwork before. However, building fieldwork into the curriculum can also come at a time and content cost. What might you need to take out of a curriculum in order to fit fieldwork in, without shoehorning it in because you feel you need to? It can be a conundrum!

When we looked at our fieldwork offer at KS3 we knew we could improve it, and in doing so we also felt that we could improve both student and staff confidence in delivering fieldwork as some sections were small with full instructions given. The resources given to staff included videos showing how to carry out the fieldwork, these were also chosen to be accessible to students too.

Over time, we have worked on improving our fieldwork provision. Table 8.2 details some of the fieldwork that is completed by our students. This is predominantly completed on-site and within a lesson or series of lessons. For context, we have three hours a fortnight in Years 7 and 8, and four hours a fortnight for Year 9. Alongside the examples in Table 8.2, students will also complete regular geographical skills practice in lessons. Interleaving the skills throughout KS3 strengthens students' knowledge and understanding overall and demonstrates to students the importance of mapwork, data recording and data presentation.

Year 7	**Biodiversity study** - enquiry question, methodology, data collection and data presentation, sampling methods	**Microclimate study** - enquiry question, hypotheses, methodology, data collection, data presentation and analysis	
Year 8	**Brazil** - virtual fieldwork - visiting Brazil using online mapping	**Climate change** - data collection, surveys and questionnaires	**Local area fieldwork** - full investigation on safety and accessibility
Year 9	**Sustainability study** - enquiry question, hypotheses, methodology, data collection, data presentation and analysis and conclusion	**Infiltration survey** - enquiry question, data collection and evaluation of methods	**Oceans and plastics** - designing and conducting online surveys

Table 8.2 Examples of fieldwork included in Kingsthorpe College's 11–14 curriculum

BIODIVERSITY STUDY

Year 7 students complete a short piece of fieldwork on biodiversity within the first five weeks of the school year. The lessons on fieldwork follow a lesson on temperate deciduous forests as part of a wider unit called 'Becoming a geographer'. In this unit, students recap some of the knowledge they should have learned at Key Stages 1 and 2 and then investigate the biome that we live in, identifying some of the characteristics that we associate with the temperate deciduous biome as well as examples of the flora and fauna that can be found there. The fieldwork that students participate in takes the knowledge learned in the classroom out into the field. Students can test their understanding of the biome and whether the knowledge they learned was correct. It also introduces students to one of our curriculum 'golden threads' – biodiversity.

Students are tasked with investigating whether our school site is an area of high or low biodiversity. For many, this will be the first piece of geography fieldwork that they will have completed. We have students join us in Year 7 from a large number of primary schools, therefore we know that the amount of fieldwork completed during KS1 and KS2 is variable. They are introduced to basic pieces of fieldwork equipment (quadrats, tape measures and species guides) and are shown how to conduct a simple survey along a hedgerow and on the school field. Our equipment is limited. When we complete this study students will swap between the hedgerow survey and the quadrat sampling on the school field. The data collection portion is carried out during one lesson, and we spend three lessons overall on this investigation.

In the first of the three lessons students are introduced to some key terminology associated with fieldwork, including primary and secondary data and sampling. We gauge the students' level of understanding with careful questioning and students write down definitions of fieldwork, primary data and secondary data. We then walk through how to complete each method before we head out to the field. We find showing students how to complete each method in the classroom is far less distracting than doing it outside. To aid non-specialists we also use two clips to demonstrate how the methods are carried out (see also chapter 23). In department meetings we also discuss how to carry out the fieldwork so that everyone who teaches a Year 7 class feels confident in being able to carry out the investigation.

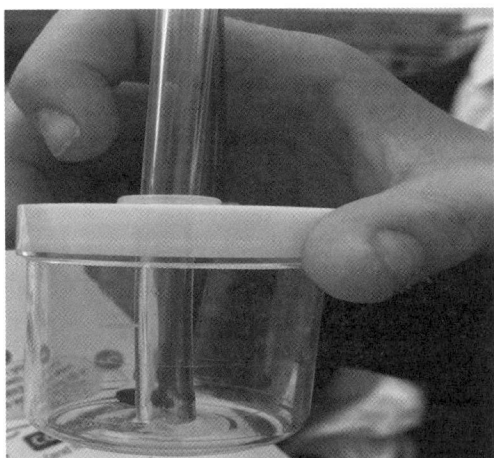

Figure 8.2 Students using a pooter to identify insect species

Students complete two methods, the first is a hedgerow survey and the second is a quadrat survey. To complete the hedgerow survey, students use a tape measure to measure a five-metre distance along the hedgerow; they then split that distance into one-metre intervals. This introduces them to the concept of systematic sampling. Using a species guide, they look up and down each one-metre interval and tally each new species they find. We also borrow pooters from science to encourage students to take a closer look at some of the insects that they find. We have purchased identification guides from the Field Studies Council, however there are lots of free guides available on the internet such as 'Discover Wildlife' and the guide from Imperial College.

To carry out a quadrat survey, students are given a 100-square quadrat and a species identification guide. They are told to place their quadrat randomly at ten sites across the school field. This introduces them to the concept of random sampling. We restrict the area in which they place their quadrats so that we can keep a close eye on them. Once they have placed their quadrat down, they need to count the number of squares each species is found in, for example if ten squares had daisies in them, we would estimate that there is 10% coverage of daisies in that quadrat. Students add this tally to their data collection sheet and then move their quadrat to the next site.

Figure 8.3 Students using a quadrat

On returning to the classroom, students are given an opportunity to discuss the methods and we ask the following questions:
1. How could we improve the methods we carried out?
2. Do we have enough data to judge how biodiverse our school is?
3. What else could we do to investigate biodiversity?

At this point, the students' knowledge of fieldwork is limited but asking them to think critically about what they have done so far is a great way to assess their understanding of the fieldwork process as well as reviewing their ability to think and talk like a geographer. Students often propose interesting fieldwork methods, and a debate usually starts which deliberates the time we have available to complete fieldwork.

The final lesson in the series is dedicated to data presentation and simple methods are used. Students present their hedgerow data on a bar chart and their quadrat survey in a pie chart. As we progress through our fieldwork curriculum, students will practise more sophisticated data presentation methods. We model how to complete a bar chart and a pie chart, talking students through each stage before getting students to complete their own. Once completed, they then analyse their data and write up their findings before answering their enquiry question. This was the students' first attempt at presenting and analysing data. Students were given four key areas to discuss and analyse.

1. Which species are found the most across our hedgerows and field? Use data to back up your points.
2. Which species are found the least across our hedgerows and field? Use data to back up your points.
3. How do plants differ between the hedgerows and field?
4. Were there any surprises?

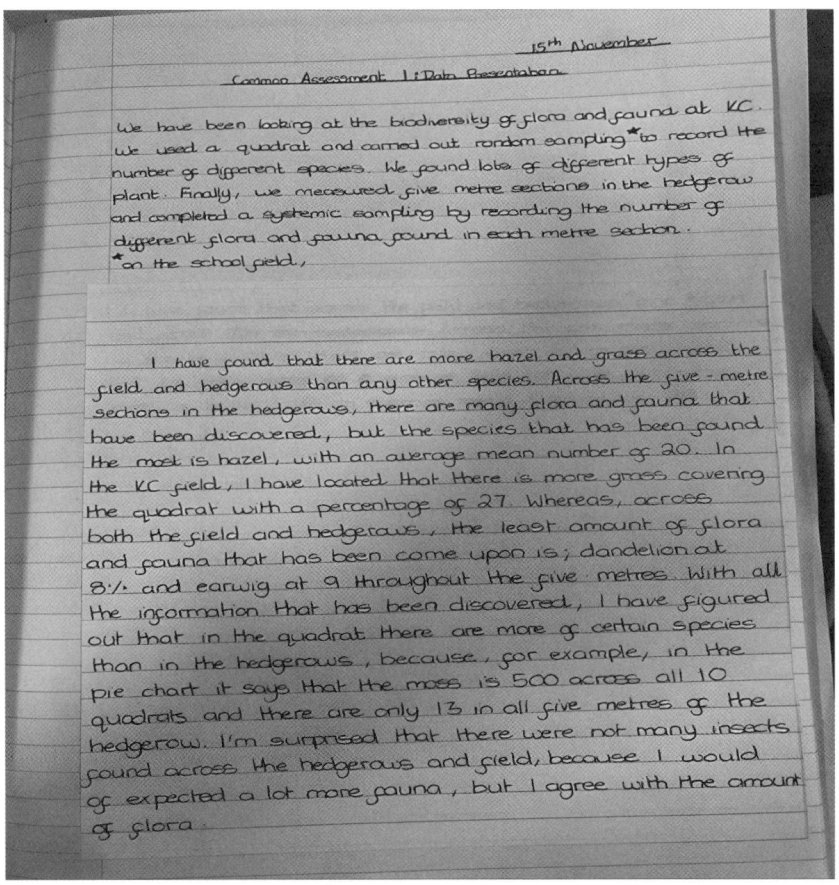

Figure 8.4 Student work, write-up of their analysis

Students were able to read the data they had collected and begin to describe the story that their data told. Giving students the opportunity to do this early in Year 7 not only lays the foundation of the importance of fieldwork but also demonstrates that geography is more than just the maps, countries and capitals that many expect. When students were asked what surprised them, we wanted to show that often fieldwork is messy and that there are unexpected results. Some students struggle with this and need a 'right' answer. We remind them that this 'messy' learning is essential in geography as we are questioning the world and how it works. The Geographical Association reminds us that 'Fieldwork – that is, learning directly in the untidy real world outside the classroom – is an essential component of geography education' (Geographical Association, 2009).

Having the right equipment can sometimes cause concern, especially when school budgets are dwindling, however there is little need for concern for most fieldwork. DIY fieldwork equipment can be created such as making quadrats from wire coat hangers or by taping metre rulers together. You may also find that your science department has quadrats, metre rulers, tape measures, thermometers and callipers which could be borrowed. You could also speak to a local university; many want to help out schools and sharing fieldwork equipment is an excellent way to develop links. They may also want to help you with fieldwork. We have been incredibly fortunate with the University of Northampton as they have joined us on fieldwork and suggested a range of ideas that we could implement.

FIELDWORK WITHOUT EQUIPMENT

Human fieldwork lends itself to surveys and questionnaires, and this often means that little is required in the way of equipment. An interesting on-site investigation to carry out is 'How sustainable is our school site?'; all this requires is some worksheets, clipboards and a pen. During this investigation, students complete: environmental quality surveys, bi-polar surveys, emotional mapping and observational studies. Students also take photographs to annotate later in the classroom.

Students complete this fieldwork in Year 9 as part of a sustainability study unit called 'Incredible Cities'. At the end of this unit, they complete an enquiry on sustainability in Dubai and Masdar and we follow this with some on-site fieldwork to find local examples of sustainable practice. 'Fieldwork enables students to relate concepts to the reality of unique places.' (Roberts, 2022) Sustainability is another of our golden threads. As this fieldwork is in Year 9, students should have a concrete understanding of sustainability as their understanding of each of the golden threads is built on each year.

We take students to four different sites around the school, and they complete a bi-polar survey and environmental quality survey at each site. They also note down their observations around sustainability; this could include the function of each site, types of waste disposal and the materials used to build each site. They also take photographs of features that they consider to be sustainable or areas where sustainability could be improved.

Figure 8.5 Completing surveys in the school courtyard

On returning to the classroom, students also complete emotional mapping, using different colours to represent their emotions in each part of the school. They are given an aerial photograph and a map of the school and four different coloured pens. They add a coloured circle to each part of the school, the larger the circle the bigger the emotion or feeling. For example, if a student really enjoyed part of the school, they may add a large green circle to express happiness, whereas a large red circle may show anger.

Figure 8.6 Students completing their emotional mapping methods

Once students have completed each of their methods, they use the information gathered to write a letter to the senior leadership team suggesting ways in which the school site can be made more sustainable. They can apply their knowledge learned in lessons to the fieldwork and the real-life application of sustainability.

This piece of fieldwork is carried out over one or two lessons and could be extended to investigate sustainability in the school's locale. Homework could also be set for students to investigate sustainability in their home neighbourhood. Results from student surveys could be compiled to build up a larger picture of sustainability in the school's catchment. Bringing the data together could also lead to an interesting GIS mapping activity (see chapter 11).

CONCLUSION

Fieldwork needn't be a worry or concern, but we should try to find opportunities for it throughout our KS3 curriculums. There are a wide range of on-site fieldwork investigations that can be carried out to help you ensure that your students are ready to hit the ground running with fieldwork at GCSE. Fieldwork is messy, it makes you question what you know, and it is dynamic and delicious. Go on, get outside and show your students just how practical geography is, and help to inspire the next generation of geographers.

> **Five reflection questions**
> 1. Does your department prioritise fieldwork at KS3?
> 2. Where are you giving students the opportunity to practise geographical fieldwork and skills?
> 3. Is fieldwork integrated into your curriculum or is it a bolt on? How do you know?
> 4. Do all of the teachers in your department feel confident in delivering fieldwork?
> 5. What is preventing you from including further fieldwork in your curriculum?

> **DIGGING DEEPER – THREE RESOURCES TO DELVE FURTHER**
> - Monk, P. (2016) Progression in fieldwork. *Teaching Geography*, 41(1), pp. 20–21.
> - Lambert, D. & Reiss, M. (2014) *The Place of Fieldwork in Geography and Science Qualifications*. UCL IoE Press.
> - Kinder, A. (2018) Acquiring geographical knowledge and understanding through fieldwork. *Teaching Geography*, 43(3), pp. 109–12.

REFERENCES

Bonnett, A. (2008) *What is Geography?* London: Sage

Department for Education (2014) The national Curriculum in England: Key stages 3 and 4 framework document. https://www.gov.uk/government/publications/national-curriculum-in-england-secondary-curriculum (Accessed: 28/05/2025)

Geographical Association (2009) *A Different View: A manifesto from the Geographical Association.* www.geography.org.uk/GA-Manifesto-for-geography (Accessed: 28/05/2025)

Roberts, M. (2022) Powerful pedagogies for the school geography curriculum. *International Research in Geographical and Environmental Education*, 32 (2), pp. 1–16.

9. USING HOMEWORK TO ADD BREADTH AND DEPTH TO THE CURRICULUM
JENNIFER MONK
@JENNNNNN_X

'Great teachers set great homework.' (Tom Sherrington, 2012)

WHAT IS THE PURPOSE OF HOMEWORK?

Parents, teachers and students have differing opinions on homework and there are many debates on the type, frequency and 'marking' (or feedback) surrounding homework. Often homework can feel like an add on, increasing workload for teachers, taking away family time for students and parents, and usually adding further pressure on everyone. Schools usually have a homework policy which dictates the frequency and time that homework should take and, to try to help students manage their work at home, there is often a rigid timetable. These restraints, along with many others, can lead to homework being set that simply creates higher workload for teachers, is of little benefit to students and can also result in a negative experience for them.

There has been a wide range of research surrounding homework. However, results vary hugely – mainly due to the number of external factors that also affect homework. For example, the Education Endowment Foundation

(EEF) Toolkit on homework suggests that generally with older children (secondary-school age), homework has a positive effect on progress of up to five months. But the key findings also suggest that homework has to be supported for those who may not have a quiet space at home and must also have a clear purpose for students. Research also suggested that homework that is clearly linked to learning in the classroom and includes feedback has a much higher impact on students' learning.

Homework as a tool can be an effective way for students to take their learning outside of the classroom. Most homework I set will be one of four things:

- Preparing for new learning – often known as flipped learning.
- Application of learning – applying something learned inside the classroom to a different context.
- Enrichment of learning – adding breadth or depth to learning.
- Consolidation of learning – making sense of previous learning.

It is important we think about the bigger (and long-term) purpose behind homework – the independent learning skills we want students to develop. We know that those students who are successful at GCSE and A-level are the students who are resilient and resourceful – two skills that can be learned through giving homework earlier down the school. Resilience is particularly important with homework as often the support (in terms of a teacher or even an exercise book with classwork in) is taken away.

IS HOMEWORK WORKING FOR ALL STUDENTS?

There are many reasons why homework could be effective (or in some cases ineffective) and it is important that, when designing and setting homework, we ensure that it is meaningful and easy to understand. We should also ensure that we offer some support for those students who may struggle to complete homework at home, otherwise those who don't have a suitable device at home or a quiet workspace may struggle to complete homework, and this can increase the attainment gap for disadvantaged students.

Further to this point, if homework is too complicated to understand or if students do not have a good understanding prior to the homework being set, then it is highly likely that the homework will be less effective.

When setting homework for students Cathy Valterott (2018, p.9), in *Rethinking Homework*, suggests that:

> *'Despite there being more diversity among learners in our schools than ever, many teachers continue to assign the same homework to all students in the class and continue to disproportionately fail students from lower-income households for not doing homework, in essence punishing them for lack of an adequate environment in which to do homework.'*

Reflecting on the above, particularly thinking about key stage 3 where teachers could have three or four classes per year group, setting different homework for different students would be an impossible task. It would add to workload and make giving feedback much harder. Instead, I would propose we think about how we can take those barriers away from students to ensure success with homework. For example, we offer two drop-in homework clubs specifically in geography (in addition to a whole-school daily drop-in homework club) with geography teachers and student leaders (GCSE students in Years 10 and 11) to support. We ensure that devices are available for students to use and that we have printed copies of reading tasks for students to take away.

HOW CAN WE USE HOMEWORK TO BROADEN THE CURRICULUM?

One of the biggest issues with organising our curriculum is all the wonderful things we want to include, but don't have time to. For example, we want to ensure we develop our students understanding of 'place' well. The 'Getting our bearings: Geography subject report' published in September 2023 suggested that:

> *'Some schools taught the concept of place very well. Where this happened, students were able to use their locational knowledge to explain some of the features of the places studied. They also considered how a range of different human and physical processes applied to the place and contrasted the place with other places that they had learned about. In these schools, leaders were very aware of the issues around presenting a 'single story' about a place, and they planned to revisit places in a range of contexts over time.'*

This inspired us to further utilise our homework as an opportunity to broaden and deepen our curriculum. Below are some examples of homework we have used to support our teaching and learning:

1. **'How is this different to …'**– inspired by the wonderful work of Will Bailey-Watson and Richard Kennett who co-created 'Meanwhile, elsewhere …' to give students the opportunity to look at other places during the same timescales as the places studied in their history lessons, I created some tasks to compare places we were studying in current geography lessons to places we hadn't yet studied. For example, when we looked at extreme weather in the UK, the homework task was to compare the UK example from our classwork with another example, either in the UK or somewhere else in the world. Students watched a video or read an article and then completed a range of tasks to support their understanding of the other places.

   The example below (Figure 9.1) shows the worksheet students used when we were learning about the climate in India and their homework was to complete tasks linked to the climate of Russia. In the next lesson, students discussed why these two places have different climates, e.g. looking at latitude.

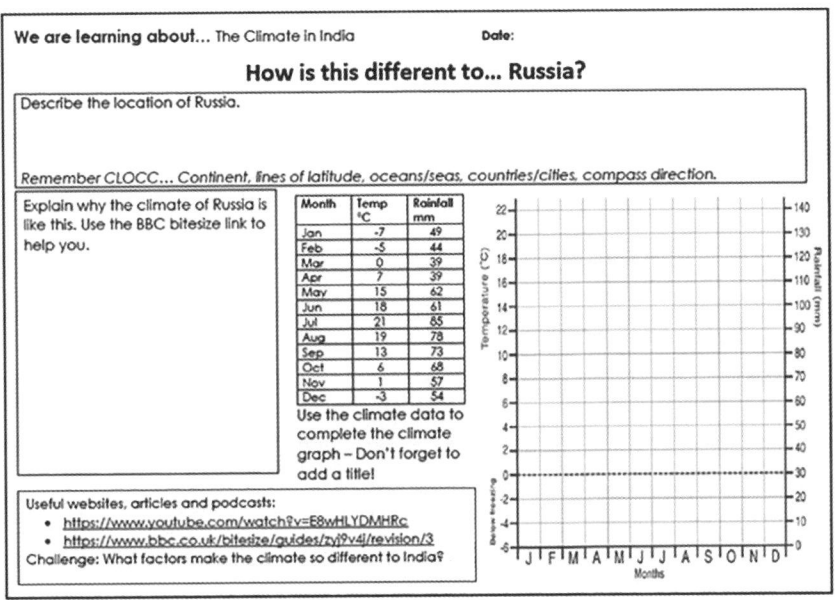

Figure 9.1 How is it different to … Russia? homework sheet

2. **'Geog your Memory'** – Retrieval practice is clearly an effective way to revisit previously learned material, making it an effective tool to use for homework. It can take on many forms outside of the classroom. One example is to set regular quizzing using a self-marking tool such as Google forms. Having a bank of these questions for various topics means students can revisit specific ones as needed or you can use them as a blanket tool to suggest next steps in terms of wider revision. This is not only manageable for staff workload but can be used across departments to encourage consistency. If all students are completing the same questions, it means you can assess misconceptions easily. We used to do this through the use of '5 a day' where I would choose five questions based on misconceptions from lesson time or from assessments and students would work through them. In the next lesson we would go over the homework and students would self/peer assess in purple pen. Choosing topics that link to the lessons currently being studied also helps with creating a schema for the students.

The 'Give me 5' retrieval grid below (Figure 9.2) aims to support students with retrieving key information about five different areas of knowledge.

Give me 5...   Date:

	1	2	3	4	5
Causes of deforestation					
Responses to Chile Earthquake (label them as immediate or long term)					
World biomes & a characteristic of each					
Examples of extreme weather in the UK					
Impacts of tropical storms (categorise them into primary or secondary)					

Challenge: Outline the primary and secondary effects of a tropical storm, use a named example. (6 marks)

Figure 9.2 An example 'Give me 5...' retrieval grid

3. **Videos/podcasts/reading tasks** – Our curriculum is based around a 'big' question and our lessons and homework aim to allow students the opportunity to be able to answer the question. As part of this, in the very first lesson of the topic, we ask students to come up with all the 'things' they need to know to answer this question. We aim to cover most of the ideas through our lesson delivery but some of the topics students are interested in or link into our topic may not be given curriculum time in the classroom. In this case, we might set homework for students to watch a video, read an article or listen to a podcast. We then use a stimulus in the following lesson to hear views and feedback on what students have learned during their homework. We encourage them to summarise their thoughts into their book at the start of the lesson – with the intention that if they do this well, it should support them in

answering our 'big question' at the end of the unit. Sometimes we may provide a worksheet or activity for students to complete while listening or watching at home. There are so many good-quality videos and documentaries to use. As with the other homework approaches we have looked at, often these video/podcast/reading homework tasks will take students to another location or place. For example, when our Year 7s study tropical rainforests, they look at sustainable management; one of the homework tasks during this module is looking at ecotourism in a rainforest in India to compare and contrast solutions to deforestation. Below is another example of an in-lesson review, following a reading task. Students read a text, then did an online Google forms quiz. They then came to the following lesson and answered some questions as a starter task. (If we include a piece of text for our homework, we also include a recording of one of the teachers reading to support students who may struggle with reading texts. We also include key terms and definitions with all our reading texts to support all students with vocabulary they may not be familiar with.)

Independent learning review: ILA Date 18th March 2024

1. What do we mean by accessible transport? (1 mark)
2. Why is improving accessibility for disabled people so important? (4 marks)

Quiz Score: ___/10   Review Score: ___/5

Figure 9.3 Independent learning review sheet

CONCLUSION

Homework has the potential to be effective and meaningful. With careful thought and consideration, I think both are easily achievable. Although some support needs to be planned in to ensure homework doesn't put any student at a disadvantage, there is clear evidence to suggest homework is worthwhile for most students.

> **Five reflection questions**
> 1. Has the purpose of the homework been made clear to students?
> 2. Does homework support teacher and student wellbeing?
> 3. Do students have everything they need to complete the task independently?
> 4. How do you move forward following misconceptions identified from homework?
> 5. How do you ensure students get feedback on their homework?

> **DIGGING DEEPER – THREE RESOURCES TO DELVE FURTHER**
> - Huntingdon Research School (10 November 2017) Homework: What Does the Evidence Say? Available at: https://researchschool.org.uk/huntington/news/homework-what-does-the-evidence-say (Accessed: 28/05/2025)
> - SecEd article: Research analysis: Getting the most out of homework – www.sec-ed.co.uk/content/best-practice/research-analysis-getting-the-most-out-of-homework/ (Accessed: 28/05/2025)
> - Finch Noyes, H, (2019) Making homework count, *Teaching Geography*, 44(3), pp 105-107.

REFERENCES

Bailey-Watson, W. & Kennett, R. (n.d.) Meanwhile, elsewhere .... https://meanwhileelsewhereinhistory.wordpress.com/ (Accessed: 28/05/2025)

Education Endowment Foundation (2021) Homework. https://educationendowmentfoundation.org.uk/education-evidence/teaching-learning-toolkit/homework (Accessed: 28/05/2025)

Sherrington, T. (2012) Homework: what does the Hattie research actually say? https://teacherhead.com/2012/10/21/homework-what-does-the-hattie-research-actually-say/ (Accessed: 28/05/2025)

Valterott, C. (2018) *Rethinking Homework: Best Practices that Support Diverse Needs*, 2nd edn. ASCD.

10. INCORPORATING GEOGRAPHICAL SKILLS WITHIN A CURRICULUM

LOUIS VIS AND MASON DAVIES
@MRVISGEOGRAPHY AND @THEGEOGTEACHER

'In an educational context, young people need to learn about geographical practice as well as learning through geographical enquiry and practice. This approach includes becoming competent at using the more practical skills, methods and approaches of geographical enquiry, as well as learning about and practising the analysis and argumentation involved in confirming how we know what we know.' (Rawling et al., 2022, p.9)

WHAT ARE GEOGRAPHICAL SKILLS?

It is through the teaching of geographical skills that students will be given the tools and techniques to understand, interpret and analyse the Earth's physical processes and human actions. Geographical skills are essential to understand the relationships between people, places and the environment.

In their 'Study Geography 2023/2024' brochure, the Royal Geographical Society (2023) describes geography as 'one of the most relevant areas of study today, providing a perspective on the world's most pressing issues.' In other words, teaching geographical skills matters because they empower students to:

- understand the world's diverse physical and human landscapes while making sense of its complexity and interconnectedness
- make informed decisions about global issues such as resource management, urban planning, environmental protection and disaster preparedness
- develop an appreciation of the world's cultural diversity, leading to a better understanding of different ways of life
- build transferable skills valued by employers in various sectors, such as urban planning, tourism, international relations, ecology or government.

Yet, when it comes to selecting skills for our curriculum, there seems to be little consensus in the geography teaching community on what these exact skills should be. When we first met to write this chapter, we agonised over which skills to pay service to given the exhaustive list of them and our limited word count. What would geography teachers find most valuable? After consulting colleagues within the wider geography community, we concluded that the five most fundamental geographical skills are:

1. **Map reading and interpretation**: The ability to understand and navigate maps using symbols, scales and keys to gather information about locations, features and distances. This means understanding the distribution of physical and human features on the surface of the Earth.
2. **Critical thinking and analysis**: Evaluating geographical information, recognising bias and drawing conclusions by using evidence from maps, data and sources.
3. **Fieldwork and enquiry process**: Using geographical enquiry to gather primary data and observations about physical processes and human actions.

4. **Interpreting data**: Analysing geographical data such as population, weather and climate or economic indicators to understand patterns and trends at different geographical scales.
5. **Communication skills**: Expressing geographical ideas and information effectively through written, verbal and visual means to different audiences.

We know that we will never meet everyone's preferences and that this list may vary slightly between schools, curriculums and individual teachers. However, we believe that these five fundamental geographical skills are key to helping students develop a deep understanding of the world and its complexities, while helping them make informed decisions about various geographical, environmental and societal issues.

This being said, we acknowledge the importance of a school's context. It is likely to, and should, heavily influence the decisions we make about which skills to include in our curriculums. The considerations are vast: prior knowledge each cohort brings from primary school, the expertise in your team, the expertise of (and skills taught by) your maths and science departments, the strength and uptake of your Department of Education programmes, your CPD budget, your CPD opportunities and attitudes towards these, the number of non-specialists teaching at Key Stage 3, and your own beliefs about the role geography plays in the lives and futures of the young people you serve. The number of factors seem somewhat endless and, even when you have taken them into consideration, your choices will need constant adaptation as cohorts change and staff come and go.

In summary then, the choices you make about what skills to include in your curriculum should be highly context specific but need to enable students to see the world through a variety of perspectives while analysing the impacts of human actions on physical processes. It is by studying geography and learning geographical skills that today's students will find the answers to tomorrow's problems.

WHAT DOES THIS LOOK LIKE WHEN PLANNING SKILLS INTO THE CURRICULUM?

When we worked together all those years ago at a school in Bristol, our department had a clear rationale for the importance of teaching geographical skills as a standalone topic at the start of Year 7. We argued that exposing students to skills early on in their curriculum journey would set them up for a greater chance of success when they encountered them again in the future. While we still see some value in this argument, our thinking has changed somewhat since then. Since parting ways and taking on new roles as Assistant Head Teacher for Teaching and Learning and Head of Geography respectively, we have been fortunate enough to read more widely about what the evidence suggests leads to more effective learning. We have distilled this learning into the context of mapping skills across a curriculum, creating three principles that we hope anyone reading this finds useful:

Learning principle 1: Prior knowledge is really important. We are more likely to learn something if we can link it to what we already know. So attempting to learn something in isolation is less effective.

Impact on curriculum planning: We should probably avoid these one-off skills topics that don't really link to anything. Instead, we should deliberately and 'planfully' embed skills so they can be regularly encountered (and subsequently linked to prior learning) across the curriculum.

Learning principle 2: Space to *nearly* forget. Forgetting is a crucial part of the learning process. So spacing our retrieval of information and encountering the same knowledge/skill across the curriculum in a planful way is more likely to support effective learning.

Impact on curriculum planning: We should probably consider retrieving and practising the same skill regularly across our 5- or 7-year curriculum in order for forgetting to take place and knowledge to be further strengthened when retrieved.

Learning principle 3: Applying knowledge in a different context is more likely to lead to deeper, more meaningful learning and longer-term retention of that learning. So retrieving this knowledge and applying it in the same context each time is less likely to lead to long-term learning.

Impact on curriculum planning: We should probably reapply/reuse the same skills across multiple topics and lesson sequences in order to develop student mastery and flexibility of those skills. Bonus alert: this will also help students to deepen their understanding of the topic they are studying too!

WHAT MIGHT THIS LOOK LIKE IN REALITY?

We now want to start with a non-example to highlight what we now think curriculum writers might want to avoid:

Non-example: Mapping skills in isolation at the start of Key Stage 3 in a topic on skills

Year 7, Topic 1: Map skills
Lesson 1: Continents and oceans of the world
Lesson 1: 4-figure grid references
Lesson 2: 6-figure grid references
Lesson 3: Height on maps
... and so on and so forth.

Six weeks later ... you've moved on to Year 7, Topic 2: Exploring urbanisation through a case study of China. You project a map of the world. How many students can locate China? How many can tell you what continent China belongs to? How many can identify the physical features of China and locate them on an OS map using grid references?

Our hunch would be not many. The knowledge they 'acquired' during Topic 1 has been forgotten. Why? It didn't mean anything. Students didn't assimilate the skills with prior knowledge or the knowledge they acquired in the next lesson. As a result, this knowledge wasn't anchored to anything in long-term memory and, unfortunately, the forgetting in this case has become permanent. So what happens now? You have to reteach the skill again (with much frustration). This has eaten considerably into your three lessons a fortnight (and now you feel even more frustrated).

How many of us have been here! Let's look at another way.

Table 10.1 Better example: Interpreting choropleth maps across a KS3 curriculum

Year	Unit and Lesson	Skill building: Potential lesson activities
7	**Country study: China** Lesson 1: China's physical features and population distribution	Interpreting choropleth map of population distribution in China
	Rivers Lesson 3: Why do so many people live next to rivers?	Interpreting choropleth map of population distribution in Egypt around the River Nile – potential comparison to population distribution in the UK vs distribution of major rivers in the UK
8	**Coasts** Lesson 8: Where should Lancashire Council build their new sea defence?	Using population distribution data to create choropleth map of Lancashire as data to support decision-making exercise on location of new sea defence
	Rainforests Lesson 4: Why are our rainforests disappearing so quickly? Lesson 5: What is happening to rainforest biodiversity and why is it a problem?	Potential choropleth map analysis of deforestation rates in the Amazon vs South-east Asia Comparison of choropleth map of population distribution in Brazil and choropleth of rainforest deforestation rates OR choropleth of biodiversity levels in 2000 vs today
	Migration Lesson 12: How has conflict in Syria led to a mass migration movement across Europe?	Changes in population distribution: choropleth of refugee destinations across Europe by country
9	**Tourism** Lesson 10: Can mass tourism ever be sustainable?	Comparison of tourist distribution across South America by country or region (choropleth) Potential comparison to rainforest biodiversity choropleths from Year 8
	Climate change Lesson 8: How is climate change impacting global migration patterns?	Choropleth map showing origins and destinations of climate change refugees by region or country Deliberate link back to migration patterns studied at the end of Year 8 – any similarities/differences/patterns we can see?

A crude example, but we hope it illustrates the application of the three principles above (i.e. 1. link prior knowledge to make meaning, 2. space to nearly forget, 3. recontextualise to develop flexible knowledge). This should mean that students:

- Grow rich bodies of knowledge that lead to meaningful learning and longer-term retention.
- Develop more flexible knowledge: skills are more easily applied to other topics/contexts – especially in further study at GCSE, A-level and beyond.
- Appreciate the value of skills in helping them to understand the world around them with greater depth and clarity.

We need to constantly encounter and practise skills, but we should also hastily add here that this isn't about shoehorning skills into topics or lessons in the hope that they fit there. Instead, this is about thinking hard about where certain skills can enhance the learning and geographical understandings across our curriculum, lesson sequences and individual lessons.

Another tip when planning skills into your curriculum comes from Ollie Lovell (2022) in his book *Tools for Teachers*, where he talks about 'sub-skill review'. This is when you pre-teach the skill that is a prerequisite for the content that is about to be taught. For example, you might quickly recap 4- and 6-figure grid references by asking students to find some symbols on a map before using these skills in context about the location of a new coastal flood defence. This should ensure that the students have retrieved the knowledge they need in order for easier, more successful application in your lesson.

WHAT COULD THIS LOOK LIKE IN THE CLASSROOM?

As mentioned above, for geographical skills to be embedded into students' long-term memory, the skills have to be incorporated into lessons, revisited and reapplied multiple times rather than studied in isolation. In the following section, we will suggest a variety of strategies that could be used within a classroom context to develop students' geographical skills.

1. Use of technology

Use online resources, apps and geographical software to provide interactive learning experiences for students. Alistair Hamill regularly shares fantastic ideas on social media on how to use platforms such as GIS to illustrate geographical processes (see chapter 11). This can often help students to better visualise key concepts in our subject while developing their map reading and interpretation skills. This is particularly useful with physical processes such as palaeomagnetism, which students always find incredibly hard to understand.

2. Field trips

Organise field trips to local geographical sites or landmarks to provide students with the opportunity to experience geography first hand (see chapter 8). In its recent report, 'Getting our bearings: geography subject report', Ofsted (2023) wrote that:

> 'Fieldwork was underdeveloped in almost all schools, as the curriculum did not consider how pupils would make progress in their ability to carry out fieldwork over time. Although COVID-19 had an impact on the number of field trips and visits taking place, fieldwork had rarely been a strong feature of the curriculum before the pandemic. Leaders had not considered how fieldwork should be taught or how pupils would learn more about how geographers carry out their work. In some secondary schools, pupils did not carry out fieldwork in key stage 3. In primary schools, field trips had often replaced geographical fieldwork. Fieldwork at key stages 4 and 5 rarely went beyond the minimum requirements of the exam boards.'

As geography teachers, it is our job to act on this. Our advice is to keep it simple. In our schools, we have decided to ensure that all fieldwork opportunities at Key Stage 3 and 4 take place within our local area and at no extra cost to students. This helps to improve inclusion and enables our students to better understand the communities they live in. In some situations, we even take students around the school site to show them hydrological processes in action or ecosystem diversity. (See also chapter 8.)

3. Global current affairs

Provide up-to-date examples of geographical events taking place around the world. Encourage students to watch the news and look for geography in the real world. I (Louis) do this through a 'Flag of the Week' display in my classrooms. Having collected flags from a very young age, I now have a large collection. Every week, I scroll through the news to find a geographical event and makes all my classes guess the flag and the story behind it. Students love this and will even recommend stories for me to read in anticipation of next week's flag. My department is also subscribed to the *Geographical Magazine* by the Royal Geographical Society and the *Economist*. Copies of these are made available in a geography library and articles will often be photocopied for students to use in class or to read for homework. This helps to contextualise geographical concepts. I also actively share geographical news stories on my social media accounts for students to read.

4. Questioning

Knowing what question/s to ask, who to question, and how to encourage in-depth answers is an important part of teaching. We sometimes write down examples of questions in the notes section of our PowerPoints. Not only is questioning an essential part of checking for understanding within your classroom, it also provides us with an opportunity to create momentum and 'dig deeper' by getting students to engage in their learning by thinking critically. Mini whiteboards (MWBs) are a super-effective way of making visible the thinking of all learners. They can be utilised to check for understanding of multiple different skills. We won't discuss how to set up the routine for MWBs here, but please check out Tavassoly-Marsh's (2021) excellent blog on this! We will however, give you a few tips for how you might want to use them to develop skills!

Mini whiteboards and multiple-choice questions

Asking a good multiple-choice question (MCQ) that includes in the possible answers the most common misconceptions and pitfalls allows us to accurately pinpoint exactly where each individual student might have gone wrong. For example:

Q: **Give the 6-digit grid reference for XXX.**
a. 126,348 – correct answer
b. 348,126 – misconception 1: up the stairs before along the corridor
c. 128,346 – misconception 2: up the stairs for the 3rd digit rather than along the corridor
d. 136,348 – misconception 3: wrongly identifying the grid square to the right

If students write their answer on their MWB, this allows us to get a reasonably accurate picture of whole-class understanding and which particular element of the skill we might need to reteach and to who!

Mini whiteboards and graph interpretation

MWBs can be used as a quick and easy way to check if students can read and interpret a graph. For example, you could project a climate graph and ask the following questions:

'On your MWBs, what is the month with the highest/lowest temperature?'

'On your MWBs, what is the month with the highest/lowest rainfall?'

Again, this can make everyone's thinking visible and can really quickly help you to identify how many students have fallen into the 'bars = temperature and/or line graph = rainfall' trap.

5. Assessment

These often require students to write lengthy and generic essays about a certain topic. Instead, assessments need to complement our curriculum. They need to become more meaningful through careful curriculum planning. What do we want students to know? What do we want students to be able to do? Ultimately, assessments should be seen as opportunities for students to showcase the different geographical skills embedded into our curriculum (see chapter 5).

6. Regular retrieval opportunities

In the curriculum section of this chapter, we talked about the three key principles for effective long-term skill retention and development. Building in regular retrieval opportunities at logical points in lessons and in lesson sequences can help us to meet all three of these principles (see chapter 19). For example:

- Using MWBs to interpret a climate graph or to answer MCQs about grid references at the start of a lesson that includes the reapplication of these skills.
- Using MWBs to interpret a climate graph or answer MCQs about grid references just before the point in a lesson where students need to reapply these skills. (Retrieval doesn't always have to take place as a quiz at the start of the lesson – sometimes it has more impact within the lesson too!)

It is worth pointing out that there is no perfect solution to teaching geographical skills. It is likely that a combination of these approaches will be needed in order to ensure successful mastery for each student in your context.

KEY TAKEAWAYS

1. There are so many skills in the geography discipline but limited curriculum time. Therefore, we need to think deeply about *which* skills are right for the context in which we teach.
2. Geographical skills are important because they help students to make sense of the world, make informed decisions about global issues, develop an appreciation of the world's cultural diversity and build transferable skills valued by employers in various sectors.
3. When planning skills into the curriculum, we need to consider prior knowledge, spacing to nearly forget, and applying skills to a variety of contexts in order to develop mastery and flexible knowledge.
4. When teaching skills, we need to consider the importance of real-world examples, technology, making thinking and application visible, and prior knowledge.

Five reflection questions

1. With limited curriculum time, which skills do we prioritise? Why are these skills the most important?
2. Is it more important to teach a wider breadth of skills or fewer skills but applied and practised more frequently?
3. What skills do students need to be competent in by point X to be able to deepen their understanding of Y?
4. Where else do these skills appear in our curriculum? Why there? Why then?
5. Do students encounter skill X often enough in a variety of contexts to make their knowledge of that skill 'flexible' so they can apply it with ease to a wider variety of different contexts?

DIGGING DEEPER – THREE RESOURCES TO DELVE FURTHER

- Mark Enser (2021) *Powerful Geography: A curriculum with purpose in practice*. Crown House Publishing.
- Mary Myatt (2018) *The Curriculum: Gallimaufry to coherence*. Woodbridge: John Catt Educational Ltd.
- Kate Stockings' web article 'Curriculum thoughts: geography curriculum'. Available at: https://www.katestockings.com/geographycurriculum (Accessed: 28/05/2025)

REFERENCES

Enser, M. (2021) *Powerful Geography: A curriculum with purpose in practice*. Carmarthen: Crown House Publishing.

Lovell, O. (2022) *Tools for Teachers*. Woodbridge: John Catt Educational Ltd.

Myatt, M. (2018) *The Curriculum: Gallimaufry to coherence*. Woodbridge: John Catt Educational Ltd.

Ofsted (2023) Getting our Bearings: geography subject report. www.gov.uk/government/publications/subject-report-series-geography/getting-our-bearings-geography-subject-report (Accessed: 28/05/2025)

Rawling, E. et al. (2022) A framework for the school geography curriculum. https://geography.org.uk/wp-content/uploads/2023/04/GA-Curriculum-Framework-2022-WEB-RES_final.pdf (accessed 01/07/2025)

Royal Geographical Society (2023) 'Study Geography 2023/24' brochure. Royal Geographical Society with IBG. www.rgs.org/schools/resources-for-schools/study-geography-supplement (Accessed: 28/05/2025)

Stockings, K. (n.d.) Curriculum thoughts: Geography curriculum. www.katestockings.com/geographycurriculum (Accessed: 28/05/2025)

Tavasolly-Marsh, J. (2021) Checking or really checking?. Durrington Research School. https://researchschool.org.uk/durrington/news/checking-or-really-checking (Accessed: 28/05/2025)

11. MAKING GIS 'PART OF THE FURNITURE' FOR THE TEACHING OF GEOGRAPHY

BRENDAN CONWAY AND ALISTAIR HAMILL
@MILDTHING99 AND @LCGEOGRAPHY

'GIS is waking up the world to the power of geography, this science of integration, and ... creating a better future.' (Dangermond, 2015)

In the quarter final of the BBC quiz show *University Challenge* in 2023, the teams were asked this question:

'Which English physician is noted for his investigations of cholera in nineteenth-century London, including the Broad Street pump study?'

Bristol University provided the correct answer 'John Snow' and won the contest.

We might not be surprised that top-calibre university students would know that answer. There is a strong argument that our geography students should also know about John Snow. Behind the question lies the most ubiquitous of geographers' tools: the humble map. Snow mapped the locations of cholera outbreaks to test his hypothesis that they were related to contaminated water. The spatial pattern became clear: the water pump on Broad Street appeared to be the epicentre of the outbreak. Snow's geospatial information gave him a 'super' vision, the ability to see what others had previously failed to. He had inadvertently laid the cornerstone for Geographic Information Systems (GIS) and spatial epidemiology.

The story of his work is an excellent case study, providing a context to develop substantive, disciplinary and procedural knowledge for geography and other subject areas such as science and history as well.

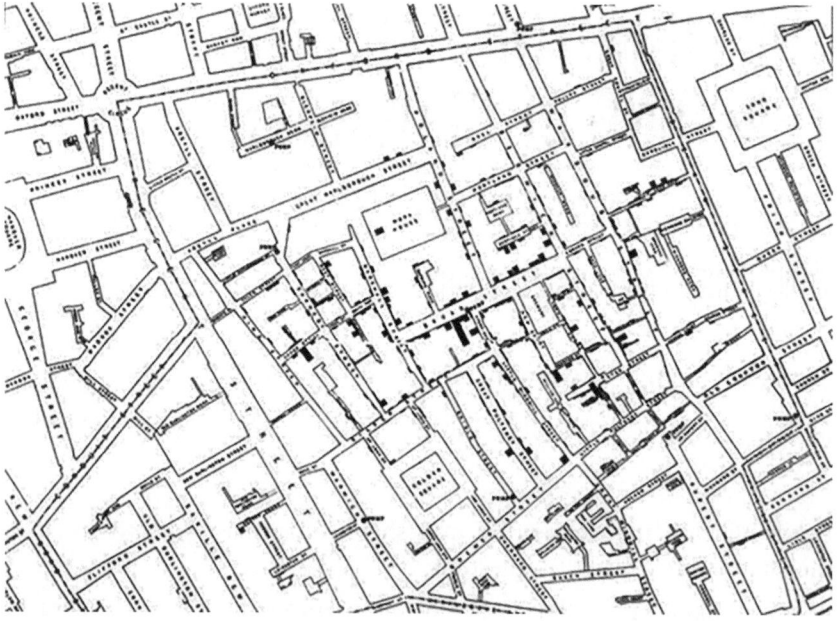

Figure 11.1 Mapping a London Epidemic (from the map made by John Snow, 1854)

London cholera epidemic 1854. (John Snow data – web map by Brendan Conway.)

In this chapter, we explore the potency of GIS to enable our students to develop this same 'super' vision, equipping them to think more deeply as geographers about today's pressing global issues.

WHAT IS GIS?

Geographical information systems (GIS) are systems used to create, manage, analyse and map various types of data. GIS makes links between data and maps.

GIS is used here as shorthand to embrace the range of 'geospatial' resources, including aerial and satellite imagery (remote sensing). A useful quick way we can talk about it is 'The Science of Where' (Dangermond, 2017).

WHY USE GIS IN OUR TEACHING?

There are several important reasons for integrating GIS into the teaching of geography which may be summarised as four types of imperative: curriculum, pedagogical, technological and vocational. It is also essential that GIS teaching fully acknowledges the three areas of knowledge: substantive, procedural and disciplinary.

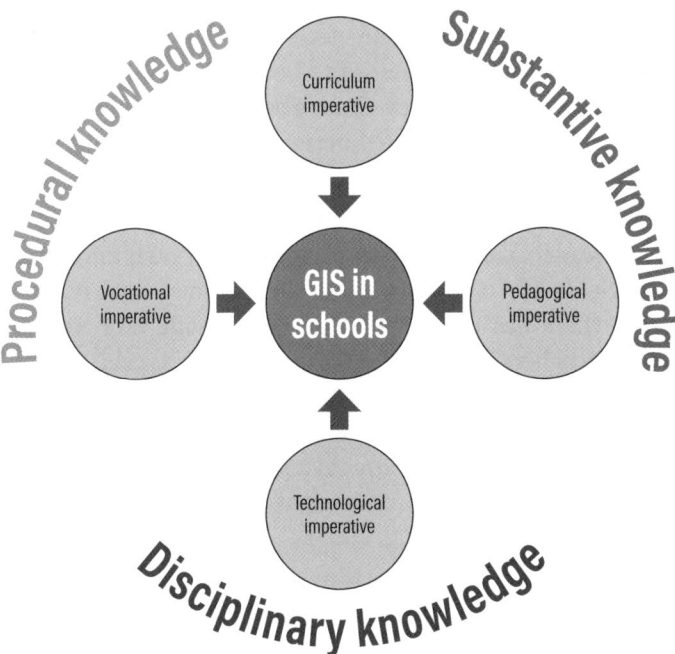

Figure 11.2 Why use GIS in our teaching? (Conway, 2023)

THE CURRICULUM IMPERATIVE

GIS has been part of the curriculum for well over a decade for all key stages in all the devolved administrations of the UK (QCA 2007, DfE 2013).

The rationale behind the statutory requirements for GIS has increased. GIS needs to be integral because it makes a distinct contribution to deeper geographical thinking, empowers students to become critical consumers of geospatial information and because GIS is increasingly how geography is *done* in the 'real world'.

Despite both statutory and geographical reasons for the inclusion of GIS, this curriculum mandate is still struggling to find expression across the sector. 'Geographic information systems (GIS) were not on most secondary schools' curriculums, despite being part of the national curriculum at this phase'. Although certain explanatory factors have sometimes been put forward, such as lack of IT access, 'this was not the main barrier to using it' (Ofsted, 2023).

An urgent review would therefore seem necessary. We hopefully offer helpful advice here, but other factors need to be considered, beyond the control of schools. For example, policy about geospatial education needs to become more effectively co-ordinated between government agencies, professional organisations, GIS providers and exam boards.

There are also questions about training and resource support. Ofsted have observed that 'leaders sometimes have gaps in their own knowledge of areas such as fieldwork, GIS and procedural knowledge' (Ofsted, 2023). In the past, there has been a tendency to provide GIS training mainly for novice teachers, quite often as one-off sessions without follow up. Perhaps it is time to pivot training towards expert teachers, including substantial follow-up coaching and mentoring at school level, commensurate with best practice for effective professional development (Mccrea, 2022) and the introduction of Intensive Training and Practice (ITAP) (National Institute of Teaching, 2023).

Ofsted's earlier 'Research review' strongly emphasised the need for integration of GIS throughout the curriculum rather than as a disconnected 'add-on' (see chapter 3). GIS needs to become 'part of the furniture', because it makes a vital contribution to students' capacity for 'spatial thinking' (Ofsted, 2021). A key consideration for schools is the learning progression for GIS. 'Sequencing geographical content is complex ... leaders need to identify precisely what pupils need to know and to sequence it clearly' (Ofsted, 2023). The GI-Learner project (Zwartjes *et al*, 2017) provided a useful foundation for consideration of progression. The Oak National Academy (2024) have taken a major step forward, providing a very carefully sequenced and practical curriculum from key stage 1 to 4 with dedicated GIS and the integration of GIS throughout all of their lesson materials, provided free to all online. This will become a much-needed benchmark for all and a model to follow.

GI Learner competencies		K7-8	K9	K10	K11	K12
1	**Critically read, interpret cartographic and other visualisations in different media** — *interpretation*	A	B	C	C	C
	A: Be able to read maps and other visualisations *Example: use legend, symbology …*					
	B: Be able to interpret maps and other visualisations *Example: use scale, orientation; understand meaning, spatial pattern and context of a map*					
	C: Be critically aware of sources of information and their reliability *Example: critically evaluate maps identifying attributes, representations (e.g. inappropriate use of symbology, or stereotyping) and metadata of the maps*					
2	**Be aware of geographic information and its representation through GI and GIS.** — *learning about*	A	B	C	C	
	A: Recognize geographical (location-based) and non-geographical information *Example: describe GPS, GIS, Internet interfaces; be able to identify geo-referenced information*					
	B: Demonstrate that geographical information can be represented in some ways *Example: employ some different representations of information (maps, charts, tables, satellite images…)*					
	C: Be critically aware that geographic information can be represented in many different ways *Example: be able to evaluate and apply a variety of GI data representations*					
3	**Visually communicate geographic information** — *produce*	A		B	C	
	A: Transmit basic geographic information *Example: produce a mental map, be aware of your own position*					
	B: Communicate with geographic information in suitable forms *Example: basic map production for a target audience - using old and new media, Share results with target group*					
	C: Be able to use GI to exchange in dialogue with others *Example: discuss outcomes like survey results/maps online or in class, referring to a problem in own environment*					
4	**Describe and use examples of GI applications in daily life and in society** — *applying*	A	B	C	C	
	A: Be aware of GI applications *Example: know about GPS-related/locational (social networking) applications including Google Earth; produce a listing of known GI applications or find them on the internet/cloud*					
	B: Use some examples of (daily life) GI applications *Example: problem-solving oriented with GI application like navigating; use an app to read the weather, environmental quality, travel planner*					
	C: Evaluate how and why GI applications are useful for society *Example: assess the functionality and use for society of a GI application (emergency services, police, precision agriculture, environmental planning, civil engineering, transport, research) and present the results*					

Figure 11.3 GI-Learner competencies

Source: Zwartjes *et al*, (2017), GI Learner project, www.gilearner.eu

THE PEDAGOGICAL IMPERATIVE

GIS has the potential to shape and influence our pedagogical planning and practice in geography classrooms. In its simplest form, it can act like a geography visualiser, bringing landscapes to life using different layers of meaning and scale which can be more challenging with static paper maps.

GIS is a major step forward in our capacity to visualise landscapes, spatial data and many of geography's big ideas – space, place, scale, interdependence and flows. Conventional 'paper' map resources continue to be enormously valuable, but they can sometimes cause cognitive overload, called 'map shock' (Blankenship and Dansereau, 2000; Tergan and Keller, 2005). Of course, this is also a potential issue with GIS as the great GIS cartographer John Nelson indicates: 'How much information can fit in somebody's working memory at one time? Are you just throwing too much at them at once?' (Nelson, 2021). Fortunately, GIS can provide ways to manage 'map shock' such as selection of different basemaps or toggling of layers to step a narrative. The capacity of GIS to visualise 3D landscapes and spatio-temporal change can make a distinctive contribution to helping students 'see geography happen', using time-enabled layers or timelapses (Conway, 2023).

The Erasmus Plus GI Pedagogy project found that 'to visualise data and improve understanding of socio-economic and territorial reality requires a change in teachers' pedagogies' (GI Pedagogy Teaching Model, 2022), particularly by using Rosenshine's Principles of Instruction (Rosenshine, 2012). For example, GIS learning is especially enhanced by using small steps, checks for understanding, modelling, scaffolding and deliberate practice in order to build robust schema and fluency.

The value of metacognition in pedagogy is well-established (EEF, 2021) and beneficial when teaching with and about GIS. When providing expositions using a geospatial resource, it is beneficial to narrate the visualisation with metacognitive commentary, e.g. 'I'm just going to click here to toggle this layer and alter the basemap'. In this way, students experience GIS procedures and terminology so that when they use GIS themselves, there is already familiarity with the procedures. For example, if using a flight tracker, students can be taught that each aircraft icon is 'carrying' large amounts of data revealed by clicking it, including the origin and destination, speed, altitude, aircraft type, bearing and current latitude and longitude (Conway and Freeman, 2023).

How can we use GIS to teach longitude?

THE TECHNOLOGICAL IMPERATIVE

Technology must always be the servant of pedagogy and not the other way round. There have been occasions when edtech has been introduced with insufficient thought given to pedagogical considerations. Fortunately for geographers, GIS providers and apps seem to be less prone to such errors.

At one time, GIS apps required hefty downloads, but most apps are now browser-based as 'web GIS' meaning that GIS is now highly accessible and considerably easier to use. For schools, GIS is largely free (ArcGIS Online; National Library of Scotland Maps) or very low cost (Digimap for Schools), including access to digital Ordnance Survey maps, historical maps, aerial imagery and remote sensing data.

A key technological imperative at work is the new capacity and speed with which any data, including 'big data', can be processed, meaning that spatial visualisation is at everyone's fingertips in an unprecedented way.

Powerful geography is enhanced by the technological power of GIS. In the tradition of John Snow, big data can now be visualised to reveal previously obscure spatial patterns, thereby contributing to sustainable development goals (UN, 2015). Examples include the 'Lives on the Line' map showing significant disparities in life expectancy across short distances in London (Cheshire, 2012) and more recently the spatial epidemiology of Covid-19 (John Hopkins, CRC 2020–23).

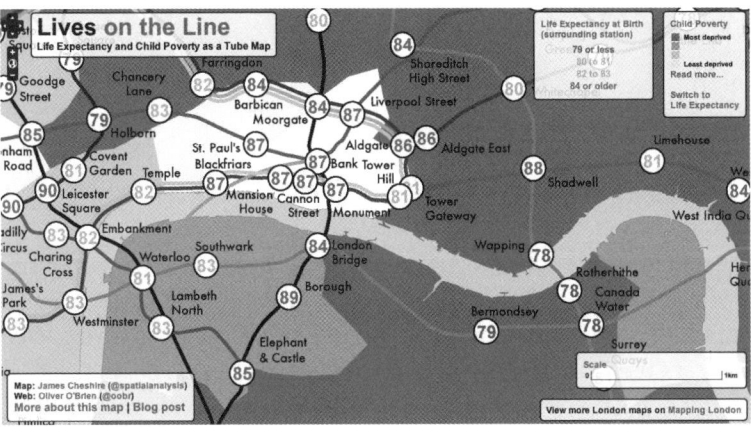

Figure 11.4 Lives on the Line map of London showing life expectancy at birth and child poverty on a Tube map (Cheshire, 2012)

Whereas students have always been able to map data in conventional ways, the layering of data has been challenging. Although this is possible using layers of tracing paper (sometimes called 'paper GIS'), web GIS makes such tasks considerably more ambitious and quicker, with more time to think about the geography. For example, students can map their own geo-referenced primary data to a web map, then very easily compare it with secondary data. With some apps, such as Esri's Survey123 (part of ArcGIS Online), this is possible in real time. New ways to access and compare historical spatial data are also becoming very accessible.

Collecting, presenting and analysing data using Survey 123.

How can we use LiDAR to go back in time to see the R Chelmer in 1706?

Some selected options are shown in Table 11.1 which are useful for teaching with or about GIS. All are accessible to school age students with the exception of ArcGIS Pro, which is for GIS experts. However, it is free to school users if required. All have capacity as 'GIS viewers', but not all of them offer the capacity to add your own primary or secondary data, which is an essential requirement for students.

A range of apps is available for outdoor activities such as walking, running and cycling which can be useful in schools, but they are not reviewed here.

Key features / WebGIS app	Basemap range	3D for relief	Route finding	Measurement tools	Spatial analysis tools	Street View	Ordnance Survey Maps	Add own data (basic GIS)	Add own data (full GIS)	Free option	Paid option
ArcGIS Online https://www.arcgis.com/	✓	✓	✓	✓	✓	✗	✓	✓	✓	✓	✓
Digimap for Schools https://digimapforschools.edina.ac.uk/	✓	✗	✗	✓	✗	✗	✓	✓	✗	✓	✓
Google Earth https://earth.google.com/	✗	✓	✓	✓	✗	✓	✗	✓	✗	✓	✗
Google Maps https://www.google.com/maps/	✗	✗	✓	✓	✗	✓	✗	✓	✗	✓	✗
Geography Visualiser https://teach-with-gis-uk-esriukeducation.hub.arcgis.com/pages/visualiser	✓	✓	✗	✓	✗	✗	✗	✗	✗	✓	✗
National Library of Scotland Maps https://maps.nls.uk/geo/explore	✓	✓	✗	✓	✗	✗	✓	✓	✗	✓	✗

Table 11.1 A comparison of selected GIS options for schools

THE VOCATIONAL IMPERATIVE

There is an ongoing explosion of demand for geospatial information in industry, government and NGOs, with a corresponding rapid expansion of career opportunities using GIS (see chapter 26). Consequently, geographers with such knowledge are very much in demand.

Whereas geographers can continue their long tradition of offering transferable skills to the workplace, in the last few years there has been a proliferation of direct employment pathways for geography specialists which simply did not exist in the past. For example, in the UK public sector, The Geospatial Commission (part of Government Digital Service) was founded in 2018 as an expert body within the Cabinet Office, responsible for framing the UK's geospatial strategy and integrating public sector geospatial activity. In the UK civil service, a rapidly growing Government Geography Profession has been set up, including a Government Head of Geography. It is vital that we make sure our students see these connections and prepare for them accordingly. As long ago as 2019, the geospatial sector was already becoming significant (see Figures 11.5 and 11.6).

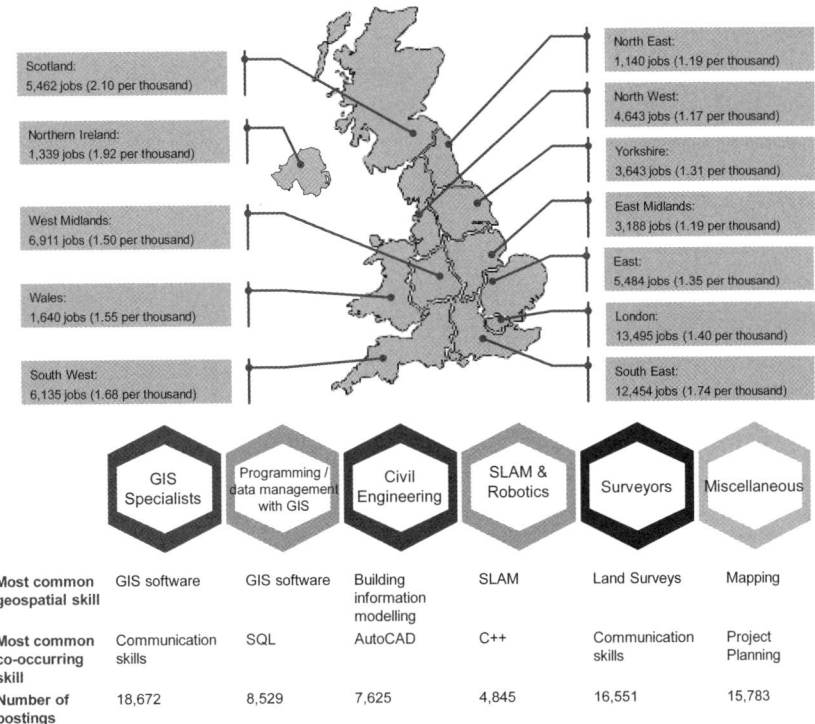

Figures 11.5 and 11.6 The demand for geospatial skills.

Source: *Frontier Economics* (2020), pages 9 and 22

The Geospatial Commission's strategy emphasises the essential role of GIS education, warning that 'The UK cannot realise the value from geospatial applications without an appropriately skilled workforce ... The pipeline of geospatial skilled individuals will depend on schools, further education and continuing professional development' with six partner bodies who 'play a key role in delivering this strategy' (Geospatial Commission, 2020; 2023). However, none of the partners have a direct role in education, so it would be useful to review this situation.

HOW DO DIFFERENT KNOWLEDGE TYPES APPLY TO GIS?

All too frequently, GIS is identified solely with procedural knowledge, but this is a mistake! 'Leaders rarely consider how to plan a curriculum for procedural knowledge in the same way as they do for substantive knowledge.' (Ofsted 2023) There is much more to GIS than developing the 'know how' to use it.

There is important disciplinary knowledge relating to GIS. The story of how we arrived at our current capacity to use GIS is a crucial aspect of geographical understanding, helping students to appreciate the pioneering work in this area so that its value for geography teaching now is properly recognised (Ofsted, 2021).

GIS also has a wealth of substantive knowledge. Students need to know how GIS works, along with the associated specialist vocabulary and concepts. Some examples would include digital layers, elevation profiles, filters, geo-referencing and remote sensing.

HOW CAN GIS BECOME 'PART OF THE FURNITURE'?

A useful approach to embed GIS in the curriculum may be to use one or more elements of the Substitution, Augmentation, Modification, Redefinition (SAMR) model to support the teacher in their on-ramp journey of using GIS. The model allows for an accessible starting point, only requiring a base level of procedural knowledge, but providing a pathway to embed GIS in the curriculum with a sense of empowerment.

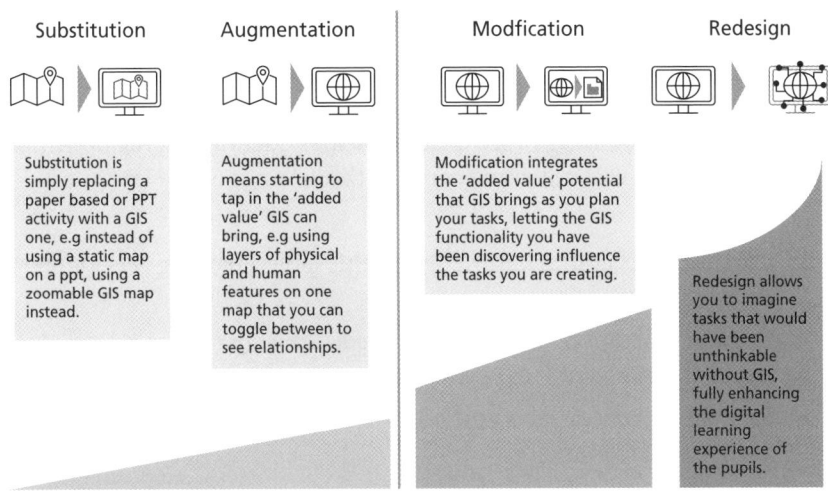

Figure 11.7 SAMR model

Source: Puentedura (2013), Hamill (2023)

SUBSTITUTION

This approach allows the teacher to become familiar with using some of the basic functionality of GIS apps in a teach-from-the-front context. Implementation could involve identifying one activity in a topic, then using a simple GIS instead of conventional resources.

Advantages:
- GIS demonstrated as an integral part of geography with introduction of GIS vocabulary.
- Some narration of tasks can help to familiarise GIS substantive knowledge, e.g. how GIS works.
- Accessible and achievable procedural knowledge; easy to incorporate.

AUGMENTATION

As confidence grows in the use of GIS, some tasks or activities can be modified. For example, a GIS layer can be introduced over a static map, including use of zooming to see different scales, identifying relationships between variables and adjusting some of the settings of the layers.

Advantages:
- Teacher's procedural knowledge is achievable, enabling development towards fluency.
- Regular narration of tasks to embed GIS substantive knowledge, e.g. wider range of GIS vocabulary.
- Students' procedural knowledge of GIS improves, supported by well scaffolded tasks.

MODIFICATION

At this stage, the teacher will bring the knowledge into reworking a topic, aware of the potential of GIS. In the example below, procedural knowledge is more advanced, but shows how students can make progress in steps rather than leaps! It may also be useful at this stage to use 'off-the-shelf' resources shared by others as models.

An example of how to teach coastal processes.

Advantages:
- GIS strongly informs curriculum planning, including strong substantive and disciplinary knowledge.
- GIS is becoming 'part of the furniture'.
- Students procedural knowledge of GIS becoming fluent, with scaffolding as appropriate.
- Subject content can be delivered that would have been difficult or impossible without GIS.

REDEFINITION

This approach allows for a complete redesign of teaching geography using GIS. It will hopefully involve students using the GIS apps themselves.

By this stage, the teacher will be able to deliver geographical content in a radically different way, opening up new routes to powerful geographical knowledge, such as a guided GIS task on Typhoon Haiyan.

Typhoon Haiyan.

La Palma volcanic eruption.

Downstream changes in the Glendun River.

Advantages:
- GIS has become 'part of the furniture'.
- Students have direct experience of GIS learning, including strong substantive and disciplinary knowledge.
- Students' procedural knowledge is fluent, enabling greater independence.
- Students have concrete experience of vocational routes using geography skills.

All teachers			Some teachers
Procedural knowledge level: **Basic** Teachers need to know how to use GIS	Procedural knowledge level: **Beginner to intermediate** Teachers need to know how to use GIS	Procedural knowledge level: **Intermediate to advanced** Teachers need to know how to use GIS to create complete more complex tasks	Procedural knowledge level: **Advanced** Teachers need to know how to use GIS to create complete more complex tasks and to have the ideas and vision to create innovative learning experiences based on GIS resources

Table 11.2 Steps for embedding GIS in the curriculum using the SAMR model. (Source: Hamill, 2023).

CONCLUSION

Hopefully the ideas outlined in this chapter will enhance the role of GIS in 'waking up the world to the power of geography' to create a better future (Dangermond, 2015).

GIS needs to become an integral 'part of the furniture' in school geography. There are persuasive imperatives from the perspectives of curriculum, pedagogy, technology and vocation. GIS learning is not only procedural. It also has significant substantive and disciplinary components.

Perhaps the most urgent points for action are the development of a coherent strategy to enhance GIS across the sector, including a significant increase in CPD support for curriculum progression.

Five reflection questions

1. Why should we use GIS in our geography lessons?
2. What GIS knowledge, understanding and skills do my students need?
3. How can we clarify ways that GIS knowledge can be substantive, disciplinary and procedural?
4. How should students' learning about GIS be sequenced and show curriculum progression?
5. What can we do to improve our own subject knowledge about GIS?

DIGGING DEEPER – THREE RESOURCES TO DELVE FURTHER

- Maps for use in Schools, National Library of Scotland. Available at: https://maps.nls.uk/guides/schools/ (Accessed: 28/05/2025)
- GIS in the Classroom – Practical Advice on Embedding GIS into your Geography Classroom (Blog Series), Tutor2U. Available at: https://www.tutor2u.net/geography/collections/gis-in-the-classroom-a-series-of-blogs (Accessed: 28/05/2025)
- Teach With GIS UK, ArcGIS Online Esri UK. Available at: https://teach-with-gis-uk-esriukeducation.hub.arcgis.com/pages/teach-about (Accessed: 28/05/2025)

REFERENCES

BBC Bitesize. (2023) Geographic information systems and choropleth maps. https://www.bbc.co.uk/bitesize/topics/zm38q6f/articles/z3rjwnb#zhgfydm2 (Accessed: 28/05/2025)

Blankenship, J. & Dansereau, D. F. (2000) The effect of animated node-link displays on information recall. *The Journal of Experimental Education*, Vol. 68(4), Summer 2000, pp. 293–308.

CCEA. (2007) Key Stage 3 Non-statutory Guidance for Geography. https://ccea.org.uk/downloads/docs/ccea-asset/Curriculum/Key%20Stage%203%20Non-Statutory%20Guidance%20for%20Geography.pdf (Accessed: 28/05/2025)

Cheshire, J. (2012) Lives on the line: life expectancy and child poverty as a tube map. https://jcheshire.com/featured-maps/lives-on-the-line/ (Accessed: 28/05/2025)

Conway, B. (2023) GIS in the Classroom Blog 8: How can we use GIS to investigate urban issues and challenges? Tutor2U. https://www.tutor2u. net/geography/blog/how-can-we-use-gis-to-investigate-change-in-human-geography-urban-issues-and-challenges (Accessed: 20 October 2023)

Conway, B. & Freeman, D. (2023) Part of the furniture: working together to embed GIS. Geographical Association Annual Conference 2023: Collaborative Geographies, 13–15 April 2023, Sheffield Hallam University. https://geography.org.uk/events-cpd/ga-annual-conference-and-exhibition/past-conferences/sheffield-2023-session-downloads/ (Accessed: 19 September 2023; Geographical Association log on required)

Counsell, C. (2018) Taking curriculum seriously. *Impact: Journal of the Chartered College of Teaching*, Issue 4.

Dangermond, J. (2015) Applying geography everywhere. Esri User Conference Plenary Session 2015. https://www.esri.com/about/newsroom/arcnews/awakening-the-world-to-the-power-of-geography/ (Accessed: 28/05/2025)

Dangermond, J. (2017) The science of where: our promise. *Arc News*, Winter 2017. https://www.esri.com/about/newsroom/arcnews/the-science-of-where-our-promise/#:~:text=The%20Science%20of%20Where%20is%20the%20science%20of%20digital%20transformation,science%20of%20insight%20and%20innovation (Accessed: 26 September 2023)

Department for Education. (2013) see: National curriculum in England: geography programmes of study. https://www.gov.uk/government/publications/national-curriculum-in-england-geography-programmes-of-study (Accessed: 28/05/2025); Geography GCSE subject content. https://assets.publishing.service.gov.uk/government/uploads/system/uploads/attachment_data/file/301253/GCSE_geography.pdf (Accessed: 28/05/2025); and GCE, AS and A level subject content for geography for teaching in schools from 2016. https://assets.publishing.service.gov.uk/government/uploads/system/uploads/attachment_data/file/388857/GCE_AS_and_A_level_subject_content_for_geography.pdf (Accessed: 28/05/2025)

Education Endowment Foundation (EEF). (2021) Metacognition and self-regulation. https://educationendowmentfoundation.org.uk/education-evidence/teaching-learning-toolkit/metacognition-and-self-regulation#:~:text=Metacognition%20and%20self%2Dregulation%20strategies%20are%20most%20effective%20when%20embedded,methods%20for%20mathematical%20problem%20solving (URL set for quote from 1:10-1:20) (Accessed: 20 October 2023)

Esri. (2019) Raster is faster but vector is corrector. https://www.esri.com/content/dam/esrisites/en-us/media/pdf/teach-with-gis/raster-faster.pdf (Accessed: 28/05/2025)

Frontier Economics. (2020) Demand for Geospatial Skills – Report for the Geospatial Commission. Geospatial Commission. https://assets.publishing.service.gov.uk/government/uploads/system/uploads/attachment_data/file/1073139/_Demand_for_Geospatial_Skills__report_.pdf (Accessed: 28/05/2025)

Geospatial Commission. (2020) Unlocking the power of location: The UK's geospatial strategy. https://www.gov.uk/government/publications/unlocking-the-power-of-locationthe-uks-geospatial-strategy#:~:text=Details,the%20use%20of%20location%20data (Accessed: 27 September 2023)

Geospatial Commission. (2023) UK Geospatial Strategy 2030: unlocking the power of location. https://assets.publishing.service.gov.uk/government/uploads/system/uploads/attachment_data/file/1162795/2023-06-15_UK_Geospatial_Strategy_2023_.pdf (Accessed: 28/05/2025)

Puertas-Aguilar, M-A., Conway, B., de Lázaro-Torres, M-L., de Miguel González, R., Donert., Lindner-Fally, M., Parkinson, A., Prodan, D., Wilson, S. & Zwartjes, L. (2022) A Teaching Model To Raise Awareness of Sustainability Using Geoinformation. Espacio, Tiempo y Forma Serie VI, Geografía 15, Universidad Nacional de Educación a Distancia, UNED. https://www.gilearner.ugent.be/wp-content/uploads/33687-Texto-del-articulo-91937-1-10-20220722.pdf (Accessed: 28/05/2025)

GI Pedagogy Toolkit. (2022) GI Pedagogy Toolkit of Innovative Pedagogical Approaches for Teaching with GIS, including the GI Pedagogy Framework and Model (IO2). Ghent University, Belgium. https://www.gilearner.ugent.be/publications-gi-pedagogy/ (Accessed: 28/05/2025).

Haywood, A. Mapping a London epidemic. National Geographic Resource Library. https://education.nationalgeographic.org/resource/mapping-a-london-epidemic/ (Accessed: 28/05/2025)

HM Government. (2020 onwards) T Level Subjects. HM Government Skills for Life. https://www.tlevels.gov.uk/students/subjects (Accessed: 28/05/2025)

Johns Hopkins Coronavirus Resource Center (CRC) (2020–23) COVID-19 Dashboard. Center for Systems Science and Engineering (CSSE) at Johns Hopkins University (JHU). https://coronavirus.jhu.edu/map.html (Accessed: 28/05/2025)

Mccrea, P. (2022) A beginner's guide to instructional coaching. Steplab. https://steplab.co/resources/papers/BP6w3bcs/A-Beginners-Guide-to-Instructional-Coaching#:~:text=%E2%80%94By%20Peps%20Mccreaandtext=Instructional%20coaching%20involves%20one%20teacher, steps%20to%20improve%20their%20practice (Accessed: 28/05/2025)

National Institute of Teaching. (2023) Intensive Training and Practice (ITAP). https://niot.org.uk/insights/What-is-intensive-training-and-practice (Accessed: 28/05/2025).

Nelson, J. (2021) What is a bivariate map? Well I'm so glad you asked... intro to part 1 of 3. John Nelson Maps, YouTube, 26 March 2021. https://youtu.be/oNwi0O4IvHc?si=jMwutvAk4VXdkh8landt=70 (URL set for quote from 1:10-1:20) (Accessed: 28/05/2025)

Oak National Academy. (2024) Oak National Academy Curriculum Resources: KS1 & KS2 geography curriculum. https://www.thenational.academy/teachers/curriculum/geography-primary/units (Accessed: 28/05/2025)

Oak National Academy. (2024) Oak National Academy Curriculum Resources: KS3 & KS4 geography curriculum. https://www.thenational.academy/teachers/curriculum/geography-secondary-aqa/units (Accessed: 28/05/2025)

Ofsted. (2021) Research and analysis: Ofsted Research review series: geography. https://www.gov.uk/government/publications/research-review-series-geography/research-review-series-geography (Accessed: 19 September 2023)

Ofsted. (2023a) Research and analysis: Getting our bearings: geography subject report. at: https://www.gov.uk/government/publications/subject-report-series-geography/getting-our-bearings-geography-subject-report (Accessed: 19 September 2023)

Ofsted (2023b) 'Education inspection framework' Available at: https://www.gov.uk/government/publications/education-inspection-framework/education-inspection-framework-for-september-2023 (Accessed: 27 September 2023)

Puentedura, R. (2013) SAMR: Moving from Enhancement to Transformation. Presentation to 2013 AIS ICT Management and Leadership Conference. http://www.hippasus.com/rrpweblog/archives/2013/05/29/SAMREnhancementToTransformation.pdf (Accessed: 28/05/2025)

Qualification and Curriculum Authority. (2007) Geography programme of study: programme of study for key stage 3 and attainment target. https://edumedia-depot.gei.de/bitstream/handle/11163/221/736434089_2007_A.pdf?sequence=2 (Accessed: 28/05/2025)

Rosenshine, B. (2012) Principles of instruction – research-based strategies that all teachers should know. *American Educator*, Spring 2012. https://www.aft.org/sites/default/files/Rosenshine.pdf (Accessed: 28/05/2025)

Tergan, S. O. & Keller, T. (eds) (2005) *Knowledge and Information Visualization: Searching for Synergies*. Berlin Heidelberg: Springer-Verlag.

West, H. & Horswell, M. (2018) GIS has changed! Exploring the potential of ArcGIS Online. *Teaching Geography*, 43(1), pp. 22–24.

Zwartjes, L., de Lázaro-Torres, M-L., Lindner-Fally, M. & Parkinson, A. (2017) GI Learner: Curriculum opportunities for spatial thinking (including GI Learner competencies). https://www.gilearner.ugent.be/wp-content/uploads/GI-Learner-competencies.pdf (Accessed: 28/05/2025)

12. USING MODELS IN THE GEOGRAPHY CLASSROOM
HANNAH STEEL
@GEOGSTEEL

'All models are wrong, but some are useful.' (George E.P. Box, 1976)

As teachers, often constrained by the limits of time, it is not uncommon for us to readily accept models at face value. As we flip through geography textbooks and seamlessly incorporate diagrams into our lesson presentations, we often embrace these models without a second thought or questioning. The saying 'All models are wrong, but some are useful', attributed to the statistician George Box, prompts us to reflect on this inherent tendency. This chapter explores how, in our role as educators, we may inadvertently adopt models without delving into their complexities, shedding light on the importance of a more nuanced understanding in our teaching practices in the geography classroom.

WHAT ARE MODELS?
Without realising, we use models all the time. They are the mental images we create to better understand, interpret and make assumptions about the complexities of the world around us. These 'mental models ... influence how we understand the world and how we take action' (Senge, 1990, p. 8). Therefore, it is important to emphasise that models are human constructs that help us better understand real-world systems and

'are used when it is easier to work with substitutes than with an actual system' (Andrew Ford, 2010).

There are four main types of modelling we often see in geography and that are used in creating a developed model for a real-world system:
- conceptual models
- interactive demonstrations
- mathematical and statistical models
- visualisation models.

CONCEPTUAL MODELS

Conceptual models serve as foundational frameworks when we're diving into the complexities of real-world systems. These models are qualitative in nature, focusing on highlighting essential connections within actual systems and processes. In our geography classrooms, we may often begin with conceptual models, particularly when introducing new concepts and ideas for the first time.

INTERACTIVE DEMONSTRATIONS

Interactive demonstrations are less often used in the geography classroom but should not be ignored as they can act as a bridge that closes the gap between conceptual models and the real world. They represent physical models for more real-world complex systems that can be scaled down and manipulated in a classroom setting. For example, wave tank demonstrations showing the impacts of coastal defences or stream tables that simulate floodplains and deltas.

Just like conceptual models, hands-on demonstrations can be a powerful tool to bring models to life and help students grasp complex concepts. For instance, when teaching about processes such as ocean acidification or the natural carbon cycle, I often show how water absorbs heat and carbon. In this demonstration (see Figure 12.1), students can see what happens when we blow carbon dioxide into water using a straw. The bromothymol blue, acting as a pH indicator, turns yellow when carbon dioxide reacts with the water, indicating increased acidity. We also show how heating the 'yellow' or CO_2-infused water releases carbon dioxide back into the atmosphere, especially when discussing oceans as natural carbon reservoirs. These demos not only let geography teachers collaborate with the science department, promoting cross-curricular learning, but they can also be referred back to during geography field trips to the coast. This helps clear up misconceptions that the ocean is 'turning into acid', even though its pH is measured at an average of 8.1. You can try this out using seawater, red cabbage juice and a pH portable probe. (See ocean acidification in a cup: https://encounteredu.com/take-action/ocean-acidification-in-a-cup.)

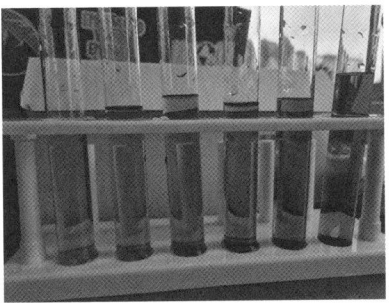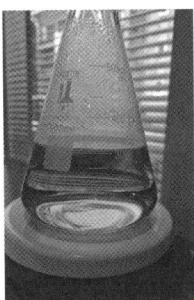

Figure 12.1 and 12.2a and b Classroom demonstration on ocean acidification and the natural carbon cycle

MATHEMATICAL AND STATISTICAL MODELS

Mathematical models encompass both analytical and numerical models, while statistical models prove valuable for discerning patterns and inherent relationships within datasets. In geography, this is a regular practice, involving the constant comparison of data and the analysis of graphs to gain insights into the surrounding world. The demographic transition model, population pyramids, choropleth maps or climate change models are some examples of these types of models.

VISUALISING MODELS

Lastly, visualising models serve as a direct bridge between data and compelling graphics or images, shedding light on the complexities of a system. This incorporates map overlays, animations, image manipulation and image analysis. These resources prove especially valuable in the geography classroom, offering spatial contexts that are often excluded in other model types. Throughout the teaching of a concept, idea or process, exposing our students to these various models becomes crucial. It is essential to be transparent about the nature of these models – acknowledging their critique-ability and their role as simplified representations of complex systems. Engaging students in discussions about the type of model being used and its creation process holds particular significance, a practice I've found particularly impactful, especially when employing visualising models such as ArcGIS maps with data overlays (see chapter 11).

HOW DO WE DETERMINE HOW USEFUL A MODEL IS?

As geographers, we may attempt to understand and interpret the complexities of real-world systems more explicitly in our classrooms through the aid of models. We may attempt to make sense of the world through geographical thinking concepts such as (Jackson, 2006):

- space and place
- scale and connection
- proximity and distance
- relational thinking.

Yet, are we applying the same 'geographical thinking' explicitly to the very models we use in our classroom? Are we using our 'geographical lens' to interpret, critique and challenge the models we use? Finally, once critiquing our models, does this change our awareness of how a system was perceived? Does the model remain justified in its use for that system, or does it open our students up to more misconceptions and a stereotyped narrative?

With time constraints in our classrooms, it can be difficult to understand the contexts and reflect on every single model we use in our geography lessons. Therefore, it is important we think carefully about the reason behind using the model: is it simply a visual aid to help geographical understanding? Is it being compared to a case study example? Or does the entire lesson revolve around the model? Something easily done when teaching concepts or processes such as the formation of waterfalls, the carbon cycle or the demographic transition model to name a few.

As curriculum makers reflecting on the effectiveness of models, the following criteria can be helpful to determine what makes a model useful in the geography classroom:

1. **Starting point:** Start by asking 'What do I want the students to know?' This helps in evaluating the essentiality of the model and its potential contribution to knowledge and understanding of the geographical concepts you are trying to teach. If a model doesn't align with these goals, or contributes to cognitive overload, consider rethinking its use.
2. **Complexity check:** Assess whether the model is too complex for the students. Consider their age and cognitive needs to recognise if the model may overload their working memory. The model may need to be scaffolded and broken down into steps before the finished product is shown to the class.
3. **Avoiding oversimplification:** Question whether the model oversimplifies the subject to the extent that it fosters misconceptions. Examine whether crucial information is omitted and deliberate on how to fill or address these gaps. Evaluate how certain models may contribute to the risks associated with presenting a single story or bias. Assess then whether the model should be avoided or used for critique and critical analysis with your students. Could they work on creating a new and 'improved' model because of their findings?
4. **Timing and integration:** Consider when and how to introduce the model. Relate it to the existing schema and contemplate its place within the curriculum, scheme of work (SoW) and individual lessons. For example, you may begin with conceptual models when introducing new ideas followed by more real-world data-driven models later in the lesson or schema to build deeper understanding.
5. **Comparison with real-world examples:** Decide on whether the model will be compared to real-world examples to highlight its limitations. Providing students with opportunities to explore these limitations through case studies and other real-world instances helps illustrate the dynamic nature of the world. Does the model manifest what we see in some real-world examples but not others? Reflect on why this may be.

6. **Creator and historical context:** Investigate who created the model and dive deeper into its historical context. Understanding the when, why, where and how of its creation becomes pivotal, influencing its relevance in the context of geography teaching. Similarly, with models that oversimplify, biases can be found when models are used outside the context they were designed for.

Numbers 5 and 6, comparing models with real-world examples and examining models in context, are where we can integrate the geographical thinking concepts through the critical analysis of the models we use in the geography classroom. It is also important to note that this is not an exhaustive list of criteria with which to choose the models we teach. It is also essential that we reflect on the bigger picture that a model is but a tool to aid in the understanding of geographical concepts.

Five reflection questions

1. What criteria are you using for choosing geographical models?
2. Where and when will the model appear in your schema? Why?
3. What is the historical and spatial context of the model?
4. How does the model create worthwhile and meaningful knowledge?
5. How will you take your students beyond the simplifications of models to create powerful and meaningful geography?

DIGGING DEEPER - THREE RESOURCES TO DELVE FURTHER

- Samingpai, B. (2023) Using physical models to aid students' understanding of coastal landscapes. *Teaching Geography*, 48(3), pp.124-26.
- Brady, B. et al. (2023) Using physical models to improve geographical learning. *Teaching Geography*, 48(2), pp.72-5. Available at: https://portal.geography.org.uk/journal/view/J9781899086010 (Accessed: 28/05/2025)
- Kit Marie Rackley. (2020) Decolonising Geography. Geogramblings. Available at: https://geogramblings.com/2020/08/01/decolonising-geography/ (Accessed: 28/05/2025)

REFERENCES

Biddulph, M., Lambert, D. & Balderstone, D. (2015) *Learning to Teach Geography in the Secondary School: A Companion to School Experience.* Routledge.

Box, G. E. P. (1976) Science and statistics. *Journal of the American Statistical Association*, December, pp. 791–799.

Brierley, G., Fryirs, K., Cullum, C., Tadaki, M., He, Q. H. & Blue, B. (2013) Reading the landscape: integrating the theory and practice of geomorphology to develop place-based understandings of river systems. *Progress in Physical Geography*, 37, pp. 601–621.

Encounter Edu. (n.d.) 'Ocean acidification in a cup'. Available at: https://encounteredu.com/take-action/ocean-acidification-in-a-cup (Accessed: 28/05/2025)

Ford, A. (2009) *Modeling the Environment.* 2nd edn. Washington D.C.: Island Press.

Geographical Association (2012) Thinking geographically: The geographical association's consultation response on the national curriculum. https://geography.org.uk/wp-content/uploads/2023/04/GA_GINCConsultation_ThinkingGeographically_NC_2012.pdf (Accessed: 28/05/2025)

Hawley, D. (2023) Are you a model geographer?. Lecture presented at Geographical Association 2023 Conference, Lecture Plus 11 (KS2–P16), Chair of GA Physical Geography Special Interest Group.

Hickel, J. (2017) *The Divide: A Brief Guide to Global Inequality and its Solutions.* London: William Heinemann.

Jackson, P. (2006) Thinking geographically. *Geography*, 91(3), pp. 199–204. https://doi.org/10.1080/00167487.2006.12094167 (Accessed: 28/05/2025)

Ortega-Becerril, J.A., Polo, I. & Belmonte, A. (2019) Waterfalls as geological value for geotourism: the case of Ordesa and Monte Perdido National Park. *Geoheritage*, 11, pp. 1199–1219. https://doi.org/10.1007/s12371-019-00366-1 (Accessed: 28/05/2025)

Roberts, M. (2013) *Learning Through Enquiry.* Sheffield: Geographical Association.

Rushton, E.A.C. (2020) The language of Geography. *EAL Journal*, 11, pp. 46–49.

Senge, P. M. (1990) *The Fifth Discipline: The Art and Practice of The Learning Organization.* New York: Doubleday/Currency.

Steel, H. (2023) All models are wrong. *GeogSteel Blog*, 13 March. https://geogsteel.wordpress.com/2023/03/13/all-models-are-wrong/ (Accessed: 28/05/2025)

Walshe, N. (2017) 'Literacy', in Jones, M. (ed.) *The Handbook of Secondary Geography*. Geographical Association, Chapter 15.

13. ORACY STRATEGIES TO IMPROVE GEOGRAPHICAL WRITING
LOUISE HOLYOAK
@LOUISEHOLYOAK

'Reading and writing float on a sea of talk'. (James Britton, 1983)

WHAT IS ORACY IN GEOGRAPHY?

Oracy uses spoken language to express ideas, build understanding and interact with others (Voice 21, 2022). Within geography, this can serve as a means for students to develop geography-specific skills and communicate their understanding of geographical concepts. When talk is planned, scaffolded and structured within the classroom, oracy can enhance critical thinking and analytical abilities, enabling students to engage with real-world geographical issues and embrace a variety of perspectives. Furthermore, the Education Endowment Foundation (EEF) found that when implemented consistently (at least three times a week), oracy approaches have a high impact on pupil outcomes and can lead to up to five months' additional progress in secondary students (EEF, 2021).

The following oracy strategies that I am going to discuss were developed by Voice 21, an organisation formed through collaboration of innovative practice at School 21, the University of Cambridge and research by the EEF. School 21 is a 4–18 school based in East London, and oracy strategies are used throughout all phases and all subjects. As the first geography teacher at School 21, I learned about these strategies and adapted them to be used to develop students' geographical thinking. I have now moved to another school where I am introducing these strategies to students who have not necessarily practised them since age four, so I have included some tentative thoughts on how to introduce oracy within the classroom from scratch.

Oracy encompasses a wide range of strategies, and it is not possible to include them all here. The focus of the strategies found in this chapter is exploratory talk. This is where learners engage 'critically but constructively' with each other's ideas (National Literacy Trust, 2012) and consensus building is valued over winning a debate. Students are supported and guided towards asking probing questions to encourage deeper thoughts on issues and where there are disagreements, detailed reasoning and alternative viewpoints should be given (Mercer, 2008). Ultimately, the purpose of exploratory talk is to aid understanding and provide alternative views to support analytical thought and evaluation, both of which are important within geography.

TALK ROLES

One way to develop effective exploratory talk in the geography classroom is the use of talk roles. These are important for structuring the talk and being explicit in the type of conversation you are expecting students to have (British Council, n.d.). Table 13.1 outlines six different talk roles which can be used to hone geographical skills and act as a verbal rehearsal for written work. In the interest of brevity and applicability to exam classes, I have conflated geographical skills and exam command words as both require students to demonstrate their understanding and application of geographic knowledge in a way that lends itself well to the stated talk roles. I will then exemplify how these can be used within a lesson.

Talk role	Description	Will say	Geographical skill/ Exam command word
Instigator	Begins the conversation or considers new topics of conversation	I would like to start by saying... I think we should consider... We haven't yet talked about... Let's also think about...	Identify/State/Name Describe Outline Examine
Prober	Explores the argument further by asking for evidence or justification of points	What do you think would be the effect of...? Why do you think...? Can you provide an example to support what you are saying?	Explain Suggest Assess Evaluate Justify Discuss To what extent Examine
Builder	Develops, adds to or expands on an idea	I agree, and would like to add... Building on that idea, I think... Linking to what X has said, I think... Yes, and also...	Describe Compare Discuss Outline
Challenger	Gives reasons to disagree or presents an alternative viewpoint	I disagree with you because... You mentioned X, but what about...? To challenge you, I think... I understand your point of view, but have you thought about...?	Assess Evaluate Justify Discuss To what extent
Clarifier	Simplifies or makes things clearer by asking short questions	What do you mean when you say...? Can you explain more about...? Does that mean...? Please can you clarify what you meant by...?	Explain Suggest Discuss Outline Examine
Summariser	Identifies the main ideas from the discussion. This could be during the discussion to help move the conversation forward, or at the end of the discussion	Overall, the main points were... The main points raised today were... Our discussion focused on... The three main things we talked about were...	Assess Evaluate Discuss To what extent

Table 13.1 Table showing Voice 21 talk roles paired with geography exam command words. (Source: adapted from Voice 21 (2022) and British Council (n.d.)).

In terms of where oracy strategies fit within the structure of a lesson, I tend to find these work best as preparation for writing tasks. Therefore, oracy would come after the 'Do now' and recap of prior learning and before independent practice. I have found that these strategies led to increased confidence for all learners, as they have had the chance to articulate their answers before committing pen to paper. This also ensures that there is a clear purpose for the talk, rather than merely using these strategies for the sake of it. Some of the following strategies are useful for introducing a topic or idea, whereas others would be more suited to reviewing content and preparing to write an extended piece on an idea. Careful planning is key to ensure that students are getting the maximum benefit.

ORACY STRATEGIES

DATA ANALYSIS: BUILDER

This can be completed in pairs or trios.

1. As demonstrated in Figure 13.1, display a graph or map on the board.
2. Partner A acts as the instigator to describe the trend or pattern shown in the figure.
3. Partner B builds on this and gives some evidence to support the trend.
4. Partner C (or A and B working together) gives an anomaly.

A variation on this can also be used to discuss the advantages and disadvantages of various data presentation methods. Partners can either share advantages and disadvantages or offer one of each.

Describe the world's agricultural production (amount of food).

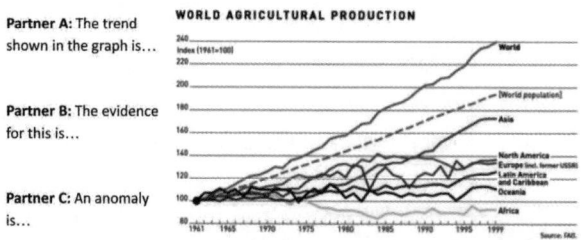

Figure 13.1 Slide demonstrating use of the talk role 'builder' to describe trends in data presentation

DESCRIBING IMAGES: BUILDER

The aim is to encourage students to look closely at an image and describe it using geographical vocabulary. This can be completed in pairs.

1. On the board, display an image and a list of vocabulary that students should attempt to use to describe the geographical features they can see. (An example is shown in Figure 13.2.)
2. Partner A will begin by describing something they can see.
3. Partner B will build on this by adding something else that they can see.
4. Repeat this several times with different images.

This can work well as a revision activity.

Figure 13.2 Slide showing how the talk role 'builder' can be used to describe images using geographical vocabulary

EVALUATION: PROBING PAIRS

Before using this strategy, it is important to set up the premise for the lesson with a statement that students will be responding to or an exam question that requires evaluation or analysis. This strategy can be used in a variety of settings: to introduce students to a new case study, build on ideas within an extended GCSE case study or as a revision technique.

1. On the board, display a fact about a case study, as in Figure 13.3.
2. Partner A will instigate by using the sentence stem (appropriate to the question being asked).

3. Partner B is required to listen carefully to A's initial response and ask a probing question to encourage them to think beyond their initial response.
(Pairs swap for the next prompt.)

---

*Exam Q: For a named **megacity**, **evaluate the success** of different **top-down** development strategies. (8)*

| The number of people taking the Mumbai Monorail is much lower than expected: only 15,000 per day and most of them tourists. |

Partner A: This indicates that the scheme **was / was not** successful because…

Partner B: What do you think would be the impact of…?
Why do you think…?
Can you provide an example to support what you are saying?

Figure 13.3 Slide illustrating use of probing pairs to evaluate the success of a particular strategy in response to a GCSE question

---

Tip: Make sure you are listening to the students' conversations and allowing adequate time for the discussion to progress beyond Partner A's initial reaction to the fact. In order for this to have the desired effect, Partner B will need time to ask the question and allow Partner A to give a detailed response.

SILENT SUMMARISER

The role of the silent summariser is to listen to a discussion taking place and take notes of the key points raised. This can be added to the probing pairs task outlined above and can serve as a useful record of the conversation.

At the end of the discussion, silent summarisers can be asked to share all or part of their notes with the class. If they are completed on mini-whiteboards, they can then be used by all three students to write up an answer or part of an answer to the question. An example of this is shown in Figure 13.4.

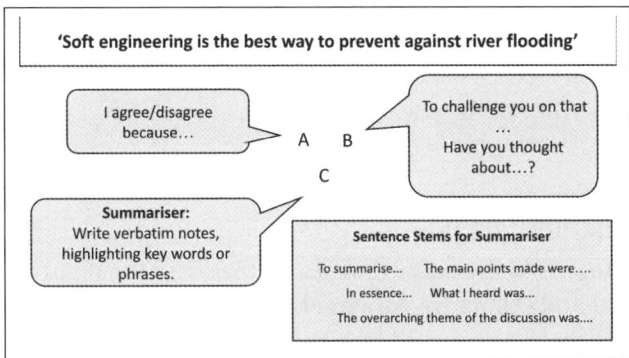

Figure 13.4 Slide demonstrating the role of the 'silent summariser'.

ASSESSING AND JUSTIFYING: RANKING IN PAIRS

This strategy works best once students have already covered material relating to a topic or case study and are now required to rank them in order to assess their significance for an extended writing piece or longer-answer exam question. This strategy can be used to support students in structuring an extended essay, where they are likely to have to compare the significance of different factors or case studies.

1. As shown in Figure 13.5, display a list of factors relating to the question.
2. In pairs, the students are required to discuss and rank the factors according to the criteria you have given them.

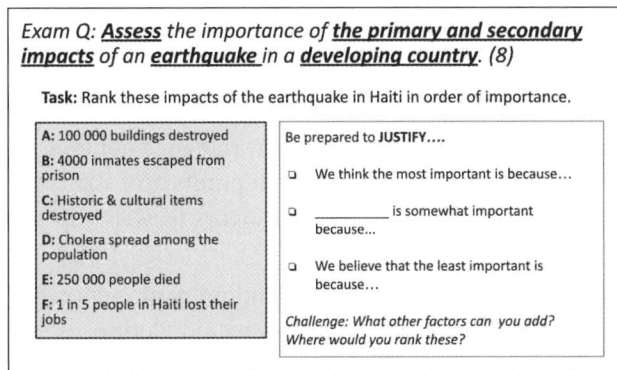

Figure 13.5 Slide showing use of ranking and justifying in order to assess

> Tip: This is also a strategy that takes longer than you might think, especially if you want students to build consensus and rank between four and six factors from most to least important.

ANALYSIS AND EVALUATION: QUADS

The aim of this strategy is to use oracy to rehearse the structure required for writing analytical and evaluative paragraphs at GCSE and A-level. In my experience, some students respond by writing everything they know about a point or case study, which is often very descriptive and does not include the type of high-level analysis required for successful answers. By breaking down a model paragraph and highlighting that several distinct skills are required, these can then be rehearsed by using different talk roles. Figure 13.6 shows an example of how to structure this.

As the name suggests, this is done in groups of four.

1. The role of the **instigator** is to essentially say a topic sentence that will be used to introduce the theme or idea of the paragraph. The **builder** will then add explanatory detail to this point.
2. The **challenger** will then introduce their point with 'However', indicating an analytical sentence is following.
3. The **summariser** will then begin their contribution with 'Therefore', thus indicating that they are evaluating the points raised by the builder and the challenger, and reaching a mini-conclusion.

ANALYSIS AND EVALUATION: WHOLE-CLASS ESSAY

A whole-class 'talking essay' can then be created using this strategy by assigning each quad a different factor or case study. Once the students have discussed in their groups, they can all say their sentence out loud. A silent summariser can be used to capture the main points and structure of the essay, which can then be distributed to students to write up the essay afterwards.

I have also used this as a revision strategy with a random name generator. I introduce an essay question, discuss what we would define in the introduction, then call on a member of the class to say the first point.

Once they have said it out loud, I generate the next name to build on their point and so on until all members of the class have contributed. This requires students to have revised adequately before the lesson and also listen carefully to their peers as they could be called on at any point to build, challenge or evaluate a point.

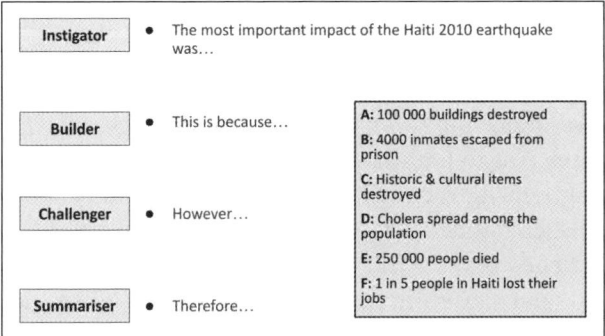

Figure 13.6 Slide demonstrating use of quads (instigator, builder, challenger, summariser) to build an analytical paragraph

Tip: It is worth considering what you will do if a student states something that is inaccurate during this task. Are you going to correct the student yourself or allow students to correct each other? What will the protocol for this be? This could include a non-verbal signal (e.g. crossing arms to indicate disagreement or holding out their thumb to indicate they want to speak) which would need to be agreed before commencing the task.

INTRODUCING ORACY

If you have never used oracy strategies in the classroom before, it can be daunting to introduce something different. My advice would be to start small, with paired talk and data response or probing pairs. Once you and your class have developed confidence with these, then think about introducing silent summarisers and expanding into quads and whole-class essays.

Off-task chat and sitting in silence are the two main challenges I have encountered when using oracy in the classroom for the first time. Despite being opposite problems, the solutions are the same. The first is to outline your expectations very clearly *before* starting the task and make it clear that you will be circulating to ensure *all* students are on task. Remember, just because it is a talk task does not mean that you should not reward/sanction students in the same way you would for written work. If a student is either not talking at all or talking about something other than what is required, they should be reminded in line with your school policy. Likewise, more importantly when introducing oracy for the first time, it is important to highlight and reward good practice for students who are engaging with the tasks. Second, think carefully about the seating plan and where students are located in the room. Students sitting next to their friends may be more tempted to continue off-task discussions from break and those sitting next to sworn enemies are not going to have productive conversations. Also consider what you will do if students are absent and are therefore missing a partner – for example, should they turn around to create a three or should they move to anyone else missing a partner?

CONCLUSION

In conclusion, oracy strategies can be used to develop geographical skills and enhance students' ability to communicate their understanding of geographical ideas. Intentional talk, which has been structured to allow students to focus on different skills, fosters critical thinking and allows students to delve into geographical issues and engage with various perspectives. By integrating opportunities for exploratory talk into your lessons and offering a clear purpose for the discussions, students will be able to prepare effectively for written tasks relating to exam command words. While introducing oracy into the classroom may present challenges at first, these can easily be overcome by setting clear expectations, providing guidance and structure, and recognising positive student engagement. As these strategies become a part of your classroom routine, they will contribute to an enriched geography education by promoting discussion, analysis and evaluation, all of which are essential in the study of geography.

> **Five reflection questions**
> 1. How can using oracy strategies enhance students' critical thinking skills?
> 2. In what ways can talk roles be used to support understanding of geographical concepts?
> 3. How can oracy strategies be best placed within individual lessons across a scheme of learning to maximise their impact on student confidence and ability in approaching written tasks?
> 4. How will you introduce oracy strategies to your classes in order to ensure productive discussions?
> 5. How might the use of oracy strategies support the overall aims of your geography curriculum?

> **DIGGING DEEPER - THREE RESOURCES TO DELVE FURTHER**
> - Voice 21 has a wide range of materials available on its website, including its oracy framework (ideas on how to develop speaking across the four strands of oracy), its oracy benchmark strategy (for introducing oracy more widely within schools) and the wider impacts of oracy within education. Available at: https://voice21.org/publications/ (Accessed: 28/05/2025)
> - British Council (n.d.) It's good to talk: oracy lesson plan. Available at: https://www.britishcouncil.org/sites/default/files/its_good_to_talk.pdf (Accessed: 28/05/2025). This resource from The British Council brings together the resources from Voice 21 and University of Cambridge to show how oracy can be introduced in the classroom, as well as developing ideas beyond those covered here.
> - For ideas on how to assess oracy in the classroom, this resource from the University of Cambridge provides a range of strategies. Available at: https://www.educ.cam.ac.uk/research/programmes/oracytoolkit/ (Accessed: 28/05/2025)

REFERENCES

British Council. It's good to talk. https://www.britishcouncil.org/sites/default/files/its_good_to_talk.pdf (Accessed: 28/05/2025)

Britton, J. (1983) 'Writing and the story of the world' in B. M. Kroll & C. G. Wells (Eds.), *Explorations in the Development of Writing: Theory, Research, and Practice*, (pp. 3–30). New York, NY: Wiley.

Educational Endowment Foundation (EEF). (2021) Oral language interventions. https://educationendowmentfoundation.org.uk/education-evidence/teaching-learning-toolkit/oral-language-interventions (Accessed: 07/11/2023)

Mercer, N. (2008) Three kinds of talk. https://thinkingtogether.educ.cam.ac.uk/resources/5_examples_of_talk_in_groups.pdf (Accessed: 07/11/2023)

Voice 21. (2022) What is oracy? https://voice21.org/what-is-oracy/ (Accessed: 28/05/2025)

14. LITERACY IN GEOGRAPHY
BETHANY ALDRIDGE
@MSALDRGEOG

'Knowledge is not static; it is shared, recycled and constructed. Knowledge must be viewed critically. Teachers must *scaffold*, or build bridges, to facilitate learning'. (Gloria Ladson-Billings, 1995)

WHY IS LITERACY IMPORTANT?

In my current setting, I serve a disadvantaged community. My students would traditionally be marginalised from geography in post-16 and higher education settings due to their ethnicity and socio-economic class (see chapter 18). In addition to this, many of them have reading ages significantly lower than their chronological age. Analysis of assessment data shows there are moderate, statistically significant correlations between reading ability and achievement in GCSE geography (GL Assessment). In a whole-school literacy drive, literacy consultant Sarah Green (@thelitcoach) works with my school and then department, to weave reading literacy into our curriculums.

There is extensive research which emphasises how important literacy is for future success in life. Reading provides opportunities to acquire vocabulary but also for learning (Ricketts et al., 2014) and by improving literacy in all subjects, educational attainment improves (Quigley and Coleman, 2021).

In the English key stage 3 geography curriculum, literacy is key to enabling students to read and understand geographical information such as maps, graphs, photographs and text. Literacy enables students to analyse, interpret and evaluate this information which is crucial to understanding and making sense of the human and physical world around us. Literacy skills are essential for students to communicate their understanding of geographical concepts and issues verbally and in writing. This is important because being able to explain and write about these concepts and issues helps students develop a deeper understanding of geography, and to be able to apply their knowledge in real-world contexts.

In Ofsted's geography subject report, it is acknowledged that many schools' key stage 4 curriculums are GCSE specifications and do not go beyond these (2023). Literacy can act as a vehicle for reading around and beyond the GSCE specification when many departments and teachers are short of time (see https://reteach.org.uk/subject/geography).

WHAT IS DISCIPLINARY LITERACY?

Disciplinary literacy is an approach from the Educational Endowment Foundation (EEF). The EEF suggest that disciplinary literacy is a vessel for teaching literacy in every subject, particularly in secondary settings. Teachers of that subject, teach students how to read, write and communicate effectively in their subject (EEF, 2021). It also acknowledges that a well-written paragraph in geography will look vastly different from a well-written paragraph in English or another subject. Similarly, the skills students will need to read in geography will look unique and different from other subjects. It also requires students to continue to be taught subject-specific language throughout their studies (Quigley and Coleman, 2021). The aim with disciplinary literacy is to build on language, knowledge, practices and communication which are relevant within a subject (O'Brien, Moje and Stewart, 2001) taught by their teacher.

In geography, disciplinary literacy might look like students reading extracts from the plethora of texts and books which are relevant to their studies. However, GCSE specifications require students to be able to 'read' geographically in other ways. This might be reading graphs, maps or images and deciphering meaning. Or looking at these graphs, maps and images and connecting inferred information to prior learning. In this chapter, I will explore how I have sought to weave disciplinary literacy into the KS3 and KS4 curriculums over the past 18 months.

INTRODUCING GEOGRAPHICAL VOCABULARY

Beck and McKeown (1985) first introduced a tiered system for vocabulary. They offer three tiers which are now mentioned frequently in education, but difficult to categorise.

- Tier 1 words are everyday words with unambiguous meanings. Students will typically hear these words at home and acquire them naturally in social situations without needing to be taught them. For example: cat, sofa, plant.
- Tier 2 words are academic words which are high frequency. In geography, these are often command words. For example: analyse, describe, examine.
- Tier 3 are low-frequency, subject-specific words. For example: place, social, flora.

Both tier 2 and 3 vocabularies will need to be taught in the classroom. In geography, these tiers also overlap frequently due to the nature of the subject. It can be suggested that 'place' and 'space' would go into tier 1 but can also be tier 3 in a geography classroom, for their meaning can vary depending on the situation (see chapter 15).

When introducing literacy, identifying the vocabulary which may become a barrier to your students, and appropriately introducing or teaching this vocabulary ensures students can access the text they are being taught. To simplify this and reduce the volume of words which need to be 'taught' explicitly, giving students a glossary or list of definitions and signposting the definition when reading can be effective. I would always recommend having the teacher read to students, as opposed to having students read aloud. This is because students should hear an 'expert reader' reading text and modelling rather than another student. It also gives you, as the teacher, the opportunity to be questioning to check for understanding while introducing vocabulary.

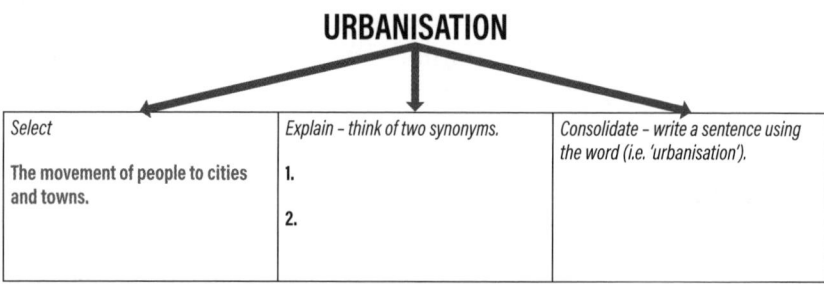

Figure 14.1 Examples of SEEC model for the word 'urbanisation'

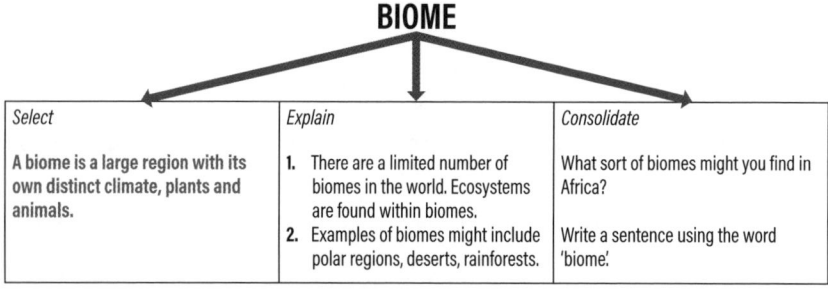

Figure 14.2 Examples of SEEC model for the word 'biome'

In my department we use Alex Quigley's SEEC model (Select, Explain, Explore, Consolidate) for initial introduction of words, particularly for tier 2 in KS3 and tier 3 in both key stages. I will typically choose a word that students need to understand to grasp the lesson content, and I will only choose one to introduce in this manner.

The 'select' portion is the selection and definition of the word (see Figure 14.1), which is as simple and minimal as possible. The 'explain' and 'explore' are two that I usually combine, and this is where students will delve into the meaning in more detail, perhaps giving examples or sometimes non-examples (see Figure 14.2). The 'consolidate' is the time for students to apply this newly gained knowledge to a task. This can vary depending on class, ability and key stage. This is a more effective means of introducing new vocabulary as it means the students are applying the vocabulary rather than passively writing down a definition.

LITERACY FROM AN EQUALITY, DIVERSITY AND INCLUSION PERSPECTIVE

Increasing equality, diversity and inclusion (EDI) is something I seek to do as frequently and as deeply as possible in my teaching. By introducing geographical texts and weaving literacy into a geography curriculum, you are able to increase the variety of knowledge and information students are absorbing.

Ladson-Billings (1995) writes about 'culturally relevant pedagogy' by which teachers must acknowledge they are part of a community of experts and learners. In the classroom, this can manifest as allowing students to be experts. However, it can also mean welcoming alternative perspectives and points of view and amplifying marginalised voices through reading (see chapter 31).

Reading from a variety of sources allows students to access information and resources which can enable them to understand and appreciate a plethora of perspectives. This can assist in improving understanding and acceptance of diversity. And on the other end of the continuum, it can also ensure that students are taught about the world in a nuanced and accurate way. For example, when teaching about ecosystems, I ensure to include text from the perspective of Indigenous people to teach students about a place. In a Year 8 scheme of work, students learned about hot and cold deserts. There are many texts written by travel journalists or outsiders learning about a place. However, for a more accurate and authentic perspective, I sought out extracts about indigenous groups in Siberia and Chile for students to gain understanding. This amplifies marginalised voices and increases the diversity of knowledge in the classroom, reinforcing Ladson-Billings' concept of a 'community of experts' (1995).

Literacy in geography is an essential tool for promoting equality, diversity and inclusion. It allows students to access knowledge and gives them the tools to express themselves and participate in critical thinking.

CHOOSING A TEXT

Text does not always need to be a traditional 'text'. Graphs, maps, photographs and infographics are all used as resources in geography and teaching students to read these is just as much an essential skill as being able to read a book chapter or extract of text at length.

Multimodal texts are common in geography and are particularly useful in lessons for students to practise demonstrating geographical understanding and applying knowledge and understanding, which are assessment objectives in the English GCSEs. This can also be done in any key stage, despite the focus being on a GCSE assessment objective. By introducing these literacy skills in key stage 3, students are building knowledge and skills essential to a strong GCSE grade.

Multimodal texts might be a group of images, texts and graphs. Their meaning is conveyed to the reader through this combination of 'texts' that geography students must be able to read and interpret. Reading a graph, map or image is still reading in geography. Giving students a piece of text like this, and using a visualiser to make inferences and connect to what they are learning or have learned, allows them to apply knowledge and understanding and increase their geographical understanding. This type of text is common in my classroom. The first time I use them with any year group, I train students how to read and make inferences. This means using a visualiser and talking through my thought process aloud; it may also include questioning as I model *how* to read this text. This questioning can prompt students to make synoptic connections between their prior learning and the information in front of them, and then together on the visualiser, annotating together. After being repeatedly exposed to multimodal texts, students become more confident in the routine. Then, I will read through the information with students and point out anything I would like them to focus on, and then students annotate and make their own connections.

Choosing text extracts from articles, magazines, books and so on can be difficult. Many texts are not immediately suitable for students, particularly in KS3. Choosing a text which is too hard can mean you as the teacher need to work much harder than the students, at which point the activity becomes redundant. Lots of text is dense and will need adapting. Typically, this means I will scan (or copy) a page or chapter into a word processor and then read through and analyse the vocabulary and importance of certain parts. I will only include text which is relevant to what my students are learning. Generally, I will rewrite the text, making it more concise and improving the readability. Checking the reading age and difficulty of a text is essential to scaffolding correctly. Websites such as *The First Word* ensure that your text is accessible to your target audience, your class.

There may be occasions when the text you are seeking does not exist, or you do not have access or are unaware of it. Sometimes an online or news article is a perfect fit for one aspect but does not cover everything, and in this case, compiling information to create a text can be a resolution. This can also increase plurality by including multiple perspectives in one text.

The voices you choose to amplify in your classroom are also an important factor to consider when choosing a text. Using a singular text for anything more than a singular perspective can be reductive. Teaching a whole scheme of work focusing on one book by one author can reduce the diversity of opinions in your classroom and lean towards biases. Oral histories are a fantastic opportunity to amplify voices in your classroom and there is less opportunity for generalisations and misconceptions. By amplifying the voices we are teaching about, the topic or case study becomes less of a 'human zoo' where we make assumptions with no real basis (see chapters 16 and 31).

Focusing on only one author or voice can put too much pressure on an author to have *all* the answers and can unintentionally silence different voices and perspectives. Using multiple voices provides students with multiple viewpoints and perspectives and adds duality or plurality. Therefore, pairing texts which explore similar or very contrasting ideas increases plurality and the quality of the information your students are engaging with.

ACTIVITIES AND COMPREHENSION

When reading aloud to students, questioning can be an effective means of scaffolding in order to ensure students can access text. Text should be challenging, and I would always want to 'teach to the top' when using disciplinary literacy in class.

Comprehension questions or activities can also support in scaffolding text to make it accessible. Text that is too challenging can put students off reading in geography, or completely go over their heads and become a pointless task.

Usually when creating a worksheet, I will include some information about the author and where the extract is from. There will usually be a definitions list too and up to six questions or tasks for students to engage with. These questions will be deliberate and have significant reasoning behind them. They will also build upon one another and the end question will be more open-ended, and in KS4 this may culminate as an exam-style question. I will attempt to include a variety of assessment objectives, particularly comprehension and application of knowledge (AO2 and AO3 in English GCSEs).

By designing a well-sequenced set of questions which build upon one another, different strands of ideas can be broken down and scaffolded for students into manageable chunks and are then metaphorically tied together by the student. By designing activities like this, you are able to set students up for extended writing activities after reading in class, reinforcing and entrenching their acquired knowledge and giving opportunity for students to demonstrate their understanding.

Five reflective questions

1. What does literacy look like in your department?
2. What skills do you want students to gain?
3. What knowledge do you want students to gain?
4. What texts do you use or include? Why?
5. Whose voice is being amplified?

DIGGING DEEPER - THREE RESOURCES TO DELVE FURTHER

- The Voice Project: https://www.thevoicesproject.co.uk/ (Accessed: 28/05/2025)
- Milner, C. (2020) Classroom strategies for tackling the whiteness of geography. *Teaching Geography*, vol. 45 (3), pp. 105-107.
- Cushing, I. (2022) Word rich or word poor? Deficit discourses, raciolinguistic ideologies and the resurgence of the 'word gap' in England's education policy. *Critical Inquiry in Language Studies*, DOI: 10.1080/15427587.2022.2102014

REFERENCES

Beck, I. & McKeown, M. (1985) Teaching vocabulary: making the instruction fit the goal. *Educational perspectives,* 23, pp. 11–15.

GL Assessment. (n.d.) Read All About It: why reading is key to GCSE success. https://camdenlearning.org.uk/wp-content/uploads/2020/03/GL-Assessment.pdf (Accessed: 28/05/2025)

Ladson-Billings, G. (1995) Toward a theory of culturally relevant pedagogy. *American Educational Research Journal,* 32(3), pp. 465–81.

O'Brien, D.G., Moje, E.B. & Steward, R.A. (2001) 'Secondary literacy in sociohistorical and cultural perspective: Understanding literacy in people's everyday school lives', in Moje, E.B. and O'Brien, D.G. (eds), *Constructions of Literacy: Studies of Teaching and Learning in Secondary Class.* Erlbaum, pp. 27–48.

Ofsted (2023) *Subject Report Series: Geography.* Ofsted.

Quigley, A. & Coleman, R. (2021) *Improving Literacy in Secondary Schools.* EEF.

Ricketts, J., Sperring, R. & Nation, K. (2014) Educational attainment in poor comprehenders. *Frontiers in Psychology,* 5, pp. 1–11.

15. WRITING LIKE A GEOGRAPHER
CATHERINE OWEN
@GEOGMUM

'The messy, complex, rich and rewarding act of writing can and should be at the heart of best practice in the classroom. The demands of extended academic writing – or 'school' writing – can be met over time, and we can close the writing gap, one move at a time.' (Quigley, 2022, p.6)

In his book, *Closing the Writing Gap*, Quigley (2022) challenges the reader to read a paragraph about a cyclone in Kolkata, then write a single sentence summary in ten words or less, writing with their non-writing hand. This is followed with questions asking the reader to reflect on how this felt (physically) and made them feel. Quigley then reviews the 'sheer array of knowledge and skill' needed to compete this task:

- Overcoming the physical discomfort of writing with a different hand from usual, resembling the challenge students who aren't able to handwrite automatically and fluently face.
- Skimming and scanning the words and connecting them to background knowledge before deciding which information is essential.
- Creating a 'clear and complete' sentence to complete the task.

Quigley likens the complexity of writing to playing a game of chess, with students having to think about many moves, including 'handwriting, word choices, spelling, paragraphing, writing for their audience, activating their prior knowledge, along with considering the purpose and genre of their writing, and more.' He acknowledges that students use many types of writing in a school day as they move from lesson to lesson, such as instructional writing for a science experiment, essay writing in history and prose writing for telling a story in English. Quigley's book aims to support teachers in developing their skills so that they can support students in developing their writing, placing great emphasis on disciplinary writing.

GENERAL ACADEMIC WRITING

We can support our students by encouraging them to improve their grasp of grammar. This could include:

- **Expanded noun phrases**: This could involve adding more adjectives to the noun to add detail or using appropriate adjectives to make a sentence more concise and accurate. For example, students may describe an 'arid, barren landscape' rather than just saying it is 'empty' or that it is 'hot and dry with not much to see'.
- **Shrunken verb phrases**: Sometimes a single verb can be used to replace a two-word verb phrases. For example, 'find where' could be replaced with 'locate'.
- **Sophisticated synonyms**: We can encourage our students to think carefully about their choice of words. For example, they could use 'precipitation' instead of 'rain'.
- **Tentative language**: When students express their ideas and opinions in geography it is often best to use modal verbs such as 'may', 'might', 'could', 'should' and 'appears to' rather than asserting a point with certainty. For example, students explaining the impact of fairtrade on farmers' lives may say that the farmers 'could use the money to send their children to school'.
- **Right-branching sentences**: These sentences start with the subject (noun) and verb, then have additional details branching off to the right. For example 'Shopping centres use grey water to be

sustainable by collecting rainwater in tanks and using this to flush toilets.'

(Source: Adapted from Quigley, 2022, pp.78–81)

We may be nervous to use these techniques if our own grounding in grammar is weak, but students will have worked on grammar at primary school and may be more ready to develop their writing than we expect.

A whole-school focus in my school has been supporting students in constructing sentences. Our wonderful Assistant Principal, Vicky Pickford, has provided CPD to improve staff confidence, along with techniques to use to improve students' writing. My department has used these ideas to create success criteria for writing to use alongside the geography criteria for each of our key stage 3 mid-cycle assessments, as shown in Figure 15.1.

We provide examples of how to use these techniques before students complete each assessment, taking progress through the key stage into account. The support provided for the task in Figure 15.1 is shown in Figure 15.2.

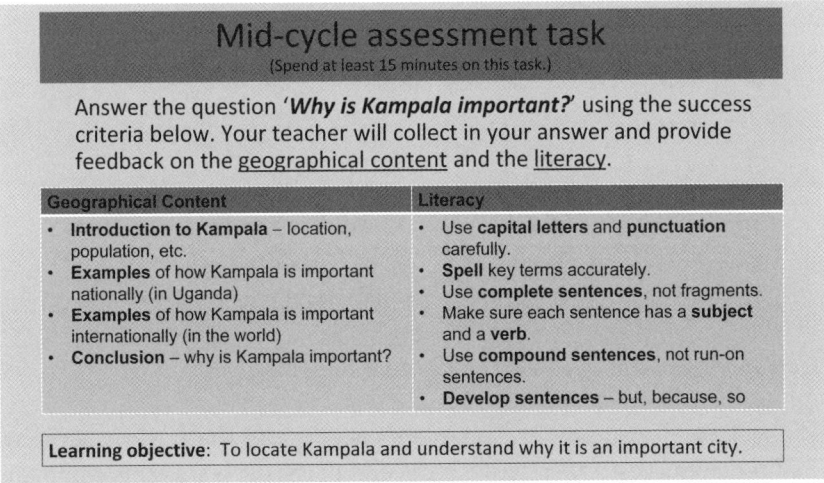

Figure 15.1 Example of a KS3 mid-cycle assessment task

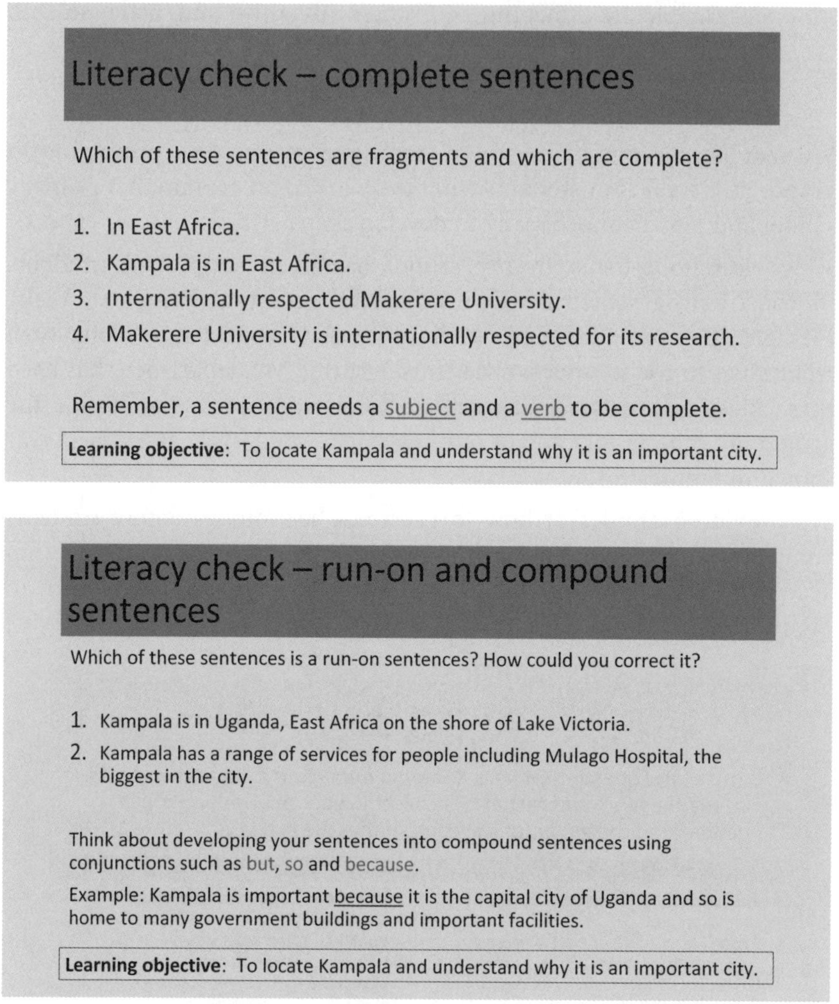

> **Literacy check – complete sentences**
>
> Which of these sentences are fragments and which are complete?
>
> 1. In East Africa.
> 2. Kampala is in East Africa.
> 3. Internationally respected Makerere University.
> 4. Makerere University is internationally respected for its research.
>
> Remember, a sentence needs a <u>subject</u> and a <u>verb</u> to be complete.
>
> Learning objective: To locate Kampala and understand why it is an important city.

> **Literacy check – run-on and compound sentences**
>
> Which of these sentences is a run-on sentences? How could you correct it?
>
> 1. Kampala is in Uganda, East Africa on the shore of Lake Victoria.
> 2. Kampala has a range of services for people including Mulago Hospital, the biggest in the city.
>
> Think about developing your sentences into compound sentences using conjunctions such as but, so and because.
>
> Example: Kampala is important <u>because</u> it is the capital city of Uganda and so is home to many government buildings and important facilities.
>
> Learning objective: To locate Kampala and understand why it is an important city.

Figure 15.2 Example of support given for a KS3 mid-cycle assessment task

Feedback on the mid-cycle assessment encompasses both the geography and literacy success criteria and at the end of the cycle students complete a similar piece of writing, giving them a chance to develop and improve their writing.

When it comes to developing sentences, using '**but, because, so**' has proven very successful in our lessons. Hochman (2009) suggests an exercise to develop students' sentences by asking them to take a short independent clause and expand it in different ways. A geographical example could be:

> The industrial estate was redeveloped.
> The industrial estate was redeveloped **because** many of the buildings were derelict.
> The industrial estate was redeveloped **but** the road names still show what the area was once known for.
> The industrial estate was redeveloped **so** new businesses moved into the area and offered jobs to local residents.

Students now often develop their sentences without prompting, not just because we have encouraged this in geography lessons, but because teachers across our school are using these techniques. (See also chapters 5 and 6.)

WRITING LIKE A GEOGRAPHER

The first recommendation from the Education Endowment Foundation's (EEF) guidance on improving literacy in secondary school is to prioritise 'disciplinary literacy' across the curriculum:

'Literacy is key to learning across all subjects in secondary school and a strong predictor of outcomes in later life. Disciplinary literacy is an approach to improving literacy across the curriculum that emphasises the importance of subject specific support. All teachers should be supported to understand how to teach students to read, write and communicate effectively in their subjects. School leaders can help teachers by ensuring training related to literacy prioritises subject specificity over general approaches.' (EEF, 2019, p. 4)

This report challenges us to consider the following questions:

- 'What is unique about your subject discipline in terms of reading, writing, speaking and listening? What is common with other subject disciplines?

- *How do members of this subject discipline use language on a daily basis?*
- *Are there any typical literacy misconceptions held by students, for example, how to write an effective science report?*
- *Are there words and phrases used typically, or uniquely, in the subject discipline?'* (EEF, 2019, p. 7)

Answering these questions helps us explore what it means to 'write like a geographer'. Quigley (2022, p. 123) focuses on the frequent need to assess and evaluate in geographical writing, with students needing to refer to evidence, offer a **balanced argument** using **tentative language** and make connections between different aspects to show depth of geographical understanding. If this overwhelms students they may respond in an over-simplistic way, failing to develop their sentences and ideas. Students may have developed misconceptions about geography from the media, which we can sometimes tackle by encouraging use of tentative language, rather than absolutes. Quigley (2022, p. 125) summarises some features of geographical writing:

- **Dense noun phrases** convey information, e.g. 'tensional plate margin'.
- **Appositive phrases** (two noun phrases next to each other) concisely describe geographical phenomena, e.g. 'Bristol, a city in the west of England, is a centre for banking and insurance.'
- **Sentence signposts** describe places clearly, e.g. 'Firstly, ..., for example ..., In the event of ...'.
- **Common use of graphs and diagrams** to represent human and physical geography.
- **Strong verbs** to describe geographical phenomena clearly, e.g. 'Rock armour protects ...'.

Digby and Holmes (2020) suggest seven areas for students to focus on so that they can write like a geographer:

1. Be able to use geographical terminology.
2. Know the key words for each topic.
3. Be able to write briefly when required.

4. Be able to develop (or extend) your writing.
5. Be able to explain causes.
6. Be able to explain consequences.
7. Be able to argue a case.
8. Be able to make decisions or judgements.

Our students won't automatically know how to do this, so we need to build explicit teaching, modelling and opportunities for practice into our curriculum. In chapter 14, Bethany Aldridge explains how she introduces geographical vocabulary using the SEEC model. Once students have been introduced to a term, they will need to keep revisiting it and using it to embed it in their memory, making it available for future writing. We can use low-stakes quizzing to recap terminology as well as modelling its use ourselves and using questions to prompt students to improve terminology used. A useful exercise is to give students an answer, such as the one in the box below, then ask them to make it better by using terminology and developing points.

Rewrite this answer and make it better.
How do animals adapt to conditions in hot deserts?
It is really hard for animals to live in deserts. They are really hot and dry. Animals have changed stuff so they can live there. Camels have fat in their humps. They have two sets of eyelids. They have pads on their knees.

....

....

....

....

....

When preparing students for exams it is important to make sure they know which key words are associated with each topic. They can use the section prompt in the exam paper to help them recall which words may be appropriate to use in answers. Students should be given checklists of key words for each topic so that they can check their understanding and self-quiz. Some may find it useful to create dual codes for each key term, creating an image to make it more memorable.

The skill of writing briefly needs attention, whether to support student in note-taking or when answering low tariff questions in exams. Sullivan et al (2021, p.74) discuss how they worked with their KS3 students to improve their note-taking skills:

> *'We crossed out connectives (e.g. it, the, and) in the text and summarised the remaining information. We used a modelling technique that is already embedded in teaching practice – I do, we do, you do – to demonstrate this effectively. This technique gradually increases the ownership of the work being completed. First the teacher talks through an example (I do), then the class collaborate to produce another example (we do), before students complete the task independently (you do). Students needed some guidance with deciding which words were important, but we observed that going back to basics with key skills boosted confidence, as students were able to achieve better notes with this scaffold. ... We began each subsequent note making lesson by crossing out connectives in text and summarising meaning to ensure students were continually building on their skills.'*

Another technique to support students' note-taking is use of Cornell notes, as shown in Figure 15.3.

Key questions/prompts	Notes
What are megacities?	Cities with populations over 10 million
Examples of megacities?	Shanghai, China; Dehli, India; Kinshasa, DRC.
Why have megacities grown so rapidly in newly emerging economies (NEEs)?	Most megacities are in Asia.
	NEEs such as China have industrialised since the 1980s. More factories were built in cities, so people moved from the countryside to get jobs in them (rural-urban migration).
Summary of notes Megacities have over 10 million people and include Shanghai in China. Cities in NEEs grew rapidly due to industrialisation and rural urban migration.	

Figure 15.3 Note-taking using Cornell notes format

This technique encourages students to organise their notes and also makes self-quizzing easy as they can cover the notes and use the key questions or prompts to quiz themselves. The summary forces them to write briefly.

Digby and Holmes (2020) stress that students often don't need to answer in detail in exams, so need to practise answering questions concisely. If they are asked to identify one factor but the question is worth two marks they will need to develop their answer – Digby and Holmes suggest using **connectives** such as 'so that' to do this, pointing out that this is also the foundation for structuring more extended writing. To develop points in extended writing they suggest adding more detail, giving a consequence and/or giving an example.

When students explain causes in geographical writing they should think through the **sequence**, creating a **chain of reasoning**. A good process to model this is the formation of a coastal stack – starting as a crack in a cliff, then being eroded to form a cave, then an arch, then a stack. Students are likely to understand this sequence – if my students are finding this tricky, I get them to model the stages using play dough, then write up what they have done. They can then apply the idea of sequencing to other answers where the sequence is less apparent.

There are several different terms used for exploring consequences in geography – effects, impacts, results – which can also be positive and negative. Students may be asked to consider these consequences in the short and long term or to classify them as social, economic or environmental (SEE). Writing frames may help students get to grips with these different challenges. Butt (2005, pp.56–57) explains that writing frames:

'support and scaffold the organisational process before students start writing and they are, therefore, encouraged to come to decisions about which facts go where when writing rather than rushing headlong into recording lots of unrelated facts in almost any order. Frames usually contain a variety of 'starters', 'connectives' and 'sentence modifiers' which give students a structure within which they can concentrate on what they want to say, rather than getting lost in the form of the writing. In certain circumstances the frame may represent students' first drafts that they can subsequently amend.'

Over time we need to plan to reduce dependence on writing frames, encouraging students to amend frames they are given and then create their own. Different students will be able to move towards independence at different times, but all need to be able to write independently by their final exams.

Being able to argue a case is an important part of writing like a geographer, with students expected to produce balanced arguments and come to conclusions. Quigley (2022, pp. 140–41) compares a balanced argument to a persuasive argument, suggesting that it is helpful for us to make this explicit to our students. His structure for a balanced argument includes:

- Introduction – explain the relevance of the topic and the competing claims in the debate.
- Evidence supporting your argument – explain your position and pose evidence.
- Evidence challenging your argument – explain counter arguments with evidence.
- Weighting up competing claims – balance competing positions and claims.
- Conclusion – summarise both sides of the argument and restate your reasoned position.

Canell et al (2018) created an excellent resource for teachers wanting to develop their students' critical thinking skills, with several of the activities also ideal for developing students' ability to argue a case. Examples include:

- **Flat chat** – students work in groups in silence to add notes to a piece of A3 paper about a stimulus (image, artefact, statement, infographic), then move around to add notes about other stimuli, building upon the notes left previously, still in silence. Once they have visited all the stimuli, they can discuss what they have learned before using this evidence to support a piece of extended writing.
- **Silent debate** – similar to flat chat but focused on students expressing their opinions in what they write down, challenging each other and developing arguments on their sheets without talking.

- **Argument frames** – these are writing frames designed to support students in producing a balanced argument with a reasoned conclusion.

These techniques are useful when students are required to make a decision or judgement about an issue. Digby and Holmes suggest that students always think carefully about the criteria they are using to make their decision and stress the importance of **justifying** their decision with evidence and reasoning.

You may have noticed that this chapter doesn't advocate using Point, Evidence, Explanation (PEE), Point, Development, Link (PDL) or such like to train students to write paragraphs. As Enstone (2017) 'tentatively concludes', using the PEE formula can lead to complacency and limit the length and quality of responses. Using the techniques highlighted in this chapter will help our students to develop their geographical writing as they progress through education, closing the writing gap between those who can fluently express themselves in writing and those who struggle, enabling all to write like a geographer.

Five reflection questions

1. What barriers do your students face when it comes to writing?
2. Do you need to brush up on your own writing and grammar skills to help you support your students?
3. Do you need to make your teaching about writing more explicit?
4. How could you use these ideas to develop students' general academic writing?
5. How could you use these ideas to develop students' geographical writing?

> **DIGGING DEEPER – THREE RESOURCES TO DELVE FURTHER**
> - Quigley, A. (2022) *Closing the Writing Gap*. Routledge.
> - Canell et al. (2018) Critical Thinking in Practice: Practice guide. Geographical Association. Available at: https://geography.org.uk/wp-content/uploads/2023/05/ITE_683_Critical_Thinking_in_practice_guide_final.pdf (Accessed: 28/05/2025)
> - Geographical Association. Writing in Geography. Available at: https://geography.org.uk/ite/initial-teacher-education/geography-support-for-trainees-and-ects/learning-to-teach-secondary-geography/geography-subject-teaching-and-curriculum/geography-knowledge-concepts-and-skills/geographical-practice/literacy-and-numeracy/writing-in-geography/ (Accessed: 28/05/2025)

REFERENCES

Butt, G. (2005) Engaging with extended writing. *Teaching Geography*, Spring 2005. Sheffield: Geographical Association.

Canell et al. (2018) *Critical Thinking in Practice: Practice guide*. Geographical Association. https://geography.org.uk/wp-content/uploads/2023/05/ITE_683_Critical_Thinking_in_practice_guide_final.pdf (Accessed: 28/05/2025)

Digby, B. & Holmes, D. (2020) Think like a geographer, write like a geographer. Geography Education Online. https://www.geographyeducationonline.org/webinars/skills-boost/think-like-a-geographer-write-like-a-geographer

Education Endowment Foundation (2019) *Improving Literacy in Secondary Schools*. London: Education Endowment Foundation.

Enstone. (2017) Time to stop PEE-ing. *NATE Teaching English*, Issue 13.

Hochman, J. (2009) *Teaching Basic Writing Skills*. Dallas, TX: Sopris West.

Quigley, A. (2022) *Closing the Writing Gap*. London: Routledge.

Sullivan, K., Thompson, H & Willis, H. (2021) Note taking: it's just writing stuff down … isn't it?. *Teaching Geography*, Summer 2021, pp.72–75.

16. CHOOSING AND USING CASE STUDIES
RAMYA RAJKUMAR
@RAMYA_SARUJAN

'It is impossible to engage properly with a place or person without engaging all of the stories of that place.' (Chimamanda Ngozi Adichie, 2009)

Geography is inherently context dependent and case studies allow us to explore the nuances with which geographical processes take place in different contexts. When building a curriculum, we often break it down into thematic topics and then sequence concepts to build upon one another. Case studies exemplify these concepts, and move them from the abstract to the concrete. This approach is successful in building conceptual schema, but what is lacking in conscious thought are the wider schemas on place that are developed. Many curricula also include some place-based units focused on a continent or country – often defined by the national curriculum specification points – which serve as in-depth case studies. But what sense of place do students leave with? And how can we consciously and deliberately support students to build a sense of place through our case study teaching?

HOW DO WE MANAGE OUR OVERALL CURRICULUM?

A - REPRESENTATION

Geography adopts a horizontal knowledge structure (Bernstein, 1999), which means that it does not integrate everything into one big picture, but instead adds together unique and different ideas that can exist side-by-side (Vernon, 2016). Consequently, some of the most difficult curriculum decisions we make come from the fact that we have to be hugely selective about what – or where – to teach, and how much depth to teach it in, as the building blocks of knowledge are less clear and there is an absence of a clearly defined sequencing strategy. This makes building suitable representation a challenging task.

We want students to leave with a balanced understanding of places. They should have a wide knowledge of many places so that they can understand broad patterns that exist, but also have a deep knowledge of places so that they understand how different physical and human processes interact within them. And this is required for multiple places, so that students can appreciate how processes interact differently in different areas. So we should aim to teach about many places in some depth and some places in lots of depth.

We also want to consider who our students are, and whether they can see themselves reflected in our curriculum, to empower them, strengthen their sense of identity and make the learning experience meaningful. After all, the curriculum is not comprised of immutable content; it is a process, and it is our responsibility to consider the classroom scale of this process (Mitchell, 2017). The following strategy helps us to review existing representation in our curriculum by assessing its balance:

1. Plot the case studies that you teach onto a world map, and their associated lesson budget. This should be done for each learning cycle (key stage), as some students will not opt to learn the subject at the next cycle.
2. Analyse your map by using the following questions:
   a) How balanced is the split between thematic and place units?
   b) Which regions of the world are underrepresented?
   c) How balanced is breadth vs. depth of case studies?

d) What impression of a place is given if it is not revisited?
e) How well do the case studies reflect the heritage of your students and their families?
3. Use your analysis to map out and prioritise amendments to your curriculum.

An example

Figure 16.1 shows an example of a Key Stage 3 curriculum, with the lesson titles, focus countries and lesson budgets plotted. The annotations list initial ideas that could be explored.

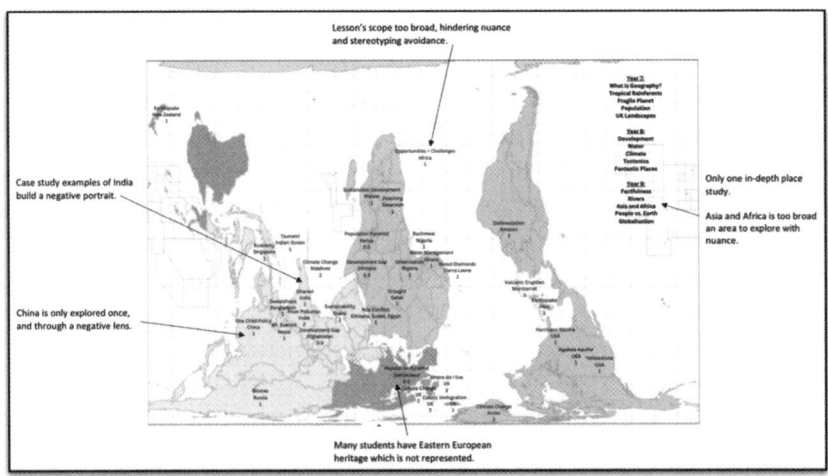

Figure 16.1 An example of a KS3 curriculum map

As well as identifying regions of the world which are underrepresented, this map highlights the lack of depth into which most case studies go, resulting in unbalanced and distorted presentations of place. This activity is a useful form of CPD that can be used within departmental meetings and discussions.

B - PORTRAIT

It is important to consider the narrative presented of each case study location and the overall portrait that is created, as this shapes students' understanding of place over time. Most case studies are taught as an exemplification of a concept within a thematic unit, so it is likely that we teach about a place only with a particular focus. For example, a case study on a hurricane may speak only of the causes, impacts and responses, and provide little about a country's history, culture or global significance, all of which are important in building an overall understanding of a place.

But if this place is never revisited, students' understanding will be skewed and incomplete, and the narrative presented will be unbalanced. For example, social studies curricula currently fail to represent Black joy – the positive experiences and developments within Black communities (Duncan, Hall and Dunn, 2023). It is dangerous to only present places through a lens of oppression and suffering because it distorts students' understanding and perpetuates curriculum violence. This is when lessons have negative impacts, which can be unintentional, on the wellbeing of students (Jones, 2020).

So we should centre marginalised voices and perspectives to support agency and autonomy in building narratives that create equity and balance (see chapter 31). And, because it is the interaction of different processes which shape place, we must build an understanding of these multiple influences through our curriculum. As Chimamanda Ngozi Adichie's concept of Single Story says, 'It is impossible to engage properly with a place or person without engaging all of the stories of that place ... the consequence of the single story is this: it robs people of dignity.' (Adichie, 2009).

Of course, we will never be able to capture the full depth and richness of places, as curriculum teaching time constraints mean that our representations will always be distorted. But we must strive for place representations that are as comprehensive as time allows; otherwise, we reproduce narratives that parallel the negative tropes pervading our wider cultural landscape.

So, what questions can we ask to ensure that we paint a more holistic portrait and build a deeper understanding of place across our curriculum?

A strategy

For each place visited, reflect on the overall portrait created by asking the questions in the mind map in Figure 16.2.

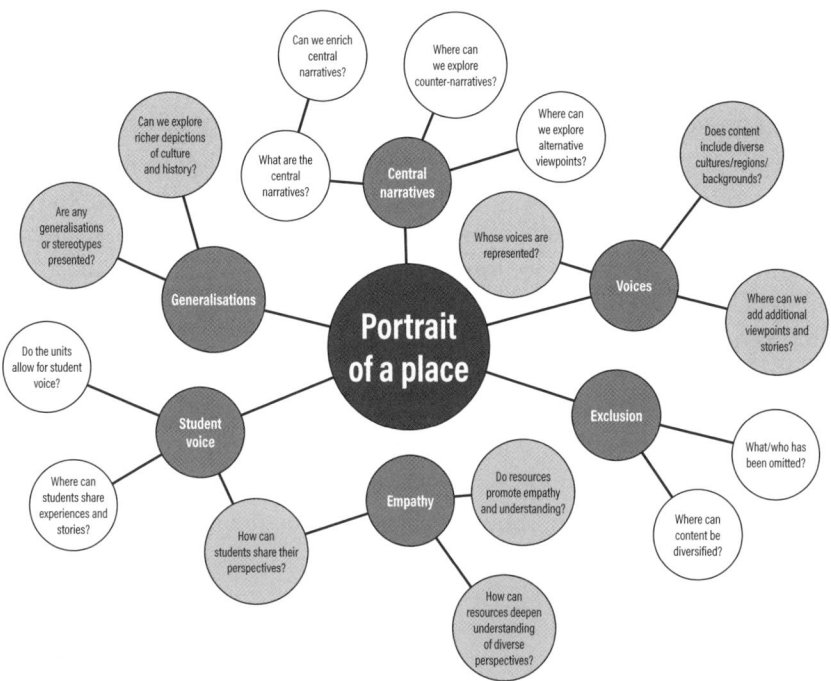

Figure 16.2 A portrait of place mind map

This helps teams to explore some of the predominant ways in which distortion manifests and create priorities for change. It may be unrealistic for each case study to reflect a balanced portrait of its location. However, it is appropriate to aim for balance across a unit or learning cycle.

WHAT DO WE INCLUDE IN OUR CASE STUDIES?

C – CONTEXTUALISATION

Understanding the history of a place deepens students' understanding of how it exists only in relation to other places (Massey, 1991). Historical context forms important hinterland knowledge for our case studies, and serves a proximal function (Counsell, 2018) towards understanding how different geographical processes interact to shape a place. An appreciation of this allows students to produce evaluations that are distinct and authentic to the places studied. Yet this is often absent in lesson resources. Consequently, our teaching offers an isolated and fragmented understanding of place.

Additionally, formal education has grouped knowledge into discrete subjects. While these inevitably overlap – and we should make better efforts to explore a more interdisciplinary approach – geography has fought to hold onto such an identity and place within the academic community. So how do we know which history is 'ours' to teach? One approach is to consider whether that history has directly shaped the way that geographical processes influence that place today, and where the answer is 'yes', it is our responsibility to teach it. Even better would be to maintain a dialogue with history departments so that we can draw upon prior knowledge learned in history and build upon this to create a richer understanding (Geographical Association).

When deciding what and how much to include, the questions shown in Figure 16.3 can guide us.

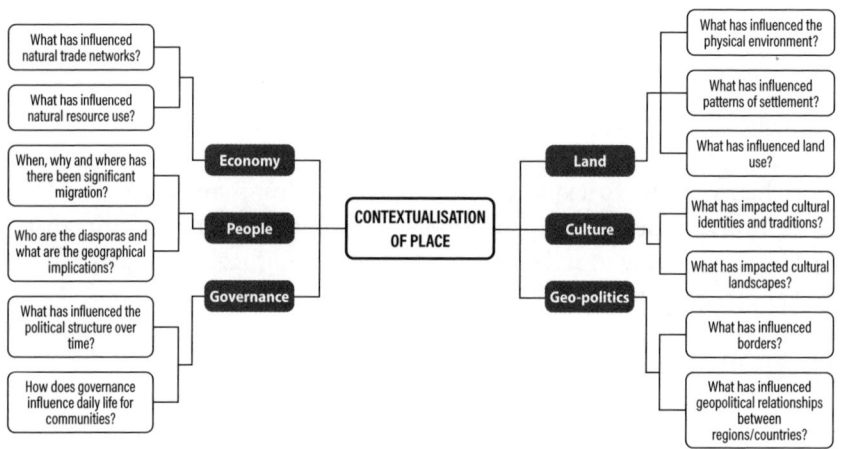

Figure 16.3 Contextualising of places

Not all case studies will neatly conform to the themes listed, and the framework is not a prescriptive list, but these questions stimulate thoughtful exploration of areas where contextualisation may strengthen a case study.

An example

Figures 16.4 and 16.5 show sections from resources discussing Hurricane Katrina (left) and the 2010 Haitian earthquake (right) which help students to understand the root causes of destruction. They also build broader schema on development, geopolitical relations and race, and consequently, a deeper understanding of place.

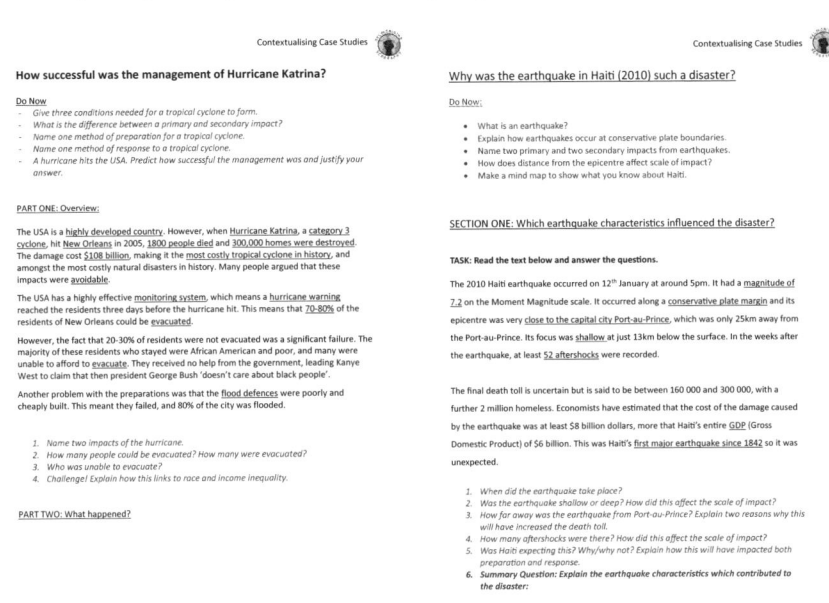

Figure 16.4 Hurricane Katrina case study

Figure 16.5 2010 Haitian earthquake case study

D – VOICES AND STORIES

For our curriculum to be both balanced and inclusive, we should include a range of perspectives, including those of our students and indigenous voices. It is important to explore ideas that are totally different from what is 'normal' and push boundaries; otherwise, we risk losing the power to think critically and creatively about geography (de Leeuw et al., 2017).

Stories allow us to introduce ideas and perspectives that transcend conventional textbook resources (see chapter 32). They connect with others and are a powerful way to express ideas about the world (Cameron, 2012). They can illustrate the ways in which geographical processes impact communities and facilitate authenticity. Given that geography is about understanding general patterns and identifying unique details, even the smallest stories can create geographic knowledge (Hofmann, 2014). They are therefore an important instrument in case study teaching.

What distinguishes stories is their capacity to capture our attention and communicate complex ideas in an accessible way. They make case study knowledge 'sticky' because their narrative structure is easy to remember (Willingham, 2009). Stories are more memorable than explanatory texts because they create a context for new knowledge, making it easier to connect new ideas to those that students already know, which in turn builds conceptual schema. Beyond their cognitive role, stories offer insight into authentic experiences, which promote empathy and an understanding of diverse perspectives. So we should use stories within case studies to narrate important concepts, ideas and perspectives (see chapter 32).

An example

Figure 16.6 is an excerpt from an aural interview which has been translated into English. It uses local knowledge to explore landscape diversity in Sri Lanka.

> So my country Sri Lanka, I am a Tamil from Sri Lanka. It is a tiny little island...in the Indian ocean, and even through it's a very small island...it has an amazing variety of vegetation and topography. Every hour as you travel across, you will see the vegetation and geographical aspects changing all the time. It's very interesting that way. So you have all these coastal areas...with beautiful, beautiful beaches.
>
> And then there are some arid areas...and further down as you leave the northern province, you get tropical rainforests...with monsoon rain, very lush green areas.
>
> And then in the central highlands, due to a volcanic eruption...you have a rise there, a mountainous region.

Figure 16.6 Transcript of an interview exploring the landscape of Sri Lanka

The depiction of landscapes through aural narrative offers imagery that supports retention. The use of authentic voice adds to the source's genuineness. Despite its divergence from a conventional explanatory text, the narrative retains all of the knowledge that would be explored in this lesson (see chapter 31).

HOW DO WE DELIVER CASE STUDY TEACHING?

E – CRITICALITY

As well as its intrinsic learning and examination aims, the geography curriculum is a powerful tool for developing critical thinking skills in students. But our common model of exploring an 'issue' through 'Causes → Impacts → Solutions' is limited, because change is not just about making small adjustments, but also about challenging the current system. Critical geography asserts that scholars can provide a profound understanding of issues, and in doing so can confront those in power, challenge prevailing beliefs, and facilitate political change (Blomley, 2006). Therefore, criticality spans broader than awareness of oneself and others in our world; it provides the tools to deconstruct and disrupt existing discourses within the discipline, and works towards social justice (Kim, 2019). Teaching critical thinking through our curriculum is, then, a useful way to begin challenging dominant thinking within secondary geography and create balance.

A strategy

The HEADSUP tool (Andreotti, 2012) is a useful scaffold for students to ask questions and critique existing resources (see Figure 16.7). The following steps outline a constructive method for implementation:

1. Create simplified questions that are accessible to your class.
2. Choose one or two questions to explore.
3. Use a 'think aloud' to walk students through your thought process as you critique your resource.
4. Choose an additional question to explore as a group.
5. Over time, build criticality skills with your students by:
    a) encouraging independence through small group discussions
    b) increasing the number of questions that you explore within one session.

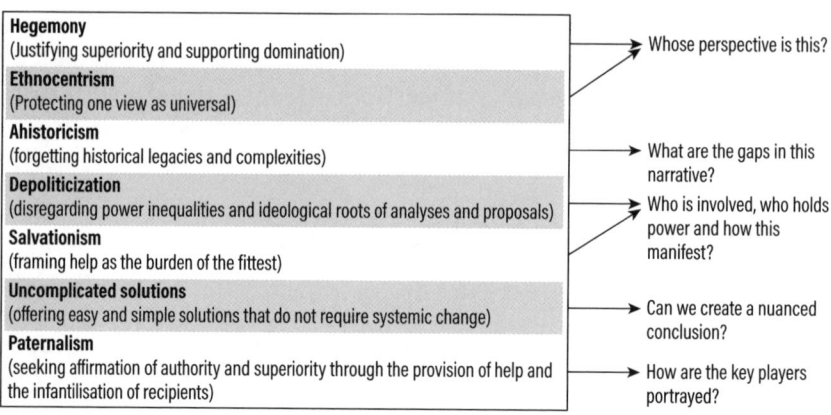

Figure 16.7 HEADSUP tool

The dialogues that we engage in with our students catalyse their development of critical thinking. While curriculum adjustments take time to implement – and time constraints within the profession pose ongoing challenges – the development of critical pedagogies alongside existing resources is possible to achieve.

F – SENSE OF PLACE

Often when we teach about places, they are presented as 'unknown' places in entirely unfamiliar locations. But this continual focus on difference perpetuates an Us vs. Them culture. Many of us teach in demographically diverse classrooms, so have a powerful opportunity to instead use places to create a culture of togetherness. Because our experience of place is not just influenced by material capital, but also by forms of power such as race, gender and colonialism, Doreen Massey (1991) argues that we should stop viewing places as static and start understanding them as processes. Places take on multiple identities, which means that each students' sense of place will – and should –differ, and teaching should explicitly explore, embrace and celebrate this idea.

In the classroom, we should empower students to build Massey's Progressive Sense of Place. And our students' own experiences of place should be brought into the classroom and used to explore and compare experiences of place, what shapes them, and consequently, co-create an understanding of place. In doing so, we can break down the Us vs. Them narrative and foster togetherness.

Students should, of course, share only what they are comfortable with, and never feel singled out, so do consider whether a given space is appropriate for cultural sharing. But where it is, this can be a powerful classroom tool.

An example

An exploration of existing knowledge of a place before embarking on a thematic-based exploration is useful. In the example in Figure 16.8, at the start of a lesson about flooding in Bangladesh, students volunteered information and perspectives on the country.

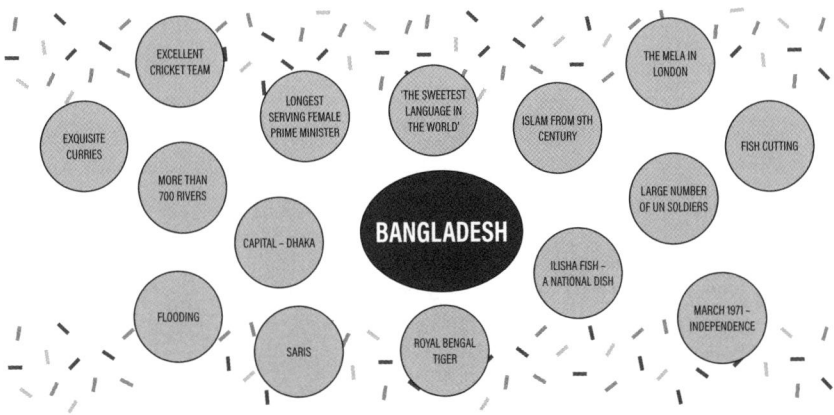

Figure 16.8 Bangladesh brainstorm

The collaborative task allowed students to understand the diverse ways in which Bangladesh is perceived, and contributed to the creation of a more comprehensive portrayal of the location.

CONCLUSION

Geography is inherently context dependent, and case studies help us to explore the nuance of geographical process in different places. When designing and implementing our curriculum, we should view the role of case studies as more than just an illustration of thematic concepts, and consider the larger perspectives on place that we want students to develop. Place is not just a location or where communities live; it is a process that is constantly evolving, and shaped by socio-economic, political and cultural processes. To reflect this, we must consciously and deliberately design case study teaching and learning to not only build knowledge, but also nurture a genuine and profound sense of place.

> **Five reflection questions**
> 1. How can you balance breadth and depth of place representation in your curriculum?
> 2. What portrait is built of each place and how can you minimise distortion?
> 3. How can you teach students to think critically about your existing resources?
> 4. Where can stories, local knowledge and authentic voice make your curriculum richer?
> 5. How do you deliberately and consciously build sense of place with your students?

> **DIGGING DEEPER - THREE RESOURCES TO DELVE FURTHER**
> - Doreen Massey. (1991) *A Global Sense of Place*. Routledge.
> - Chimamanda Ngozi Adichie. The danger of a single story. TED Talk. Available at: https://www.ted.com/talks/chimamanda_ngozi_adichie_the_danger_of_a_single_story?language=en (Accessed: 28/05/2025)
> - Teaching About a Place? Stop and Think First! Decolonising Geography. Available at: https://decolonisegeography.com/blog/2021/04/teaching-about-a-place-stop-and-think-first/ (Accessed: 28/05/2025)

REFERENCES

Adichie, C. N. (2009) The danger of a single story. https://m.youtube.com/watch?v=D9Ihs241zeg (Accessed: 28/05/2025)

Andreotti, V. de O. (2012) Editor's preface 'HEADS UP'. *Critical Literacy: Theories and Practices*, 6(1), pp.1–3.

Al Jazeera English (2011) *Haiti: After the Quake l Al Jazeera Correspondent*. (online) https://m.youtube.com/watch?v=XP2V-0WqcgM (Accessed: 28/05/2025)

Bernstein, B. (1999) Vertical and horizontal discourse: An essay. *British Journal of Sociology of Education*, 20(2), pp.157–173. doi: https://doi.org/10.1080/01425699995380.

Blomley, N. (2006) Uncritical critical geography? *Progress in Human Geography* 30(1), pp.87–94. doi: https://doi.org/10.1191/0309132506ph593pr.

Burnett, L., Brack, S. & Anderson, S. (2021) Teaching about a place? Stop and think first! - decolonising geography. https://decolonisegeography.com/blog/2021/04/teaching-about-a-place-stop-and-think-first/ (Accessed: 28/05/2025)

de Leeuw, S., Parkes, M.W., Morgan, V.S., Christensen, J., Lindsay, N., Mitchell-Foster, K. & Russell Jozkow, J. (2017) Going unscripted: A call to critically engage storytelling methods and methodologies in geography and the medical-health sciences. *The Canadian Geographer / Le Géographe canadien*, 61(2), pp.152–164. doi: https://doi.org/10.1111/cag.12337.

Cameron, E. (2012) New geographies of story and storytelling. *Progress in Human Geography*, 36(5), pp.573–592. doi: https://doi.org/10.1177/0309132511435000.

Counsell, C. (2018) Senior Curriculum Leadership 1: The indirect manifestation of knowledge: (A) curriculum as narrative. (online) the dignity of the thing. https://thedignityofthethingblog.wordpress.com/2018/04/07/senior-curriculum-leadership-1-the-indirect-manifestation-of-knowledge-a-curriculum-as-narrative/ (Accessed: 25 October 2023)

Resources - Decolonising Geography. https://decolonisegeography.com/resources (Accessed: 28/05/2025)

Duncan, K.E., Hall, D. & Dunn, D.C. (2023) Embracing the fullness of Black humanity: centering Black joy in social studies. *The Social Studies*, pp.1–9. doi: https://doi.org/10.1080/00377996.2023.2174926.

Geographical Association (2023) Making links with other subjects. https://geography.org.uk/ite/initial-teacher-education/geography-support-for-trainees-and-ects/learning-to-teach-secondary-geography/geography-subject-teaching-and-curriculum/curriculum-and-curriculum-planning/making-links-with-other-subjects/ (Accessed: 28/05/2025)

Hofmann, R. (2014) Narrating spaces. Innovative entries to (school) geography. *European Journal of Geography*, (online) 5(1), pp.70–80.

Jones, S. (2020) Ending Curriculum Violence. *Teaching Tolerance*. https://www.learningforjustice.org/magazine/spring-2020/ending-curric (Accessed 26 October 2023)

Kim, G. (2019) Critical thinking for social justice in global geographical learning in schools. *Journal of Geography*, 118(5), pp.210–222. doi: https://doi.org/10.1080/00221341.2019.1575454.

Massey, D. (1991) A global sense of place. *Marxism Today*, pp.24–29.

Mitchell, D. (2017) 'Curriculum making', teacher and learner identities in changing times. *Geography*, 102(2), pp.99–103. doi: https://doi.org/10.1080/00167487.2017.12094017.

National Alliance to End Homelessness (2021) Homelessness and Black History: Poverty and Income. https://endhomelessness.org/blog/homelessness-and-black-history-poverty-and-income/ (Accessed: 28/05/2025)

Vernon, E. (2016) 'The structure of knowledge'. *Geography*, (online) 101(2), pp.100–104. https://www.jstor.org/stable/10.2307/26546723 (Accessed: 28/05/2025)

Willingham, D.T. (2009) *Why Don't Students Like School?: A Cognitive Scientist Answers Questions About How the Mind Works and What it Means for the Classroom*. San Francisco, Ca: Jossey-Bass.

17. SUPPORTING STUDENTS WITH SEND IN GEOGRAPHY

AMY CUSHING
@AMY_GEOG

'The inclusion debate is long-running and reminds us that inclusion is a complex concept that is difficult to apply in practice. It is complex because rather than labelling pupils or categorising them, it requires you to understand them all as individuals with their own talents, hopes and aspirations.' (Biddulph et al, 2015, p.166)

WHY SUPPORTING STUDENTS WITH SEND MATTERS

A student has a special educational need and/or disability (SEND) if 'special educational provision' is required for them (Gov.uk, 2020, p.93). Most teachers will be familiar with a wide range of strategies that can be implemented to support students with SEND. Teachers will know that they have a responsibility towards supporting students with SEND as part of the teacher standards and as part of the SEND code of practice (Gov.uk, 2020). Aside from standards and codes, we know that these are some of our students who need us the most in the classroom.

Government published statistics for the 2022/23 academic year show that 4.3% of all students in England have an educational health care plan (EHCP). These plans are given to students with high needs and include increased funding for schools to meet those needs. These statistics also show that 13% of students in England have special educational needs (SEN) support in school but no EHCP. This accounts to over 1.5 million young people. These two statistics have increased each year since 2016. We know that actual needs will be higher than these reported government figures (see chapter 18). While there is debate on the usefulness of the 'SEND' label, it can aid the selection of effective pedagogical strategies (Gross, 2023: Sobel and Alston, 2021).

This chapter aims to consider the specific challenges that students with SEND may experience in geography and then presents a three-question approach to aid the selection of appropriate pedagogy. The strategies suggested are intended to be manageable in terms of teacher workload and able to be implemented in busy classrooms where 1-2-1 support is not always possible. This chapter draws on examples from my school which is a state secondary school that also provides specialist provision for a small number of students with moderate learning disabilities (MLD).

This chapter is not attempting to provide a specialist's knowledge of SEND needs. It is recommended that research is carried out by teachers with regards to the needs of students they teach. A good starting point is to speak with the special educational needs co-ordinator (SENDCO) at your school. What is offered in this chapter is the view of a current head of geography who has worked closely with SENDCOs and also has a background in teaching English as a foreign language.

SPECIFIC CHALLENGES FOR STUDENTS WITH SEND IN GEOGRAPHY

There are numerous teaching and learning publications that cover the challenges students with SEND experience and suggest strategies for support. However, what can often be omitted is the specific challenges faced in individual subjects. The 2023 Ofsted geography review contains a small section on supporting students with SEND and acknowledges that there is little research available on the specifics of this in geography (Gov.uk, 2023a). The following is a list of some of the challenges that students with SEND may experience in the geography classroom.

1. **Knowledge and skills:**
    - Amount of knowledge, specialist vocabulary and place examples.

- Ability to comprehend another's situation and infer opinion.
- Range of written, numerical and cartographic skills.
- Use of figures including colour, size and complexity.
2. **Fieldwork and fieldtrips:**
    - Meeting send need, including medical needs, outside of school.
    - Data collection tasks.
    - Working in groups.
    - Working in unknown settings.

UNDERSTANDING NEED AND PLANNING SUPPORT

Teachers know of strategies that can be implemented to support students in their classrooms, for example, I give sentence starters to students with dyslexia. When you are planning to meet the needs of each of your students with SEND, ask yourself the following three questions:
- What is the student's need?
- How will you support their need?
- How does this strategy support their need?

It is the last of these three questions that makes teaching staff further their understanding of SEND and enable them to say not only *what* they are doing but *why* it helps. These questions were developed from an approach used by a SENDCO to encourage heads of department to think more deeply about their departmental SEND offer. The following are examples of learning needs often found in our geography classrooms.

EXAMPLE 1: STUDENT WITH A SPECIFIC LEARNING DIFFICULTY (SPLD)
- What is the student's need? *My student has a specific learning difficulty and struggles to start written sentences when we practise extended writing in geography.*
- How will you support their need? *I prepare sentence starters, plan on mini whiteboards or verbally develop a sentence starter with this student.*
- How does this strategy support their need? *The mini whiteboard allows the student to try and rehearse their answer. Verbally working with the teacher gives them confidence in their ideas.*

EXAMPLE 2: STUDENT WHO HAS COLOUR VISION DEFICIENCY (CVD)

- What is the student's need? *My student has colour vision deficiency (CVD) and cannot distinguish between red and green, which makes it difficult to use many figures. This includes exam materials.*
- How will you support their need? *I speak with them to ask which colours they can differentiate between on a particular resource. I point to the key or legend on the figure, and we add labels to correspond to what they can see.*
- How does this strategy support their need? *The student can use the teacher as a prompt as they would be entitled to in the GCSE exam.*

EXAMPLE 3: MAINSTREAM CLASSROOM WITH SOME STUDENTS WITH MODERATE LEARNING DIFFICULTIES (MLD)

- What is the student's need? *My school is a regional hub for students with MLD. Students attend geography lessons as part of their mainstream curriculum and find the vocabulary load challenging. A teaching assistant will be present in the classroom, which could contain three students with MLD in addition to SEND and unregistered need in the classroom. Therefore, there is a high level of need in the classroom and a traditional 'go over and help' is not practical.*
- How will you support their need? *I use support worksheets developed to enable students to independently access the same learning as the rest of the class. An example which supports the start of a lesson is shown in Figure 17.1.*
- How does this strategy support their need? *The worksheets mirror whole-class learning tasks but contain extra cognitive prompts. An additional benefit is that they can be useful for support staff to refer to in lessons, especially if provided with suggested answers.*

Title: What are high and low energy coasts?
Date:

1. Today – What adjectives (describing words) would you use to describe the coast shown in the photo?

2. Key word – what do the words destruct and construct mean?

3. Last topic – What could be causing the weather in the photo?

Figure 17.1 Sample support sheet for students with specific or moderate learning difficulties

Source: Cushing (2023)

These worksheets are not intended to predispose the set tasks and limit students, as is warned by Myatt (2020). Instead, they support a preidentified need in a large classroom by 'putting in place cognitive support so that everyone can learn the same thing in much the same way' (Enser, 2021). It is imperative that teachers have the same high expectations and that these should not be limiting and should still provide 'intellectual stimulus and challenge' (Biddulph et al, 2015). Worksheets come with a workload demand, but this can be attained with a department approach.

EXAMPLE 4: STUDENT WHO HAS ENGLISH AS AN ADDITIONAL LANGUAGE (EAL)

While many schools still refer to students having EAL, there is an argument towards referring to students as bilingual on the basis that it is 'broader and more inclusive' (Conteh, 2006, p.3).

- What is your student's need? *The amount of content, keys words and case studies are a particular challenge.*
- How will you support their need? *Develop core geographical vocabulary using pictures, translation, contextual use, verbal use and recall.*
- How does this strategy support their need? *They develop their geographical vocabulary in English by connections to their first language. Theories of second language acquisition stipulate that closed practice and verbal activation will support the storage of new vocabulary to long-term memory (Conteh, 2006).*

Again, this approach brings with it additional workload. The use of a universal instruction list has proved to have success in developing geographical language and student independence in my department as well as being manageable in terms of workload. It requires only the key words, a closed practice activity and five questions to be changed each lesson. The instructions can be translated into a student's first language.

Geography lesson instructions

1. Write the date and title in your book and underline it.
2. Complete the do it now.
   a) Write 5 words from last lesson in English.
   b) Write a sentence about the photo on the board in English.
   c) Write a question about the photo on the board in English
3. Write the new words for the lesson in your book. Write a translation for your language. Draw a picture of the word.
   a) a river
   b) a flood / to flood
   c) a problem
   d) people
   e) money

4. Complete these sentences with the words in task 3.

   When a river .... it can cause problems. There are problems for .... For example, people's houses are damaged and it costs ....

5. Answer the questions below
   a) What is a flood?
   b) Where do floods happen?
   c) Why do floods happen?
   d) What problems do floods cause for people?
   e) Who do they affect?

6. Read the text and answer the questions (teacher to give translated part of lesson).

7. At the end of the lesson, rewrite the 5 new words from the lesson and their meaning.

Figure 17.2 Tasks and instructions for a geography lesson

SUPPORTING STUDENTS DURING FIELDWORK AND FIELDTRIPS

Fieldwork is an essential component of geography (see chapter 8) but can present specific challenges for students with SEND and for staff planning fieldtrips and residentials. Always share a list of students being taken offsite with relevant staff in your setting, including the SENDCO, safeguarding lead, pastoral team and medical team. It is imperative to speak with staff in school. In some cases, professionals in the local authority can also provide information. I have previously met with the visually impaired (VI) support team in my local area, educational psychologists and NHS staff to enable me to meet high SEND on fieldtrips and residentials. Some of the following strategies are quick to implement adjustments, some may require consultation with relevant professionals.

What is the student's need?	How will you support their need?	How does this strategy support their need?
Student has a visual impairment (VI) which could include limited distance vision, difficultly with outdoor bright lights, difficulty reading small text and/or difficulty distinguishing colours.	Use a booklet that is the correct text size, font and colour according to need. Use a phone to record data by voice. Ensure that the student has support with walking in unknown locations. Contact local VI team for further advice.	The student can access written instructions, collect and record data.
Student is hard of hearing and finds it more difficult to hear over background noise, for example in an urban area.	Necessary verbal instructions are planned and added to the booklet.	The student can use the written text to compliment spoken instructions on the day.
Student has ADHD tendencies and may find it difficult to follow a long set of instructions, for example how to collect data.	Preparation work in the classroom using prerecorded videos of staff completing fieldwork. Use of fieldwork equipment in the classroom to rehearse.	The student practises the fieldwork skill step by step which aids retention.
Students with autism can find the unknown challenging.	Share the 'plan for the day' with students in advance.	The student knows what is happening step by step. On residentials, this could be repeated the evening before each day.
Students with autism can feel overwhelmed, particularly in crowds such as urban areas and transport hubs.	If travelling through an airport or large transport hub, student may have a sunflower lanyard (Hidden disabilities, 2023).	This signals that the individual may need extra support. In an airport, the individual is entitled to using a fast-track lane.
Student has experienced trauma and may struggle with large groups.	Speak with them to ascertain their coping strategies. Student may want to board the bus first/last. If possible, identify a safe space at each site.	This helps the individual with feelings of being unsafe in large groups.
Student has attachment difficulties and may struggle with unknown adults.	Student to work with known member of staff, this may be an additional 1:1. If the student has an EHCP and 1:1 support at school, this support should be upheld off site.	This helps with feelings of being unsafe, it allows a staff member to employ coping strategies if needed while maintaining the required number of staff:student ratio for the trip to continue.

Table 17.1 Supporting students for fieldwork

With specific reference to fieldwork, it is noted that 'many adjustments to support disabled learners will benefit all participants' (JISC, 2016). This is often the same for any support applied in the classroom and will constitute quality first teaching and high expectations for all (Biddulph et al, 2016; Gov.uk, 2023). It is hoped that further discussion and research can increase our knowledge of specific challenges in geography for students with SEND and how best to support students.

> **Five reflection questions**
> 1. What do you consider to be the challenges in geography for students with SEND?
> 2. For a specific student you teach, what are their needs? How will you support them? How do these strategies help?
> 3. How do you work with and support teaching assistants in your classroom?
> 4. What discussions about students with SEND happen in your department meetings?
> 5. What support do you have in place for fieldtrips and learning outside the classroom?

> **DIGGING DEEPER - THREE RESOURCES TO DELVE FURTHER**
> - Colour blind awareness. Available at: https://www.colourblindawareness.org/
> - TDA. (2009) Including students with SEN and/or disabilities in primary geography. Available at: https://dera.ioe.ac.uk/id/eprint/13792/1/geography.pdf (Accessed: 28/05/2025)
> - TDA. (2009) Including students with SEN and/or disabilities in secondary geography. Available at: https://dera.ioe.ac.uk/id/eprint/13793/1/geography.pdf (Accessed: 28/05/2025)

REFERENCES

Biddulph, M., Lambert, D. & Balderstone, D. (2015) *Learning to Teach Geography in The Secondary School. A Companion to School Experience.* 3rd ed. London: Routledge.

Conteh, J. (ed) (2006) *Promoting Learning For Bilingual Pupils 3–11.* London: Paul Chapman Publishing.

Enser, M. (2021) Are there any new ideas in teaching? *TES Magazine*, 9 July. https://www.tes.com/magazine/archived/are-there-any-new-ideas-teaching (Accessed: 28/05/2025)

Gov.uk. (2020) Special educational needs and disability code of practice: 0 to 25 years. https://assets.publishing.service.gov.uk/government/uploads/system/uploads/attachment_data/file/398815/SEND_Code_of_Practice_January_2015.pdf (Accessed: 28/05/2025)

Gov.uk. (2023a) Special education needs in England. https://explore-education-statistics.service.gov.uk/find-statistics/special-educational-needs-in-england. (Accessed: 28/05/2025)

Gov.uk. (2023b) *Geography subject review*. https://www.gov.uk/government/publications/subject-report-series-geography (Accessed: 28/05/2025)

Gross, J. (2023) SEND: A label worth having? *TES Magazine*, 25 October.

Hidden disabilities. (2023) Our sunflower is for everyone with a hidden disability. https://hdsunflower.com/ (Accessed: 28/05/2025)

JISC. (2016) Between a rock and a hard place – making fieldwork accessible to disabled learners. https://accessibility.jiscinvolve.org/wp/2016/10/03/fieldwork/ (Accessed: 28/05/2025)

Myatt, M. (n.d.) Death by differentiation. https://www.marymyatt.com/blog/death-by-differentiation (Accessed: 28/05/2025)

Sobel, D. and Alston, S. (2021) *The Inclusive Classroom. A New Approach to Differentiation*. London: Bloomsbury.

18. TEACHING 'DISADVANTAGED' STUDENTS
CATHERINE OWEN
@GEOGMUM

'We need a rethink; we need to find out what every child can offer, what we need to change, and how we can work together.' (L. Elliot Major, quoted in Weale, 2023)

WHAT IS DISADVANTAGE?

The UK government defines disadvantage by saying: *'it uses eligibility for the Pupil Premium as a guide: children who have been eligible for free school meals during the past six years, children who are in care, and children who were previously in care but left in particular circumstances such as adoption.'* (Long and Bolton, 2015).

However, we need to look beyond this definition to consider children who face other barriers to education but don't qualify for Pupil Premium. What if they could qualify for free school meals but haven't claimed them? Does it make a difference if children are persistently disadvantaged or 'dip in and out' of disadvantage? Deciding who is disadvantaged is complex – it is a value-laden judgement and may be best made by those closest to the school, who know what type of 'disadvantage' students face (NGA, 2018).

WHAT EVIDENCE IS THERE OF DISADVANTAGED STUDENTS FACING INEQUALITIES IN EDUCATION?

In March 2022 the UK government Education Committee released a report highlighting:

- regional disparities in learning loss, with some disadvantaged students up to eight months behind
- schools with the most disadvantaged students being ten times more likely to have 'a class worth of severely absent pupils'. Severe absence was defined as missing over 50% of sessions.

This is reflected in GCSE results, with the KS4 disadvantage gap index widening to 3.95 in 2022/23 in comparison to 3.84 in 2021/22, reaching the highest level since 2011.

The Royal Geographical Society (2018) reported that GCSE geography entries in England were lower for disadvantaged students than for non-disadvantaged students, but that this gap had narrowed from 14.4 percentage points in 2013 and 2015 to 11.1 percentage points in 2018, probably because of schools being encouraged to promote English Baccalaureate (EBacc) subjects due to Progress 8 measures.

This suggests that disadvantaged students are more likely to fall behind their peers in terms of learning, more likely to be absent from school and less likely to study GCSE geography.

DISADVANTAGED, DEPRIVED OR DIFFERENT?

The first UK professor of social mobility, Lee Elliot Major, has suggested that there is a mindset in education leading to working-class children being seen as inferior. He proposes referring to students as 'under-resourced', arguing that the term 'disadvantaged' leads to unconscious bias and low expectations (Weare, 2023).

Functionalist sociologists such as Hyman (1967) and Sugarman (1970) may have set the stage for this mindset in education, claiming that children from working-class backgrounds faced cultural deprivation as their families were less well able to prepare them for school (Pullinger, 2020). Bernstein (1971) proposed that working-class students communicated in a restricted code because of the conditions in which they were raised, disadvantaging them in schools, which use an elaborated code. Critics queried these sociological ideas, suggesting that maybe working-class students were different, rather than deprived. We can also ask these questions – are our 'disadvantaged' students different rather than disadvantaged? Is it the education system which is causing them to be disadvantaged?

An example of an area where working-class students could be seen through either a deficit lens or as different is that of using 'proper' English. How does insisting on a 'proper' way of talking and writing affect our students in geography lessons? Cushing (2022) argues that Ofsted 'judgements about what is "good" English are based on the language practices of the white middle classes. These ideas are embedded in schools. They perpetuate the belief that what is "correct" is the language used by the most powerful members of society.' We want our students to be able to communicate their geographical thinking well, but should we be thinking about difference in terms of the way students communicate, rather than deficit?

Elliot Major also comments upon exam papers testing students with questions related to middle-class pursuits such as trips to the theatre and skiing holidays, providing an advantage to some students, unrelated to their proficiency in the subject being assessed (Weale, 2023). We need to reflect on how this situation could arise in our geography lessons and assessments, for example the whole pre-release for the AQA GCSE Geography paper 3 in 2023 was about cruise tourism! Elliot Major suggests a 'deep listening campaign' to help teachers understand the communities their students come from; this approach may be of particular benefit to geography departments.

The 'Against the Odds' report from the Social Mobility Commission (2021) found that UK schools were increasingly spending the extra funding they received for pupil-premium students on initiatives aiming to boost their 'cultural capital'. This concept was originally used by Pierre Bourdieu to refer to the high culture associated with the upper class but is now used in English schools to refer to social and cultural knowledge which can help a student make progress. An example is that a geography student who has read widely about the world, travelled extensively and made visits to museums and zoos is likely to have a lot of knowledge to supplement that learned in the geography classroom. However, Elliot

Major feels that students are being made to become 'middle-class clones' in order to succeed in school (Weale, 2023). Again, we need to strive to find balance in our geography curriculum. We can support students in developing their cultural capital through our lessons and fieldwork, but we can also recognise the importance of cultures beyond high culture. For example, some of our students will bring cultures from other places to our lessons while some will have great depths of sub-cultural capital as they know what is happening within their youth group (Thornton, 1995).

GYPSY, ROMA AND TRAVELLER STUDENTS – AN EXAMPLE OF A DISADVANTAGED GROUP

Friends, Families and Travellers is a group which supports Gypsy, Roma and Traveller (GRT) people and works to end racism and discrimination against them. They point out that GRT children are often bullied in school, have lower levels of attainment and are more likely to be excluded. They say that young people from these groups 'feel that their culture and way of life is not recognised or affirmed within the education system. For example, Gypsy, Roma and Traveller histories are not covered in the curriculum and many other aspects of the curriculum fail to consider the lives and experiences of Gypsy, Roma and Traveller pupils.' (Friends, Families and Travellers, accessed 2022).

Bristol City Council (2022) has identified a range of factors which can affect the academic achievement of GRT students, including:

- Disrupted and/or different educational experiences.
- Educational disadvantage of their parents.
- Social, economic, health and cultural reasons.
- Transportation and accommodation issues.
- Dispersed extended family demands.
- Lack of cultural sensitivity within the education system.
- Racism in employment sector.
- Lack of role models.
- English as an Additional Language (EAL) issues.
- Refugee and asylum seeker issues.

As educationalists, we can work to improve cultural sensitivity within the education system, but awareness of these other factors can also prompt us to make other improvements to benefit GRT students and others.

Wilkins et al (2009) state that there are 'no inherent reasons why a child from a Gypsy, Roma or Traveller community should not achieve as well as any other child.' They suggest that schools concentrate on providing high-quality teaching alongside specialist interventions. Areas identified as particularly important include:

- Safety and trust – a school gaining a reputation for being caring and understanding can have a long-term effect within a Traveller community.
- Respect – this needs to be a two-way process, with schools communicating expectations but also being flexible and inclusive so that cultural differences can be accommodated.
- Access – this could be as simple providing a school uniform or could involve providing distance learning for a student who is travelling.
- Flexibility – in terms of curriculum, a topic-based or work-related curriculum may help keep students motivated.
- High expectations – if students and their parents feel they are able to be successful, they are more likely to stay in school.
- Partnerships – schools, children and their parents all need to work together.

We need to be aware that Gypsy, Roma and Traveller students are different from each other. An example is that Roma migration occurred because of marginalisation and a desire to escape poverty, so isn't related to nomadic traditions, meaning that symbols such as caravans and horses don't represent Roma children's culture (Matras et al, 2014). Geography teachers need to be sensitive to Roma experiences of marginalisation in lessons about migration; we also need to recognise that Roma tend to organise their lives in extended kinship groups and have strong values of honour, which may affect the way they see different geographical issues. Using Elliot Major's idea of a 'deep listening campaign' could help us to understand communities such as the Roma and support their young people in achieving educational success.

GEOGRAPHY WHICH EXCLUDES

We can make it hard for some of our students if we use language which they find hard to follow, for example using tier 2 and 3 vocabulary without defining it and explaining its use in the current context.

If we take an ethnocentric approach, for example only teaching about white European perspectives and experiences, we can leave some students feeling excluded, undermining their self-esteem (Coard, 2005).

Some activities may make students uncomfortable, for example fieldwork involving payment or planned without considering the needs of individuals. We also need to be conscious of the examples we use, in case our students won't be familiar with them. An example is that a popular GCSE textbook uses mangetout as an example of one of Kenya's main food exports – the author knows that very few of her students know what this is, so brings a pack into that lesson for taste testing.

Think about your own context – what else would you add?

GEOGRAPHY WHICH INCLUDES

The Geographical Association (2022) reminds us that 'As a geography teacher, you can help to break down barriers and reduce ignorance and prejudice ... In geography lessons, (this) diversity can be used as a platform to provide meaningful interactions between students. Encourage learners to explore their own identity, discuss their ideas and address sensitive and controversial issues.'

One approach is to think about issues which could arise for each of the topics to be studied with a class in advance. Table 18.1 shows some of the author's thoughts in relation to her Year 10 students, who study the AQA GCSE geography specification.

Table 18.1 Possible issues arising from our Year 10 topics

Urban issues and challenges	We study Bristol and Kampala for our two case studies. Some students will know Bristol well, others won't. When looking at opportunities in Bristol, I need to be sensitive because some students won't have had the chance to go to the theatre, etc. Bristol has clear association with the trafficking of enslaved people in the past, which needs to be tackled head on, as does the treatment of migrants from the Commonwealth who settled in the St Pauls area and historic events such as the Bristol Bus Boycott. During the lessons about Kampala it will be important to avoid misconceptions about life in a low-income country (LIC).
The living world	Students who have travelled further are likely to have a better idea of how biomes vary around the world; others will benefit from seeing examples using video clips, etc.
Resource management	Many students live in rented housing, so in lessons about reducing energy use we need to take their lack of choice about some elements into account, i.e. insulation. Talking about food waste will be a sensitive issue for students whose families are using the local food bank.
UK landscapes	Using the local coastline as an example makes the coasts topic more relevant to our students, but students who don't travel outside the Somerset Levels may not know much about rivers, so will need to be shown videos and examples to develop their understanding.
Fieldwork investigations	Students must understand that they won't be excluded from fieldwork because their family can't afford the voluntary contribution. Some may be nervous about travelling to Bristol and Lyme Regis as they are unfamiliar with these places – this may come from home and so phone calls to reassure parents/carers may be needed. Students must be given the opportunity to discuss any concerns so that we can adapt the fieldwork to meet their needs.

I have worked with my department to identify factors which are particularly important in supporting disadvantaged students in achieving good GCSE outcomes in geography, coming up with the factors shown in Figure 18.1. Several of these factors are also being prioritised at a whole-school level.

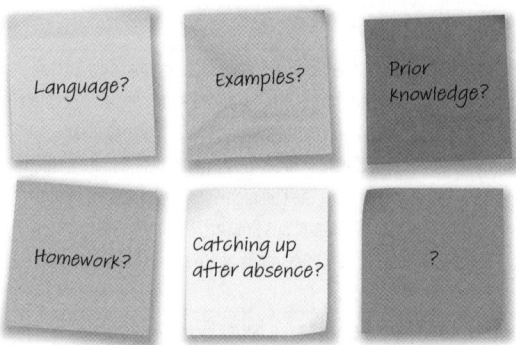

Figure 18.1 Factors identified as affecting attainment of disadvantaged students in the author's context

The school has moved to knowledge organiser-based homework in Year 10, so that all students have access to the resource they need for success. The department reviews the language we use in lessons and the examples we study ahead of each topic and have reviewed our KS3 curriculum to make sure it provides the prior knowledge students need as a foundation for GCSE. Attendance is a major concern; we have made stickers to put in books when students are absent so that they know what they missed, hold a weekly after-school session for students who need support to catch up and have topic overview sheets for students who have missed large chunks of a topic.

The question mark in Figure 18.1 reflects our awareness that we need to keep reflecting upon what we are doing, which could lead to awareness of further factors to consider. We don't have the answers but can always aim to improve our teaching of our 'disadvantaged' students.

> **Five reflection questions**
> 1. What factors may cause students to be 'disadvantaged' when learning geography in your context?
> 2. How could this affect their attainment in geography?
> 3. What can you do to support these students in achieving good outcomes in geography?
> 4. How does this fit in with your school-wide strategies for supporting students?
> 5. How could a 'deep listening campaign' help you to understand the needs of your students?

> **DIGGING DEEPER – THREE RESOURCES TO DELVE FURTHER**
> - Atkins (2023) Class Teaching – Find the bright spots. Available at: https://classteaching.wordpress.com/2023/06/28/tackling-educational-disadvantage-in-geography/
> - Elliot Major, L. and Briant, E. (2023) *Equity in Education: Levelling the Playing Field.* John Catt Education Ltd.
> - Geographical Association. Inclusive geography teaching (geography trainers). https://geography.org.uk/ite/inclusive-geography-teaching/ (Accessed: 28/05/2025, members only resource)

REFERENCES

Bernstein, B. (1971) *Class, Codes and Control: Volume 1 - Theoretical Studies Towards a Sociology of Language.* London: Routledge & Kegan Paul.

Bristol City Council. Gypsy, Roma, Traveller Team: Education. https://www.bristol.gov.uk/files/documents/3591-schools-with-grt-pupils/file

Coard, B. (1971) Making black children subnormal in Britain. *Integrated Education: Race and Schools,* 9(5), 49–52.

Cushing, I. (2022) Ofsted has been dictating what "proper English" is – here's why that's a problem. The Conversation. https://theconversation.com/ofsted-has-been-dictating-what-proper-english-is-heres-why-thats-a-problem-176742 (Accessed: 28/05/2025)

Education Committee. (2022) Disadvantaged pupils facing 'epidemic' of educational inequality. UK Parliament, Committees. https://committees.parliament.uk/committee/203/education-committee/news/161687/disadvantaged-pupils-facing-epidemic-of-educational-inequality/ (Accessed: 28/05/2025)

Friends, Families and Travellers. Education. https://www.gypsy-traveller.org/our-vision-for-change/education/ (Accessed: 28/05/2025)

Geographical Association. https://www.geography.org.uk/Creating-an-inclusive-geography-classroom#2 (Accessed: 28/05/2025)

Long, R. & Bolton, P. (2015) Support for disadvantaged children in education in England. UK Parliament, House of Commons Library. https://commonslibrary.parliament.uk/research-briefings/sn07061/ (Accessed: 28/05/2025)

Matras et al. (2014) Roma pupils at Bright Futures Education Trust (Gorton South, Manchester): background, preliminary observations, engagement strategy. The University of Manchester. http://migrom.humanities.manchester.ac.uk/wp-content/uploads/2017/11/MigRom_Bright-Futures_report_Dec-2014.pdf (Accessed: 28/05/2025)

NGA (2018) https://www.nga.org.uk/News/Blog/June-2018-(2)/The-catch-22-with-defining-disadvantage.aspx

National Statistics. (2023) Academic year 2022/23, Key stage 4 performance. https://explore-education-statistics.service.gov.uk/find-statistics/key-stage-4-performance-revised (Accessed: 28/05/2025)

Pullinger, J.T. (2020) *Sociology: An Introduction and Beyond*. Cambridge: Cambridge Academic. Royal Geographical Society (2018) https://www.rgs.org/geography/key-information-about-geography/geographyofgeography/report/geography-of-geography-report-web.pdf/

Social Mobility Commission (2021) Against the odds Achieving greater progress for secondary students facing socio-economic disadvantage. https://assets.publishing.service.gov.uk/media/60dc34c88fa8f50aad4ddb0a/Against_the_odds_report.pdf (Accessed: 28/05/2025)

Thornton, S. (1995) *Club Cultures: Music, Media, and Subcultural Capital*. Fishers, IN: Wesleyan Publishing House.

Weale, S. (2023) Warning over unconscious bias against working-class pupils in English schools. *The Guardian*, 3 October 2023. https://www.theguardian.com/society/2023/oct/03/warning-unconscious-bias-working-class-pupils-schools-england

Wilkin, A. et al. (2009) Improving educational outcomes for Gypsy, Roma and Traveller pupils – what works? Department for Children, Schools and Families. https://www.theeducationpeople.org/media/4276/improving_the_outcomes_for_grt_pupils-4.pdf (Accessed: 28/05/2025)

19. RETRIEVAL PRACTICE IN GEOGRAPHY

JENNIFER MONK
@JENNNNNN_X

'When we think about learning, we typically focus on getting information into students' heads. What if, instead, we focus on getting information out of students' heads?' (Agarwal, Roediger, McDaniel and McDermott, 2020)

MEMORY

'Retrieval Practice is the act of recalling previously encountered information into working memory, or conscious thinking. Brief spurts of Retrieval Practice help students solidify information in their long-term memories, and, importantly, understanding is not learning until it is encoded in long-term memory.'
(Lemov, 2023)

Retrieval practice has felt like a bit of a 'buzz word' in education over the past eight or so years. I, like many other teachers, have trialled a range of different strategies to encourage students to recall information previously learned. I am sure, like me, you have had a lesson where students can't recall information you taught them previously. As frustrating as it is, the idea that we will remember everything we learn is impossible, especially if we have limited knowledge of a concept to 'attach new knowledge to' – this is the theory of 'schema' which I will revisit later in this chapter.

Although the discussion in classrooms seems to have been fairly recent, the idea of working and long-term memory goes back to Ebbinghaus' Forgetting Curve in 1885.

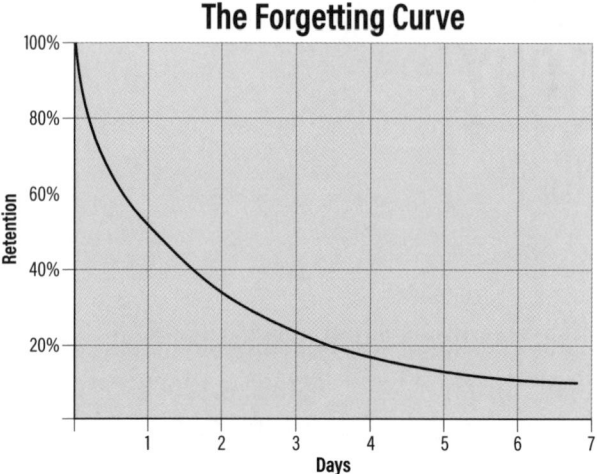

Figure 19.1 Ebbinghaus' Forgetting Curve (1885)

The forgetting curve is used frequently among researchers and teachers alike. The curve itself suggests the rate at which information is 'lost' following a lesson; it suggests that seven days after a lesson was taught the students retain less than 20% of the information given. When looked at so simply, this is quite demoralising. A few key takeaways from Ebbinghaus' findings include the idea that we forget most of the information received fairly quickly, hence the steep drop in the curve at the beginning. There are further suggestions that other factors can change the rate at which we forget information, such as whether the topic is of personal interest, how meaningful the content is to the learner and other factors linked to how the learner is feeling at the time (Busch and Watson, 2019). There are, however, some criticisms of the forgetting curve itself, mostly due to the fact that Hermann Ebbinghaus tested his own memory to develop his theory.

More recent research, for example by Watson and Busch (2019), on similar strategies to test memory have found similar results. The idea of retrieval and distributed (spaced) practice were 'found to be very effective for improving long term memory'. By revisiting material at regular intervals following the 'first teaching', we slow down the time it takes for students to forget the material and therefore they are more able to retain the information for longer and, in theory, are able to access it quicker when asked.

SCHEMA

Bjork and Bjork (2019) suggest that 'we need to forget things in order to deepen our understanding of a concept' and every time we forget something and then bring it back to our working memory from our long-term memory, we actually make it easier to retrieve in the future. This is where the theory of 'schema' comes in.

Schema is the invisible structure that organises our thinking; as we take in new information, we connect it to other things we know. This could be a belief or something we have experienced. Those who have much narrower experiences, or 'know less', find this harder to do as they don't have the prior knowledge to connect new knowledge to. Donovan and Bransford (2005) take this even further and suggest that 'the reason experts remember more is that what novices see as separate pieces of information, experts see as organised sets of ideas'. Ensuring that we map out the themes/topics across our curriculum will then help to build a schema in students' long-term memory by enabling them to relate new content to previously learned knowledge in a systematic way (see chapter 3). By teaching students where the links are and how one topic links to another, we can deepen their understanding and, in theory, help students to remember knowledge more effectively.

Brewer and Treyens (1981) took this idea one step further and experimented to see how schema affects memory of places. They carried out an investigation where people were asked to sit in a room that resembled an office for 30 seconds; some of the usual things you would find in an office were in there, along with some abnormal items, for example a frisbee and a wine bottle. Some items you would expect to find in an office were also missing, for example books and a lamp. Once the participants had left, they were asked to remember what they had seen. Many noted they had seen books (although none were present) which suggested that many people's schemas of 'office' include books. This is relevant to us in geography as many of the topics we discuss require students to have some sort of schema in mind. For example, if students have never seen a river it's unlikely they will have a detailed schema for a river, which might then mean they can't imagine how erosion would work. When we move on to look at erosional landforms, they will then find them harder to visualise or understand.

According to Ausubel (2000), meaningful learning happens when the learner chooses to relate new information to the knowledge he/she already has. This depends on various different things such as the quantity and quality of the organisation of prior knowledge held by the learner. It is of course possible to learn new information without making meaningful connections, but there are some suggestions that this learning (known as 'rote learning') is less secure. If we as teachers focus mainly on learning facts and not allowing students to create these more detailed schemas by developing their critical thinking skills, we are making it much harder for them to recall and retrieve information.

I think it is also useful to mention here the gap between 'non-disadvantaged' children and those classed as 'disadvantaged' (see chapter 18). While this is not always necessarily true for all, it is likely that students with a much smaller experience of the world we live in, for example those who haven't left their local area, haven't visited a beach or seen a river, will find it hard to build a detailed schema about some of the concepts their peers may already have. The same can be true for those students who are interested in geography and read for pleasure around the subject. They too will have a more complex schema for some of the concepts we study, more so than those who haven't read the same books or visited the same places.

BUT WHAT DOES RETRIEVAL PRACTICE LOOK LIKE IN MY CLASSROOM TODAY?

I imagine that most teachers reading this book regularly use retrieval practice in their classroom, setting quizzes, using homework to recap learning, providing revision-based activities or through the use of exams and assessments. I also imagine most of you probably think of retrieval practice as an assessment tool. However, Agarwal, Roediger, McDaniel and McDermott (2020) suggest that actually we should think of it as a 'learning tool' and when we think of it in this way and not an assessment tool, it becomes far more useful at supporting student learning.

The strategies I share below are probably not all new ideas to you, but I hope that by understanding a bit more about memory and schema you can think more deeply about how you chose the type of activity or question you use, which means you can then create a more successful approach to retrieval practice in your classroom.

Examples of retrieval practice in geography that could be utilised in your lessons:

1. Simple retrieval questions (from previous lessons) – could be short-answer or multiple-choice questions.
2. 'Geog your memory' – where the letters of 'Geog' stand for Geographical knowledge, Earlier this month, Older than a month, Geographical skills.
3. Odd one out – give three or four terms – students must identify the odd one out and say why they have chosen this term.
4. What can you see – using a photograph/graph/map as a source.
5. Find and Fix or correct my errors.
6. Last lesson, last week, last month/term, last year.
7. Where am I? – location of places on a map.
8. Definitions of key terms.

Over the past five or six years I have put a lot of thinking into how we retrieve information from our students and how we can ensure the time spent on retrieval practice gets the most out of our students. Previously, I used the first five to ten minutes of most lessons to focus on students activating prior learning. An example of a standard retrieval practice task would have been one like below, where students aimed to get as many 'points' as they could by accessing knowledge previously learned. Students would be given five minutes to score as many points as they could, writing the answers into their books. We would then discuss some of the questions in detail and I would share quick answers to all of the questions so students could self-assess their answers.

How many points can you score in 5 minutes?

Explain why China needed a population policy?	What is fertility rate? Why might it decrease as a country develops?	Why is the Scottish Highlands a sparsely populated area?	How would a population pyramid be different in stage 3 and stage 5 of the DTM?
What is urbanisation? Why did it impact the population density and distribution in the UK?	Draw a sketch of a population pyramid in an LIC. Explain the structure of the population.	Define infant mortality rate? What is the difference between that a child mortality rate?	The One Child policy had many effects, what are the problems left facing China today?
What is the demographic transition model? How does it show the link between population change and development?	Describe the difference between sparsely and densely populated. Can you name 3 examples of each in the UK.	Explain why HICs have a lower birth rate and LICs. Try to include more than one reason.	Describe the growth of world population in the last 150 years. Try to include figures where possible.
Last lesson - 1 mark	Last week - 2 marks	2 weeks - 3 marks	3 weeks - 4 marks

Figure 19.2 Example of a starter task

While there is nothing wrong with this task, I think there are almost too many questions to effectively challenge misconceptions and ensure all students in the classroom know the key information needed for the lesson. Instead, over the past few years, I have taken a more individual approach for each class and, rather than using a single starter for all of my Year 8 classes (like above), I use my assessment of their learning from the last few lessons, as well as summative assessments and their baselines, to choose specific questions which I know all/some students are unsure of.

In my department we have now moved to a 'last lesson, last week, last term/month and last year' approach, which we think has been more effective in ensuring students are revisiting material regularly. We have then taken this further with personalised questions, so as I mentioned previously, the questions you see from one classroom to the next are not the same. This has meant that our assessment for learning reflects on topics over time, but also, we are thinking deeply about the knowledge needed for that particular lesson. For example, when teaching about hazard risk in our tectonic hazards unit, one of the starter questions was linked to defining 'densely populated'. Not only did this recap learning from a previous topic, but it also meant that students understood this term ahead of considering it as a factor influencing hazard risk in the current lesson.

The example in Figure 19.3 was a starter for a lesson on the 2018 Indonesian earthquake. I have used a mixture of knowledge which is essential to the lesson – factors affecting hazard risk and GDP per capita (as we used this to compare to other places and other earthquakes) and knowledge that is less relevant, on fashion consumption. Students completed this in their books, I questioned students to check knowledge and students self-assessed in purple, as they do every lesson.

⏮ Do now: Recap

1. **Last month**: What factors affect hazard risk?
2. **Last term**: How can we reduce our fashion consumption?
3. **Last year**: What is GDP per capita?

Figure 19.3 Example of a starter task

By thinking carefully about the prerequisite knowledge needed for the lesson, as well as some other linked knowledge, not only am I guaranteeing to support students with some of the ideas in the lesson, it also allows me to see how much support they may need at certain points during the lesson. There are, however, some issues with this – namely that some topics may never be revisited, for example where place examples have been used. If that place isn't revisited, then how would you revisit that knowledge? My suggestion here is to try to use these questions when revisiting either similar places or opposite places to compare/contrast. For example, if previous knowledge relates to environmental challenges in Rio de Janeiro, you could retrieve this information when looking at another South American country or you could compare it to a HIC city to look at the different challenges faced.

The way in which we check this retrieval of knowledge is also vitally important to its success. If students are writing it down, then you may struggle to check carefully all of the students in the class. One suggestion here is that you could instead use mini whiteboards or, if you have technology available, use online quizzing software to quickly get feedback. A possible pitfall here is that there is often not an opportunity for students to ensure they know why they were wrong and, over time, this may lead to further misconceptions which will need to be revisited.

There are, of course, some exceptions to the idea of always using knowledge relevant to the lesson – and some topics may never be revisited, this is where professional judgement is important. The revisiting of 'knowledge' could be more skill based, for example we could include a map task where students label places on a world map. This would quickly show you place knowledge and students can easily correct their map once complete, using an atlas. Again, you could make this specific to your lesson in that places may be chosen carefully to see if students are aware of the place you will be looking at, as well as challenging misconceptions of other locations.

We have also used our home learning to retrieve knowledge. For example, in our Year 8 unit on development, students completed a piece of home learning on China and its growth as a newly emerging economy (NEE). Within this home learning, students revisited prior knowledge on industry and urbanisation which they had learned in Year 7. They also learned about how population control has led to uneven development and even some of the impacts of an ageing population. When we study China, during our population unit later in Year 8, they will revisit this home learning and compare it to other population policies to see the similarities and differences.

> 'Practice is an integral part of effective teaching; ensuring pupils have repeated opportunities to practise, with appropriate guidance and support, increases success.'
> (ITT Core content Framework and Early Career Framework, DfE, 2019)

> **Five reflection questions**
> 1. How do you ensure all students in your classroom are involved in retrieval practice?
> 2. Do you ensure tasks are low-stakes to encourage motivation and success?
> 3. How do you choose questions to ensure high challenge is evident?
> 4. Do you assess and challenge misconceptions when retrieving prior knowledge? How could this be done successfully?
> 5. How could you plan for retrieval in the curriculum to ensure as much content as possible is revisited?

> **DIGGING DEEPER – THREE RESOURCES TO DELVE FURTHER**
> - EEF Cognitive science approaches in the classroom. Available at: https://educationendowmentfoundation.org.uk/education-evidence/evidence-reviews/cognitive-science-approaches-in-the-classroom (Accessed: 28/05/2025)
> - Limbada, A. (2020) Effective learning strategies: Making knowledge stick. Impact, Chartered College of Teaching, September 2020. Available at: https://my.chartered.college/impact_article/effective-learning-strategies-making-knowledge-stick/ (Accessed: 28/05/2025)
> - Jones, K. (2019) *Retrieval Practice: Research and Resources for Every Classroom.* John Catt.

REFERENCES

Agarwal, P.K., Roediger, H.L., McDaniel, M.A. & McDermott, K.B. (2020) How to Use Retrieval Practice to Improve Learning. Washington University in St. Louis. https://pdf.retrievalpractice.org/RetrievalPracticeGuide.pdf (Accessed: 28/05/2025)

Ausubel, D.P. (2000) 'The nature of meaning and meaningful learning' in *The Acquisition and Retention of Knowledge: A Cognitive View.* Springer, Dordrecht, pp 67–69. https://doi.org/10.1007/978-94-015-9454-7_4 (Accessed: 28/05/2025)

Busch, B. & Watson, E. (2019) *The Science of Learning: 77 Studies That Every Teacher Needs to Know.* Abingdon, Oxon: Routledge.

Bjork, R. A. & Bjork, E. (2019) Forgetting as the friend of learning: implications for teaching and self-regulated learning. *Advances in Physiology Education*, 43(2), pp. 164–67.

Brewer, W. F. & Treyens, J. C. (1981) Role of schemata in memory for places. *Cognitive Psychology*, 13(2), 207–30. https://doi.org/10.1016/0010-0285(81)90008-6 (Accessed: 28/05/2025)

Department for Education. (2019) ITT Core content Framework and Early Career Framework. https://assets.publishing.service.gov.uk/media/6061eb9cd3bf7f5cde260984/ITT_core_content_framework_.pdf (Accessed: 28/05/2025)

Donovan, S. & Bransford, J. (2005) *How Students Learn: History, Mathematics, and Science in the Classroom*. Washington, DC: National Academies Press.

Lemov, D. (2023) Turning understanding into memory: Steve Kuninsky's retrieval practice. Teach Like a Champion. https://teachlikeachampion.org/blog/teaching-and-schools/turning-understanding-into-memory-steve-kuninskys-retrieval-practice/ (Accessed: 28/05/2025)

20. USING INSTAGRAM AND AN IPAD TO ENGAGE STUDENTS
MEG PICKEN
@MISSPICKEN ON INSTAGRAM, @PICKENGEOG ON TWITTER

'I don't know how to revise!' (Every student, ever)

THE USE OF SOCIAL MEDIA IN THE TEACHING PROFESSION

At a time when 89% of 10–15-year-olds admit to going online every day and almost half of children claim to spend three or more hours online on an ordinary school day (ONS, 2021), the opportunity to harness this time and allow students to engage in useful, translational learning is huge.

The idea for my own Instagram page started in 2021. Receiving the government bursary for geography teachers, I realised I could finally afford (and justify) the purchase of an Apple iPad to 'help develop my teaching in the classroom'. This excuse quickly snowballed once I extended the spending spree to an Apple Pencil, started playing around with the functionality and realised the scope of this equipment had seemingly limitless avenues of opportunities to develop. The first app I found on the App Store, recommended for 'aesthetic note-taking' was GoodNotes and from just downloading this one app, my love of revising was reignited. From the outset of my training, I knew I wanted students to arrive to my lesson knowing how their PowerPoints would look and be delivered and for them to leave my lesson talking about how much they loved my slides. So, while initially I used the iPad to draw diagrams which were used throughout these slides, I knew I wanted to create a space for students to access these outside of the classroom.

Instagram provides us with one example of how social media platforms can help us engage our students. In this chapter, I will discuss why I use social media and give practical tips about how to use Instagram as a teacher. I will explain how I have used this resource for revision and to promote competitions. I will tackle difficult issues such as how much of yourself you should reveal on this platform and how we can stay safe online.

1. GOODNOTES

Through the use of GoodNotes and being able to create my own notes, I realised I could share these with students via a medium the majority would almost definitely be engaging with daily. I spent a few days exploring the Instagram pages already established and decided I wanted to have a page which was pleasing on the eye (which would attract students, encourage them to interact with my content and also convince them to stay), but also one that engaged students in the learning they were exploring in lessons, as well as news stories and life skills that would take geography beyond the classroom and weave into the fabric of their own lives and the context of the real world. To that end, the page began to follow the sequential format of a post about content, a post about skills or 'geography in the news', followed by a post about a case study and then that cycle would repeat (see Figure 20.1).

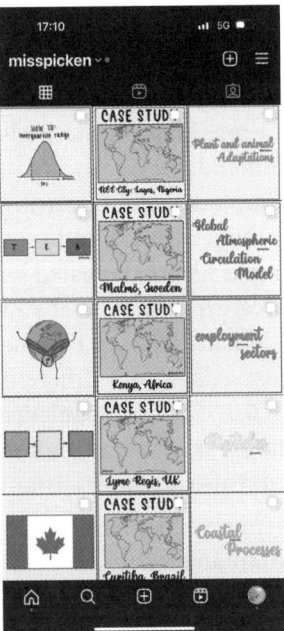

Figure 20.1 My Instagram page

Another function of GoodNotes is the ability to create your own diary. I've always personally struggled with buying diaries for school and never find they entirely meet my requirements, or they have a lot of fuss and I recognise that I struggle to be organised at the best of times, never mind if I start trying to track how many cups of water I've had in a day. Therefore, using the GoodNotes app, I draw my timetable out and can keep track of everything all in one place, without needing umpteen scraps of paper per week.

I have created a notebook for each topic and whenever we plan an exam question or agree a set of success criteria, I write this in the notebook, under the visualiser for the students to use. Then, I can screenshot this and upload it to ClassCharts (a package used for attendance, rewards, setting homework, etc.) for students to access and complete at home. These notebooks can also be made available to teaching assistants, which serves as an easy way for them to access the answers and to see how I would prefer a piece of work to be set out in exercise books.

2. CANVA

Canva is a lot like the modern-day PowerPoint. Brands and advertising companies will often use Canva to create their Instagram posts due to the large range of eye-catching design ideas and presentation methods available. This is definitely an app I would recommend, especially if you don't want to handwrite any of your content. It does take a lot of getting used to, especially as somebody who is fully indoctrinated into the Microsoft Office family, but the final product can parallel that created by a graphic designer. You can see examples of the features available if you visit my Instagram page and click any of the calendared posts. For the 2023 Year 11 cohort, I attempted to create one post a day about different aspects of the GCSE course and knew I needed to use something quicker than handwriting everything, so this was my chosen method and some of those posts ended up looking really nice, due to this software.

At the point of writing this, Canva does offer premium membership to educators, if you sign up with your school email and verify through that account.

3. SHOUTOUTS

There are many educators on Instagram, using their accounts for a variety of purposes and, for me, this enhances my experience as a teacher and allows me to draw inspiration from a multitude of sources. There are large companies like Tutor2U (@tutor2ugeog) who share exam tips and revision strategies, as well as Discover the World (@discovertheworlduk) who promote fieldwork and opportunities to travel with geography. There are individual teachers like Mr Vis (@MrVisGeography, author of chapter 10) who shares news articles and hosts 'flag of the week' to inspire and ignite the enthusiasm of students and widen their understanding of place in the context of current affairs. Another educator is K. Peppin (@mrsgeographyblog) who notably posts lots of fantastic tips for teachers on their blog, but recently has been hosting a book club, where books with geographical themes are reviewed and then recommended on their stories. Other educators like Mrs Jones (@bwsgeography) celebrate the work of their own school, highlighting fantastic work, fieldtrips and outstanding geographers. Sam Bentley (@sambentley), although not strictly in the education sector, shares good news stories, which can then in turn be shared with students to expose them to entrepreneurship, solutions and positivity in a world shrouded with problems and cynicism. Whether the Instagram account is designed to support other teachers, promote products or support a love of the subject we teach, there is an incredibly high number of accounts available to indulge in. From the outset, I knew my purpose was to engage students in geography beyond my classroom, but also to create a space where other teachers can magpie my ideas and, of course, allow me to do the same, in order to improve the standard of education for the young people sitting in front of me every day.

REVISION

If I had a pound for every time I heard a student say 'I just read through the revision guide' or 'I don't know how to revise', I think I could afford my annual holiday to Iceland. As a student, I was that person who had every colour pastel highlighter, transparent sticky notes to copy diagrams and spent hours creating aesthetic notes, never to be used again. While I am not saying there wasn't any value in this and I did achieve high grades as a result of my hard work, as an educator, I recognise that much of this was a colossal waste of my time and there are faster, more purposeful methods of revision which can activate the power of recall much more efficiently. How then, can we get students to revise, especially for those that wouldn't dream of colour-coding key terms or daring to create a flashcard? Enter: Instagram.

As a school, we have signposted a number of revision strategies for students, detailing how these methods ought to be used and how students can utilise the finished product for years to come. One of my favourite pieces of research to be posed in recent years, explored whether our attention span (particularly that of the younger generation who were never given the opportunity to struggle without Google) is growing shorter (Subramanian, 2018). The study found that while quantitatively attention span is not changing, the number of advertisements, messages and notifications a person is inundated with at any given moment is increasing. In a world of ten second reels and two-hour 'get ready with me' vlogs condensed into just 60 seconds, we have a very limited window before students metaphorically swipe off and move onto the next best thing. I, myself, am guilty of opening a video on 'X', realising it is longer than three minutes and deciding not to bother, as I have better things to do ... like watch 18 mind-numbingly, non-educational videos on TikTok through the same timeframe. We can accept that this is the case, that social media has eroded our students' attention spans, but we can also play along with this and capitalise on their willingness to engage, just on their terms.

Through the next section, I will discuss three main ways of using Instagram as a source of revision. Firstly, creating posts for students to learn from. Secondly, through the use of reels or short videos. Finally, being able to test their understanding through low-stakes multiple-choice quizzing on your Instagram story.

1. POST IT ON YOUR 'GRID'

For those of you that haven't used Instagram before, there are two main functions of the app; to post on your grid and to post on your story. Posting on the grid will be visible by all who visit your page and they will stay there unless you archive or delete them. Posting on your story is temporary (lasting 24 hours) and can be seen as more 'informal' than a post on your grid. For example, if I were going to post a photograph of my coffee, I would post that to my story, whereas I would post a picture of me at the rim of the Grand Canyon on my grid – I, personally, would prioritise this as a permanent feature of my page for visitors to see.

Being able to post multiple slides/pictures to your grid in one post means you can go into detail about a particular theme or concept and students can easily access this from your main feed. In the past, I have chosen to handwrite my notes and hand-draw my diagrams on GoodNotes; however, as aforementioned, Canva offers a great way of creating an aesthetic, consistent and eye-catching design for those who want to make something a little quicker or aren't feeling too creative.

An example of a grid post that was received particularly well was one I made on the interquartile range (IQR). Aside from percentage increase, this is one of those mathematical concepts that is sure to get students huffing and puffing. I created a series of slides detailing what the interquartile range is, an example of how to work this out with an even dataset and with an odd dataset and then finally, a slide with a number of examples for students to have a go at themselves (see Figure 20.2). Advertising this on my story, I then told students there would be prizes and achievement points given to anyone who gave this a go. While this was aimed at KS4, I was surprised to see a number of KS3 students arrive at my classroom door ready to claim their achievement points. I often find IQR can draw out defeatist attitudes in students, so having them buy into a post on Instagram and show me their working out in their Notes app on their phones from when they were waiting for the bath to run makes me feel like perhaps social media can help make a positive difference to our young people's lives.

Now, try some yourself:

The United Nations measured the income of different countries.		Alfie wanted to see how many people visited the local museum.		The United Nations measured the life expectancy of different countries.	
USA	$63,200	Day 1	52		
Portugal	$33,500	Day 2	11		
Haiti	$2,840	Day 3	23	China	78
Chad	$1,440	Day 4	6	UK	81.3
Nigeria	$4,970	Day 5	72	Qatar	76.5
Uruguay	$22,400	Day 6	27	Afghanistan	64
Tonga	$6,390	Day 7	15	Mali	62.6
		Day 8	36	Lesotho	52.6
		Day 9	12	Japan	85.1
		Day 10	44		

@misspicken

Figure 20.2 The final slide of my post about how to work out the interquartile range, where students were invited to try a few problems for themselves

2. MAKE A REEL OR SHORT VIDEO

My creativity with using reels for revision is fairly limited and, realistically, I don't quite think I've got the voice for captivating audiences – aside from Jeff Stelling, you don't often hear Hartlepudlians frequenting TV and radio. So, if you want to see examples of this being done well, @MrVis and @SamBentley are great accounts to start with. That being said, this is an area of Instagram I do intend to get better at and have got some ideas to hone my skills further down the line. At the moment, if you visited my page you would find lots of reels, but most of them are limited to showcasing my travels and adventures around the UK and internationally.

As discussed previously, short, snappy videos and reels would be best received and can help to place the theory into reality. I had created a video recently at Seaburn, where, after teaching coastal processes, I spent some time pointing out the processes of erosion along the coastline and explaining how the features and landforms observed were created. I'd hoped this would be well-received by those students who perhaps could not travel to a beach littered with stacks and stumps but was surprised to be in receipt of a number of photos and videos from students who had tried to recreate my video or show me where they thought hydraulic action would be taking place along their stretch of beach at Redcar, Marske or Saltburn.

3. LOW-STAKES MULTIPLE-CHOICE QUIZZES

One of the brilliant features of Instagram stories are multiple-choice quizzes and question boxes. Used by celebrities to collect responses to what content their fans want to see from them (e.g. a make-up tutorial or fast-fashion clothing haul), used by musicians to find out which venue they should add to their world tour, used by me to find out if students can identify a photograph of a levee or offer an explanation for why Byker (Newcastle) has a high density of budget-friendly convenience stores and charity shops (See Figure 20.3).

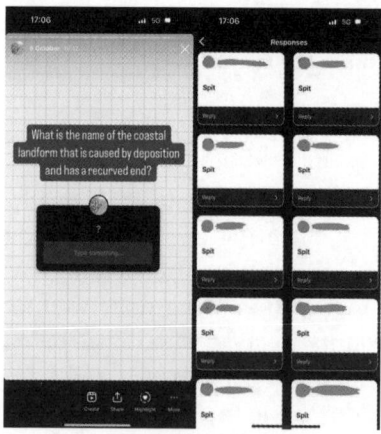

Figure 20.3 A free text quiz question

If I have posted new content, I will then host a quiz on my stories the following day. From this, I can encourage students to have a quick flick through the post, before attempting a quiz. For many of you who are avid users of Instagram, you will know how many hours are spent watching stories and how easy it is to engage in a poll or quiz if presented with one (see Figure 20.4). My thought-process was, if students are going to be on stories anyway, why not get them to 'accidentally revise' while they are there?

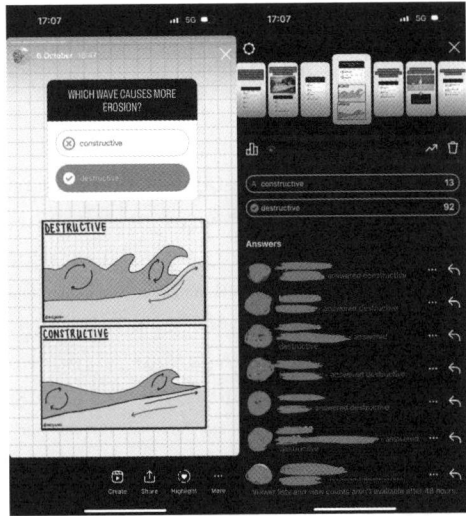

Figure 20.4 Question presented to students. Instagram allows you to break this down into the numbers of people who voted for each option and then underneath you can view how each individual cast their vote

In my opinion, this function of Instagram is single-handedly one of the most powerful tools an educator can use to engage students. This high interaction of students can then be celebrated, either by giving them a GDPR-friendly 'shout-out' on Instagram or giving them achievement points to be recognised as part of school-wide recognition.

COMPETITIONS

Students love having their work recognised and celebrated, we know this, it is well-recognised within available literature. Students particularly love the idea of you giving them a 'shout-out' on a public forum. Many students will go out of their way to complete tasks for you to simply write 'Jack in Y11 has absolutely smashed the quiz tonight – well done you!' on your Instagram story for their friends to see.

Photography competitions seem to go down really well with students. Previously, I have held half-term photo competitions to encourage students to go out and explore their local geography, or fieldwork photography competitions to encourage KS4 to document their day in a river in September, so they have content to reminisce on when revising for the fieldwork section of their exam. This can become really powerful and competitive when you raise the stakes and ask their peers to vote on the best photograph to win a prize.

In 2021, I ran a COP27 poster competition, encouraging students to submit a poster which could be used during a protest or march. This was all advertised on Instagram for the students in my school and after only being in the school for five weeks, I was so excited to receive over 60 entries, which we could then celebrate and reward students for. Perhaps this is testament to how goal-orientated our young people can be and the promise of featuring on an Instagram post is a real highlight for students.

INTRODUCING ... 'ME'?

Even as I write this section, I am acutely aware that this will be a divisive topic and some may whole-heartedly disagree with it; however, I have, personally, found showing students elements of my personal life a really rewarding and useful way of engaging them and building a strong rapport with a broad spectrum of students of differing behaviours, abilities, interests and ages. On my Instagram page I like to share photographs of my GeogDog, Oreo, and the adventures we get up to. I like to create and share reels of places I have visited on holiday or in our local area, to inspire students about opportunities that await them, whether that be when stepping out of their front door, or an aeroplane to a faraway continent. I like to applaud students' work and progress with 'shout-outs' and travel to watch them play for their local teams at the weekend, congratulating them publicly on my stories. If you like to show students that you really care about them, social media can be a fantastic method of reaching a high proportion of your cohort, with relative ease.

Many students like to start or end a lesson asking about something I have shared on my profile; I have had one parent ask about how to access a local waterfall that their son wanted them to visit as a result of a post; one student entered the room animated as a post about the taiga on my story led to them watching a Ben Fogle documentary about thawing permafrost and the consequences of a warming planet; one permanently excluded (PEX) student still engages regularly in quizzes I post to my Instagram story, even though I don't believe they even made physical contact with a BIC the full year I taught them – to ride on the coat-tails of Hans Rosling, I call this the 'accidental learning hypothesis', or the trick-them-into-engagement manoeuvre. I have found, giving students an insight into my life (albeit heavily censored) and allowing them to relate to parts of my personality has had a strong impact on the way they interact with me in the classroom.

STAYING SAFE ONLINE

Following on from my previous point, and to appease those of you reading who may have rather strong feelings of opposition, it is important to know a hassle-free and easy online presence with students can only be made possible if steps are taken to safeguard oneself against risk. School policies will differ, and advice can only be given from my own experience and what I have found to work, although I am aware of other schools and accounts that seem to have more lenient guidance. Here are a few of the top tips I can endorse and that Instagram facilitates in their settings:

- **Turn off direct messages** – this one is obvious, but giving students the ability to privately message you, external to systems governed by the school is not advisable nor permitted in many organisations.
- **Turn off comments** – similarly to the previous point, being seen to engage with individual students, even when this is in a public forum can be deemed inappropriate.
- **Don't follow students** – of course, students should be able to follow your page, but you should not be in a position to have access to their posts or stories.
- **Minimise the number of photographs/selfies you put on of yourself** – allowing students to access some elements of your

private life may be something you wish to share and control, however allowing students to access photographs of you, or pictures some may deem inappropriate, could put you in an uncomfortable position. Additionally, due to the rise of photo-based apps like Snapchat, many students are fluent with the skills to edit, so avoid giving them material to manipulate.

- **Avoid posting photos of students with their face in (unless this is an official school account and you have permission)** – alongside the usual rules of GDPR, your Instagram account will not fall under the umbrella of school policy and so any posting which allows the identification of students would be considered a breach and wholly inappropriate.
- **Don't tag student accounts or include surnames in your stories or posts** – this would direct other users to under-age users' profiles and assuming you are not vetting who is permitted to follow you, this could introduce unknown adults or children to your students' profiles, potentially facilitating inappropriate contact.

Five reflection questions

1. Which social media users would you give a shout out to for their contribution to geography?
2. Do you use social media to support students in learning and revising geography? Why/why not?
3. What new information have you gained about using Instagram from reading this chapter?
4. Look at the figures used in this chapter. How effective would these be for your students?
5. What is your school's policy on using social media to support learning and revision?

> **DIGGING DEEPER - THREE RESOURCES TO DELVE FURTHER**
> - Explore how you can use Canva to create resources for your classroom – https://www.canva.com/education/ (Accessed: 28/05/2025)
> - Find out more about using Instagram for revision – https://www.tes.com/magazine/archive/why-instagram-ultimate-gcse-revision-tool (Accessed: 28/05/2025)
> - Look at the revision resources for geography from Tutor2U – https://www.instagram.com/tutor2ugeog/?hl=en-gb (Accessed: 28/05/2025)

REFERENCES

Stripe, N. (2021) Children's online behaviour in England and Wales: year ending March 2020. Office for National Statistics. https://www.ons.gov.uk/peoplepopulationandcommunity/crimeandjustice/bulletins/childrensonlinebehaviourinenglandandwales/yearendingmarch2020#:~:text=Data%20from%20the%2010%2D%20to%2015%2Dyear%2Dolds'the%20internet%20at%20least%20daily (Accessed: 28/05/2025)

Subramanian, K.R. (2018) Myth and mystery of shrinking attention span. *International Journal of Trend in Research and Development*, 5(3), pp. 1–6.

21. LEADING A GEOGRAPHY DEPARTMENT
JO PAYNE
@_JOPAYNE

'Subject leaders are gatekeepers of standards and innovation; they are the leaders closest to the classroom.' (Jones, 2017)

INTRODUCTION

Leading a geography department is an aspiration for many geography teachers when they embark on their career. It enables you to put into practice the best of what you have encountered over the preceding years in the classroom, shaping a vision into a reality for your team of teachers and the next generation of young geographers in your school.

This chapter draws upon nearly 20 years of experience in the classroom, 17 of which have been spent as a head of geography in three secondary schools in Devon. These experiences have been in contrasting environments, both geographically, in terms of urban, rural and coastal settings, and philosophically, in terms of a multi-academy trust with a common curriculum and schools where there are contrasting perspectives on the teaching of geography. These settings come with a range of challenges and opportunities when leading geography as a discipline, but there are certainly a core set of responsibilities that fall upon any subject leader. Some of these are summarised by Kitchen (2017): 'quality geography departments display four characteristics – strong subject leadership, valuing student perspectives, a culture of innovation and sound subject knowledge.'

Let us explore five important aspects of subject leadership from a strategic perspective. It is very easy for department meetings to be derailed by the delights of the stationery order, discussing which students are frequenting the faculty detention and plans to captivate the masses on open evening. The challenge is that these procedural aspects of the role can often detract from the wider role of leading your team. This includes pursuing your shared intent through curriculum design, the subject-specific CPD to support it, and the development plan that helps you to achieve this over time.

KNOW YOUR TEAM

Geography departments can be made up of an eclectic group of educators with a wide range of experiences and backgrounds in geography, or indeed geography-related disciplines ranging from geology to politics. The unique group of individuals who make up the team may hold the same or a range of perspectives on what geography education should 'be' or 'do'. Rawling (2000) explored these ideas as curriculum ideologies, as shown within Bustin's (2018) *Geography Review* article entitled 'What's your view? Curriculum ideologies and their impact in the geography curriculum'.

Table 21.1 Ideological tradition and their use in geography departments

Ideological perspective	How they might be useful in the department
Utilitarian/informational: Education about 'getting a job'; focus on basic skills and information.	Work with the person responsible for careers education. Edit resources to highlight connections to possible careers in geography-related areas. Make connections with higher education institutions from a careers perspective.
Cultural restorationism: Cultural heritage; traditional areas of knowledge and skills which engage them to fulfil roles in society and the workplace.	Develop topics related to the geography of the UK. Map where the geography curriculum meets the requirements for spiritual, moral, social and cultural development (SMSC) and British values. *NB: This ideology needs careful managing against an imperative to decolonise the curriculum and include a diverse range of voices.*
Liberal humanist: Worthwhile knowledge passed from one generation to another as preparation for life. Rigor, big ideas and intellectual challenge.	Map the disciplinary and substantive knowledge across the curriculum. Make connections with higher education institutions to find opportunities for lectures or subject-knowledge updates. Engage with the Young Geographer of the Year competition through the GA.
Progressive educational: Personal development bringing maturity to individual students. Using academic subjects to develop the skills and attitudes to be independent.	Lead fieldtrips across the key stages ranging from school-based to international. Work with the outdoor education team to offer out-of-lesson opportunities for learning.
Reconstructionist: Agent to change society, encouraging children to challenge existing knowledge. Less focus on academic, more focus on issues and being socially critical.	Lead on co-curricular opportunities such as the model United Nations project or climate change action. Finding opportunities for decision-making activities within existing topics. Researching a range of voices and perspectives on issues in the curriculum with a decolonising geography lens.

Source: Adapted from Rawling (2000) in Bustin, R. (2018)

Where there are differences in ideology among your geography team, it can lead to tension if you are not all 'singing from the same hymn sheet'. Equally, a diverse team with a range of viewpoints about the teaching of geography can give your department a more holistic feel. It is worth finding out what drives geography team members as this might create opportunities for them to contribute to an aspect of the department that you might not focus upon due to your own vision and ideology.

SHARED VISION AND 'INTENT'

Gardner (2021) wrote about the importance of involving all teachers in a geography department in the strategic thinking. Knowing your team means using each teacher's geography subject knowledge and expertise to be able to create a shared vision of curriculum intent.

In its simplest definition, intent is the knowledge and skill that are gained at each stage. The intent is the golden thread through the core of any subject's curriculum. A cursory search of intent statements on school websites can reveal an array of comprehensive paragraphs to explain what geography departments hope to achieve through their implemented curriculum. This can easily become an unwieldy intent statement that sits in a strategy document, largely detached from the lived experience of teachers and children in the classroom. It certainly is not one that teachers can communicate with any degree of accuracy. Instead, be concise. What are the three key objectives of your curriculum? What is the 'one sentence statement' that captures your intent? For us, we could not see further than a statement from 'A different view: a manifesto from the Geographical Association' (2009, p.5) in which it says 'Geography underpins a lifelong 'conversation' about the earth as the home of humankind.'

Gleen (2020) wrote an article for *Teaching Geography* which explored how they reworked their key stage 3 curriculum after spending time reaching a shared understanding of the intent that underpins it. The article describes how they discussed what they felt was important as a department and what they wanted students to have experienced, learned and practised by the end of Year 9. After that, the team were able to look more critically at the curriculum to see where they were currently meeting their intent.

Be careful that you don't 'throw the baby out with the bathwater'. It is more than likely that you may be making small tweaks to topics to match the intent, or changes in sequencing, as opposed to a wholesale replacement of a curriculum.

DESIGNING AND REVIEWING YOUR CURRICULUM

Once the vision is in place, the department can start to create a key stage plan containing the building blocks which will support students' progress towards achieving the shared vision (Gardner, 2021).

Curriculum rarely stands still, particularly in a dynamic and ever-changing subject like geography. The curriculum evolves with every development in disciplinary knowledge or every world event that offers a different perspective on an issue. A good example would be the teaching of climate change and the role of global agreements in Year 8. Changes to the sequencing of our curriculum a few years ago resulted in the topic being delivered at the same time as the Conference of the Parties (COP) gatherings. Fortunate, I know. We therefore review and update the resources annually, reflecting on previous conferences while following present events in real time.

Some curriculum design decisions will be based on observations you make at a local level, with the children you teach. The AQA GCSE Geography specification includes the requirement for students to know seven strategies used to close the development gap. This can be a challenging concept, particularly for children in an isolated rural community in Devon, where concepts on a global scale can seem abstract. To provide appropriate building blocks towards KS4, we introduced:

- Top-down and bottom-up aid projects to a topic on development for Year 7
- the advantages and disadvantages of tourism in Thailand in Year 8
- transnational corporations and investment within a topic on globalisation in Year 9.

Kitchen (2017) refers to the broader challenges that impact the curriculum, including new initiatives, changes to staffing, tweaks to schemes of work and broader curriculum change. Departments need to be dynamic. Therefore, developing a culture of innovation, and being able to respond robustly to change, is important.

When subject leaders and their teams plan the curriculum, they are initially answering three simple questions:

1. What are we teaching?

2. Why are we teaching it?
3. Why are we teaching it then?

When addressing these questions, we keep one eye on the key stage 3 national curriculum. Regardless of whether your school is an academy or not, the geography curriculum needs to at least be as ambitious as the national curriculum. We would, therefore, be foolish to ignore its content entirely. There may be very specific reasons as to why you omit some aspects of the national curriculum and indeed go beyond it in other areas. During a department meeting, we mapped the national curriculum against our curriculum to ensure coverage and decided where the differences existed and the rationale for those decisions. These were some of the explanations:

> **Example 1:** In a previous school we did not extensively teach about glaciation as we did not have local glaciated landscapes in the south-west of England that we could draw upon. We did have a range of coastal and fluvial landscapes to use as local examples. We do, however, explore cold environments in Year 9.

> **Example 2:** Geopolitics and superpowers are not stated in the national curriculum, but was a key theme within our KS5 specification, so we included the core concepts within a Year 9 topic. Part of our intent was to assume that all children who joined us in Year 7 would continue to study geography at A-level and we would therefore prepare them accordingly.

Once the curriculum is in place, schedule time in department meetings to regularly review it. Kinder and Owens (2019) included a Venn diagram to aid departments in evaluating their curriculum (see Figure 21.1). Allocating time on the agenda to revisit the curriculum on a termly basis helps to keep the curriculum implementation aligned with the intent, which ultimately has the desired impact on student progress. If you know the dates of department meetings from a whole school calendar, individual questions from the Venn diagram could be scheduled for discussion strategically across the year.

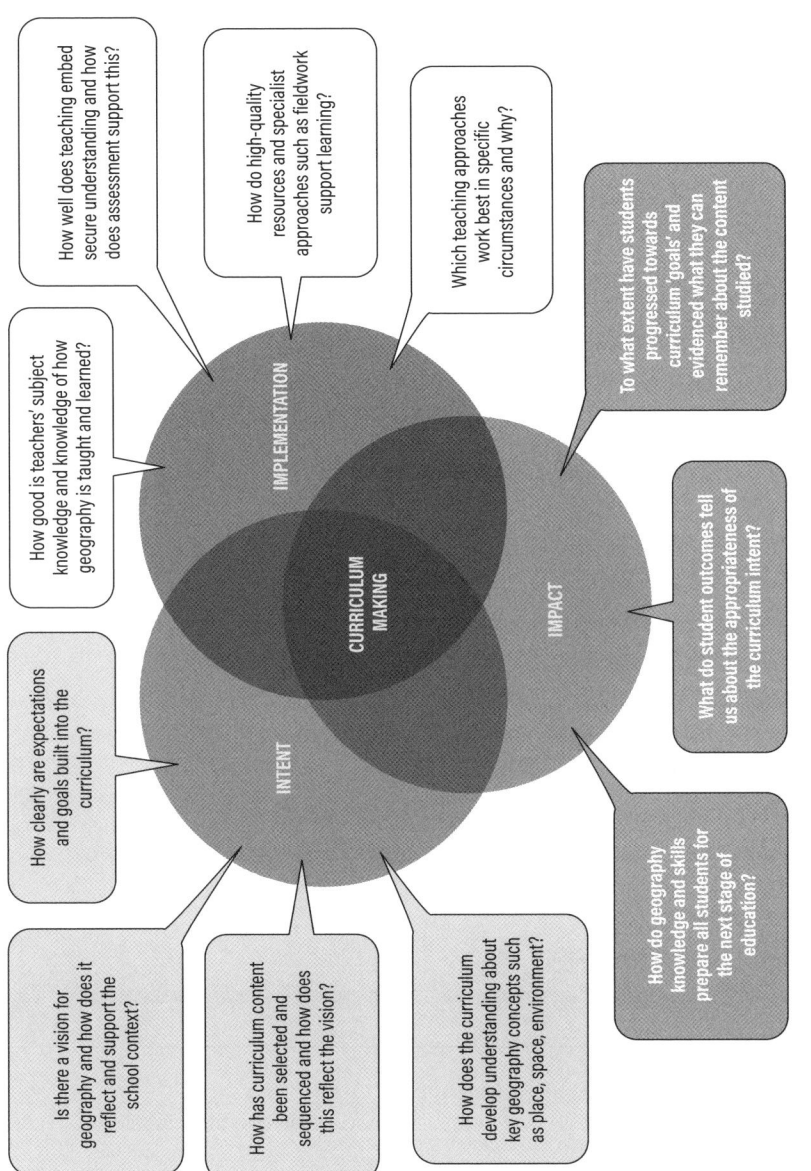

Figure 21.1 Key questions link intent, implementation and impact around the notion of curriculum making

Source: Kinder and Owens (2019), © Geographical Association, reproduced with permission of the Licensor through PLSclear

THE IMPORTANCE OF DISCIPLINARY PROFESSIONAL DEVELOPMENT

In a Chartered College *Impact* magazine article, Gatward (2020) explored the challenge of gaining powerful subject knowledge in school when CPD is often a generic approach for all. The article states that 'schools that have the poorest pupil outcomes or inspection results are often schools where this generic approach to CPD is used.' (Cordingley et al, 2018).

Within the school environment, this can only really be capitalised on when subject specialists are in one place: the department meeting. Here are some activities you could engage with as a team during this time:

Ideas for building subject-specific CPD
• Keep in contact with previous colleagues.
• Join the Geographical Association (including their social media channels).
• Contact your local university/college outreach department.
• Read about (and listen/watch) your subject.
• Work with other local departments.
• Use the evidence to highlight the importance of subject CPD with your leadership team.
• Read the publications that come with your subject association membership, e.g. *Geography* and *Teaching Geography*.
• Subscribe to resources such as *Wide World* magazine to support with procedural and substantive knowledge.
• Attend GCSE and/or A-level feedback session with the exam boards.
• Join any local geography teacher networks (often organised through teaching schools).
• Read extracts of geography-related books (both fiction and non-fiction) and discuss synoptic links or how it could be embedded in the curriculum (e.g. *The Girl with the Louding Voice*, *Africa is not a Continent*, *When the Rivers Run Dry* to mention only a few).
• Read the GCSE and A-Level examiners' reports.
• Keep up-to-date with government publications, e.g. *Getting your bearings geography subject report* (Enser, 2023).

Table 21.2 Ideas for building subject-specific CPD

Source: Adapted from Gatwald (2020)

WRITING AN EFFECTIVE DEVELOPMENT PLAN

Any professional working in a school will appreciate the pace of an academic year. We can set out with the best of intentions during the much revered 'gained time' after the rest of a summer holiday, keen to make the changes that would lead the department to be better and more successful than it was the year before. However, the day-to-day running of the department can be all-consuming if we allow it to be. To ensure that 'the most important thing remains the most important thing', a clear, achievable and accountable development plan is required.

'... in a development, working as separate individuals is a mistake because it means people pulling in different directions and tripping each other up as they do so. For a ... department to move rapidly and meaningfully it isn't possible for it to have loads of different priorities because, inevitably, all of them will be eroded. Better for a team to decide what they're working on together and do fewer things better'. (Newmark, 2018)

The development plan enables subject leaders to keep taking the temperature of the department and to regularly refocus everyone's attentions on the shared vision and how you intend to continue to improve as a department. The structure in Table 21.3 might be useful here.

Table 21.3 Typical content of a development plan

Development focus	How and why has this been identified?	Steps to be taken	Who is responsible?	Progress Term 1	Progress Term 2	Progress Term 3	What will success look like?

Documents such as this are ideal when shared as a live document with the whole team (e.g. through Google Docs or One Drive). This enables all those accountable for different aspects to edit and make changes easily throughout the academic year. It can also be the basis of conversations between the geography subject leader and the person who line manages the geography department.

CONCLUSION

The most successful geography departments are continually reflecting on what they could improve and how. Members of the team engage with their subjects (often but not solely within department meetings) and recognise that it is their expert knowledge which is ultimately the best resource in the classroom. A culture of developing one's disciplinary knowledge is an important model for any young person to witness. This knowledge and skill of the team shapes the collective curriculum intent and design. One final thought, as stated in *Huh: Curriculum conversations between subject and senior leaders* by Myatt and Tomsett (2021, p.26):

> 'Curriculum development is an ongoing process; it is not going to be finished – ever. There is no end point, there is no end product and I think you need to enjoy that process and learn from that process, taking people along with you.'

FIVE REFLECTION QUESTIONS

1. What does a successful geography department look like?
2. What are the views and ideologies of your team, and how can you use them to their full potential?
3. How can you ensure that there are opportunities for your team to receive disciplinary (subject-specific) professional development?
4. What are you teaching it, why are you teaching it and why are you teaching it then?
5. What are the priorities for your department development plan?

DIGGING DEEPER – THREE RESOURCES TO DELVE FURTHER

- Bustin, R. (2018) What's your view? Curriculum ideologies and their impact in the geography curriculum. *Geography Review*, Summer 2018, pp.61-63.
- Kinder, A. and Owens, P. (2019) The new Education Inspection Framework – through a geographical lens. *GA Magazine*, Autumn 44:3 pp.97-100.
- Gardner, D. (2021) *Planning your coherent 11-16 geography curriculum: a design toolkit*. Sheffield: Geographical Association

---

REFERENCES

Bustin, R. (2018) What's your view? Curriculum ideologies and their impact in the geography curriculum. *Geography Review*, Summer 2018, pp.61-63.

Cordingley P, Greany T, Crisp B et al. (2018) Developing great subject teaching: Rapid evidence review of subject-specific continuing professional development in the UK. Wellcome Trust. wellcome.ac.uk/sites/default/files/developing-great-subject-teaching.pdf (Accessed: 23 June 2020)

Gardner, D. (2021) *Planning your coherent 11–16 geography curriculum: a design toolkit.* Sheffield: Geographical Association.

Gatward, L. (2020) Developing subject knowledge and pedagogy for pupil success. *Impact*, Chartered College. https://my.chartered.college/impact_article/building-subject-specific-cpd-opportunities/ (Accessed: 28 October 2023)

Gleen, J. (2020) Geography's intent: developing your curriculum. *Teaching Geography*, 45(3), pp.108–110.

Jones, Mark (ed.) (2017) *The Handbook of Secondary Geography*, Sheffield: Geographical Association.

Kinder, A. & Owens, P. (2019) The new Education Inspection Framework – through a geographical lens. *GA Magazine*, Autumn 44:3 pp.97–100.

Kitchen, R. (2017) The qualities of a strong geography department. SecEd. https://www.sec-ed.co.uk/content/best-practice/the-qualities-of-a-strong-geography-department/ (Accessed: 28 October 2023)

Myers, M & Tomsett, J. (2021) *Huh: Curriculum Conversations Between Subject and Senior Leaders.* Curriculum Leadership: A conversation with Claire Hill, (pp.19–26) London: John Catt Educational.

Newmark, Ben (2018) Keeping the most important thing the most important thing. https://bennewmark.wordpress.com/2018/09/17/keeping-the-most-important-thing-the-most-important-thing/ (Accessed: 23 October 2023)

Rawling, E. (2000) Ideology, politics and curriculum change: reflections on school geography 2000. *Geography*, 85(3), pp. 209–220.

22. SUPPORTING GEOGRAPHY EARLY CAREER TEACHERS

HANNA GOWLING
@HANN AG____

'Where culture is right teachers will develop to better meet their students' needs.' (Weston and Clay, 2018, p.71)

Having a new colleague join the department is an exciting moment; it changes the dynamic, forces reflection, and through this, can enhance the curriculum, pedagogy and overall practices. Early career teachers (ECTs) bring a unique enthusiasm, revitalising the department with newfound energy and a fresh perspective, even as they navigate the initial nerves that come with their first teaching post. It is a privilege to lead, support and develop them, providing an environment for them to flourish and develop, and ultimately ensure they deliver a high-quality student experience.

The Early Career Framework (ECF) aims to improve early career support for teachers by offering a package of professional development that includes in-school mentoring, training and self-study materials, of which the latter two can be supplied by external providers. Its aims have been welcomed by most in the teaching profession, but it has not been without its issues, with criticisms around workload, lack of subject-specific support and repetition of initial teacher education (ITE) (Ford et al, 2022).

This chapter aims to highlight opportunities for ECTs and heads of geography to enhance the development of ECTs within their department. It will also suggest strategies which aim to mitigate elements of the criticisms levelled at the ECF.

BUILD SUBJECT KNOWLEDGE

In a comprehensive examination of the research on effective teaching, Coe et al (2014) found that a teacher's subject knowledge and understanding of how students engage with that subject hold the most robust evidence for impacting student outcomes. This assertion is particularly poignant in the context of geography education, where its multidisciplinary nature demands breath while also requiring a deep understanding of its various dimensions. Educators with a strong foundation in geography subject knowledge are better equipped to inspire their students, connect real-world issues to the curriculum and foster critical thinking skills, all of which are vital for the development of well-rounded, geographically literate citizens (Brooks 2010; Enser, 2018). As the importance of geography education in fostering global awareness and environmental stewardship continues to grow, the need for educators with robust subject knowledge becomes increasingly paramount. This will ensure that the next generation is well-prepared to address the complex challenges of our interconnected world.

The experience of ITE can vary significantly, not only in terms of the chosen route (e.g. PGCE, School Direct, SCITT, Teach First) but also regarding the acquired knowledge and teaching experiences. This diversity can pose challenges for ECTs, particularly in subjects like geography where curriculum models differ between schools. For instance, an ECT might have no prior experience teaching common topics like tectonics or rivers but could have taught coasts extensively at both placement schools and across key stages. Moreover, geography teachers bring diverse academic backgrounds, encompassing environmental science, planning, geology, anthropology or sociology. When combined with the fact that only 2% of mentors and 4% of ECTs have access to self-study materials tailored to their specific subject or phase (Ford et al, 2022), it becomes especially vital for mentors, departments and school leaders to offer opportunities and support for ECTs to enhance their subject knowledge.

Signposting the best reading to enhance subject knowledge is a great start. A textbook is a suitable place to start in building that subject knowledge but, as Enser (n.d.) highlights, they have their limitations; they can be out of date, narrow in terms of place and case study knowledge, and often presenting a simplified or a single story, particularly around human issues. To address this, encouraging or providing opportunities for your ECT to engage with subject knowledge beyond the textbook will be invaluable. The Geographical Association and the Royal Geographical Society both offer a range of resources from podcasts, on-demand lectures and written material. However, just signposting extra reading to improve subject knowledge is a missed opportunity – we forget that mentors and other experienced teachers are ourselves resources for our ECTs to tap into, particularly about subject pedagogy (Hattie 2012; Hughes 2021). Taking the time to discuss common misconceptions and how you approach and explain particular concepts or topics can be invaluable in developing an ECTs practice. In addition, blogs such as 'How I teach … global atmosphere circulation model' from @GeographyTom (Highnett, 2019) can be really useful to help articulate your points and structure how you do this.

GIVE FEEDBACK

An integral part of the mentor's role is providing feedback. Many of the ECF frameworks require a comprehensive instructional coaching model. Instructional coaching is when a coach, often a mentor for ECTs, identifies a change needed to enhance student learning in the classroom (Goodrich, 2021). They then guide the teacher (ECT) to make those changes through discussion and coaching, alongside modelling and deliberate practice to support the teacher in achieving this. This approach has had a significantly positive impact on ECTs since the high-stakes half-termly observation of the NQT process (Hughes, 2021; Kraft et al, 2018), allowing for the support of incremental development, observation of a range of classes, and support for various curriculum topics.

The process of instructional coaching can be daunting for a mentor. Observers need to consider, among other aspects:
- What is the most impactful area to focus on?
- How can I explain what I saw? And what the impact was?
- How can it be improved?
- What is the area of highest leverage?
- What is the best way to give this feedback?
- How can I model this?

Engaging in mentor training provided by your ECF provider is crucial, as they will provide examples and support to aid your learning in instructional coaching. If you feel the need for further support, consider requesting joint observations with a department head, induction tutor or senior leader to help you calibrate and provide precise, meaningful feedback on the most impactful areas. This will also aid consistency if you have a whole-school instructional coaching model running in parallel with the ECF.

MODELLING

An often-overlooked aspect is the opportunity for modelling in the deliberate practice element of instructional coaching. This is primarily because it can feel quite awkward to conduct in a 1-2-1 setting and takes precious time to do correctly. Often, it is skipped in favour of instructing ECTs on what they need to improve and how to do it. Consider whether we would leave our students without a model for an exam question or for how to collect beach profile data. Providing novice teachers with the opportunity to experiment with new knowledge, skills and ways of thinking, while receiving more support and feedback than actual classroom practice, is essential (Fletcher-Wood, 2018). It is a lower-stakes approach that offers further opportunities for development (Deans for Impact, 2016; Weston and Clay, 2018). The most effective way to implement this is to allocate time during your mentoring sessions and host these sessions in a classroom, ideally within the ECT's own environment, so that practice becomes an integral, integrated and natural part of your time together (Weston and Clay, 2018). Moreover, it is essential to encourage ECTs to allocate their additional time to observing and learning from experienced teachers, rather than dedicating it entirely to additional planning. Observing experienced colleagues can offer invaluable insights and serve as models for effective teaching practices. It also offers further opportunities to demonstrate and model an aspect of teaching rather than explain, which can be quite hard to put into words (Didau, 2020). However, to reduce overload and ensure maximum impact, ensure that the ECT has a clear focus or question to consider or is accompanied by an experienced teacher to allow them to narrate what they are seeing.

POSITIVE FEEDBACK

Positive feedback is paramount for ECTs and serves multiple purposes. Firstly, precise praise fosters a sense of accomplishment and boosts confidence. When ECTs receive specific commendations for their practice, it not only acknowledges their efforts but also highlights their strengths, reinforcing their commitment to professional growth. Secondly, positive feedback plays a pivotal role in developing practice. Constructive comments on what ECTs are doing well, in conjunction with targeted suggestions for improvement, provide a roadmap for refining their teaching skills. This iterative process of feedback and adjustment is fundamental for the ongoing development of effective teaching strategies. Finally, support through positive feedback creates a nurturing environment. By recognising and celebrating achievements, mentors and colleagues encourage ECTs to persevere and continually enhance their teaching methods. Such support is crucial for their journey towards becoming proficient educators (Hattie and Timperley, 2007).

GAPS IN THE ECF

Undoubtedly the ECF has provided a brilliant framework and a more level playing field for ECTs, but it is not without its problems (Ford et al, 2022). The main providers vary in structure, some of which include subject sessions, some of which deliver CPD as cross-phase or cross-subject. As a mentor or head of department, make sure you are aware of what the ECT is learning about in those sessions and support them to link their learning with their school experience. Allow them to share their learning with you and ask questions to help them connect their learning with their in-school experience. Utilising more of a coaching-style approach here could be powerful in that you are encouraging them to own their development, but you will need to pitch this based on their experience (Weston and Clay, 2018). You could also go a step further, if appropriate for your ECT, and support them to deliver an aspect of their learning during department CPD.

Common feedback from ECTs is that the framework duplicates knowledge from their ITE (Teacher Tapp, 2022). While this can be frustrating and also unusual when you have been in an environment where you are constantly learning new knowledge, it does not mean it is not useful. Just like our students, we struggle with forgetting, so revisiting key research around behaviour, questioning, assessment, etc. is important. The opportunity to engage in conversations with like-minded professionals offers the chance to be reflective and challenged professionally, and the ECF seminars and days offer fertile ground to do this, which may not be so readily available later in their careers. Promoting an awareness of this and encouraging ECTs to make the most of this opportunity is part of the mentor role. As mentioned earlier, many ECF programmes lack subject-specific pedagogy which means supporting an ECT to implement non-domain-specific knowledge gained through the ECF makes it easier to integrate learning into their lessons (Coffey et al, 2011).

FIELDWORK

Fieldwork is considered by most to be an integral part of geography. It can be both exciting and anxiety-inducing for ECTs to lead lessons in the field and be responsible for supporting students in a new environment. Remember the context is that some ECTs may have undertaken a degree and/or their A-levels with limited personal fieldwork experiences due to COVID-19. Experienced teachers will be well aware of the pitfalls of fieldwork and how to support students to get the most out of the day. Make sure that you build in time to share logistical aspects such as the route, timings, behaviour support systems in the field, what to do with the first aid kit, what to do in an emergency, and the key knowledge and takeaways students need to leave with. Also share the pitfalls and how to deal with them, e.g. if it rains, use pencils not pens, make the most of sheltered areas for instruction giving, and take the time to model data collection techniques. Much of this knowledge is hard-earned, and whatever you can do to impart your experience will ease those nerves and allow for the best possible outcomes for students. Just like in the classroom, if you can, offer a shadowing experience or a pre-trip visit first (Weston and Clay, 2018, see also chapter 8).

CONCLUSION

In this chapter, we have explored the significant potential for growth and development that comes with the arrival of an ECT in your department. Embracing their unique perspectives and providing the right support can reap significant benefits for the department and enhance students' learning. In conclusion, the journey of an ECT is a collaborative effort, and we must guide and support them as they embark on this exciting educational adventure.

> **The experience of an ECT:**
> Throughout my PGCE I was referred to the ECF continuously, so that when it came to my first meeting with my mentor, the terminology was familiar, if not repetitive. I was aware of criticisms of the ECF but also under the impression that it was the foundation for my journey to becoming a qualified teacher, and therefore a necessity. I have been fortunate enough to receive regular instructional coaching from my mentor. Though at times restricted by the framework, the structure has allowed me to focus on particular areas of development in my subject knowledge and skills. The ECF has provided a clear guide of what I should be able to do by the end of the two-year ECT programme. The expectation to begin with behaviour, for example, will be useful to many and seen as a necessary foundation for ECT's training, but needing to provide evidence of development in all areas of behaviour before moving on to a focus on subject knowledge was restrictive and, at times, unrealistic to my personal development. Applying the framework to areas of development for the department as a whole has become a way of making it relevant and specific to day-to-day teaching at my school. My individual targets therefore become collaborative in nature, as other colleagues work towards the same or similar objectives. I have absolutely benefited from the rigorous structure of the ECF, not only in finding my place within the department but professionally through achieving set objectives to develop my skills and knowledge as a teacher.
>
> Stephanie Bird – 2nd-year ECT

> **Five reflection questions**
> 1. How can we ensure that new ECTs are effectively supported in developing their subject knowledge?
> 2. What strategies can mentors and experienced teachers employ to provide effective feedback to ECTs?
> 3. What specific steps can be taken by mentors and heads of department to bridge the gap between ECTs' ECF learning and their in-school experiences, allowing them to connect and apply their learning effectively?
> 4. In what ways can experienced teachers and mentors help ECTs make the most of fieldwork experiences?
> 5. How can you help ECTs develop beyond the ECF programme?

> **DIGGING DEEPER – THREE RESOURCES TO DELVE FURTHER**
> - Deans for Impact. (2016) Practice with purpose: The emerging science of teacher expertise. Available at: https://www.deansforimpact.org/tools-and-resources/practice-with-purpose-the-emerging-science-of-teacher-expertise (Accessed: 28/05/2025)
> - Weston, D. and Clay, B. (2018) *Unleashing Great Teaching: the secrets to the most effective teacher development.* London: Routledge.
> - Haili Hughes. (2021) *Mentoring in Schools: how to become an expert colleague.* Carmarthan: Crown House Publishing.

REFERENCES

Berliner, D. C. (2015) *Practice with Purpose: The Emerging Science of Teacher Expertise.* Teachers College Press.

Brooks, C. (2010) Why geography teachers' subject expertise matters. *Geography*, 95, 143–148.

Coe, R., Aloisi, C., Higgins, S. & Major, L. E. (2014) What makes great teaching? Review of the underpinning research. The Sutton Trust.

Coffey, J., Hammer, D., Levin, D. & Grant, T. (2011) The missing disciplinary substance of formative assessment. *Journal of Research in Science Teaching*, 48(10), 1109–1136.

Deans for Impact. (2016) Practice with purpose: The emerging science of teacher expertise. https://www.deansforimpact.org/tools-and-resources/practice-with-purpose-the-emerging-science-of-teacher-expertise (Accessed: 28/05/2025)

Didau D. (2020) *Intelligent Accountability: Creating the Conditions for Teachers to Thrive*. London: John Catt Educational Ltd.

Enser, M. (2018) *Making Every Geography Lesson Count: Six Principles To Support Great Geography Teaching*. Carmarthen: Crown House Publishing.

Enser, M. Maintaining your subject knowledge. https://my.chartered.college/early-career-hub/maintaining-your-subject-knowledge/ (Accessed: 29 October 2023).

Fletcher-Wood, H. (2018) Designing Professional Development for Teacher Change. https://s3.eu-west-2.amazonaws.com/ambition-institute/documents/Designing_Professional_Development_for_Teacher_Change_-_Harry_Fletcher-Wood_1.pdf (Accessed: 29 October 2023)

Ford I., Allen B. & Wespieser K. (2022) Early Career Framework One Year On. https://www.gatsby.org.uk/uploads/education/reports/pdf/2022-10-early-career-framework-tt-gatsby-final.pdf (Accessed: 29 October 2023)

Goodrich, J. (2021) Blog. When we talk about Instructional Coaching, what do we mean? Powerful Action Steps.

Hattie, J. (2012) *Visible Learning for Teachers: Maximizing Impact on Learning*. Abingdon: Routledge.

Hattie, J. & Timperley, H. (2007) The power of feedback. *Review of Educational Research*, 77, 81–112.

Highnett T. (2019) How I teach... Global Atmospheric Circulation Model. https://teamgeography.wordpress.com/2019/11/10/how-i-teach-global-atmospheric-circulation-model/ (Accessed: 30 October 2023)

Hughes, H. (2021) *Mentoring in Schools: How To Become an Expert Colleague*. Carmarthen: Crown House Publishing.

Kraft, M.A., Blazar, D. & Hogan, D. (2018) The effect of teacher coaching on instruction and achievement: a meta-analysis of the causal evidence. *Review of Educational Research*, 88(4), 547–588.

Teacher Tapp. (2022) Early Career Framework Report. https://teachertapp.co.uk/app/uploads/2022/10/2022-10-Early-Career-Framework-TT-Gatsby-Final.pdf (Accessed: 29 October 2023)

Weston, D. & Clay, B. (2018) *Unleashing Great Teaching: The Secrets To the Most Effective Teacher Development.* Abingdon: Routledge.

Willingham, D. (2009) *Why Don't Students Like School?* San Francisco: Jossey-Bass.

23. SUPPORTING NON-SPECIALISTS TO TEACH GEOGRAPHY

JENNIFER MONK
@JENNNNNN_X

'Every team should strive not just to improve what it does, but to learn more, to know more.' (Crome, 2023)

WHAT IS A SPECIALIST?

The Department for Education (DfE) use a proxy measure of 'specialism' defined as those having a relevant post A-level qualification; CPD and qualifications at A-level and below are not covered. They state that 'the vast majority of hours taught in England to pupils in Years 7–13 in most subjects are taught by teachers with a relevant post A-level qualification. In November 2015, the respective proportions were 88.9% for all subjects and 89.0% for Geography.' (DfE 2016). This number of non-specialist teachers is predicted to be much higher in 2023/2024 due to the teacher recruitment crisis currently ongoing, and the fact that geography teacher training targets have not been met for the 2022/2023 nor the 2023/2024 academic year.

In September 2023, Ofsted released their latest geography subject review which reviewed their research visits by HMI and Ofsted inspectors to 50 schools between December 2022 and May 2023. They stated the following:

> 'Many secondary schools are struggling to recruit specialist geography teachers. In most secondary schools, at least some lessons are taught by non-specialists. In a few schools, the majority of key stage 3 lessons are taught by non-specialists. These teachers are often not supported well enough to teach geography. They are not able to provide the same kind of rich explanations as their specialist colleagues, and are less able to identify and address misconceptions.' (Ofsted, 2023)

Prior to the report, the Geographical Association along with the Royal Geographic Society and many school leaders, had voiced concern over the recruitment of geography teachers being a significant national issue. Often key stage 3 is taught by non-specialists which can lead to misconceptions not being addressed and possibly leading to bigger misconceptions later down the road. In some schools, effective support for non-specialists was seen, but for many there was a lack of support and training. Time is always cited as a reason for this issue and the recruitment crisis along with 'non-specialist' teachers having maybe less available time due to teaching multiple subjects can be a concern. However, time spent supporting non-specialists regularly should lead to better outcomes for the classes they teach.

Note that some of the strategies discussed in the chapter will also support specialist geography teachers and may also be relevant to working with trainee teachers and ECTs.

SHARING WORRIES

Since I began teaching in 2012, the geography department I have taught in or led has always had at least one non-specialist. This year I have four teachers who have never taught geography before but who have at least two lessons of geography per week on their timetable. Over the last few weeks of the academic year I spoke to these teachers and asked them their worries. Most of their concerns were linked to them 'not being good at geography', 'not understanding the content' or, their biggest worry, 'knowing if student answers are right or wrong'. There were also some concerns over ensuring students were appropriately challenged and making sure that the students weren't being disadvantaged because they were not taught by a 'specialist geography teacher'.

BUILDING KNOWLEDGE

'Previous research shows pupils make less progress when they have a teacher that does not have a formal teaching qualification; is newly qualified; less experienced; without a degree in the relevant subject; and when teacher turnover at their school is high.' (Allen et al, 2016)

Further research has shown that schools serving lower-income communities are more likely to have teachers with all these characteristics. This suggests they face greater recruitment difficulties in hiring staff and offers one explanation as to why there continue to be substantial and persistent inequalities in educational outcomes between students from disadvantaged and more privileged backgrounds.

We need to remember that varying experience can come into play here; none of my non-specialists this year have studied geography past GCSE and two of them didn't even study geography at GCSE, so their knowledge of the subject is from a very long time ago. I think it's also safe to say that *how* geography is taught in schools today is very different from when I was at school in the early 2000s.

However, there is an assumption here that non-specialists may be lacking in knowledge because they haven't studied geography at GCSE. This may be untrue. Some non-specialists may have a specific interest in geography, they may have undertaken a Subject Knowledge Enhancement Course or they could have a different lived experience from other teachers which may bring specialist knowledge to the lessons they teach. So maybe, in some cases, we need to change our thinking and instead think of the wider knowledge they may bring.

Part of my role as head of department has meant that when expanding our team this year, I tried to see the new experiences and knowledge the team would share as positives. Prior to September, I took time to meet with staff individually to discuss our curriculum and look at strengths and weaknesses in their subject knowledge, give them time to share their concerns and ensure they had time to observe lessons so that they could get a feel for geography in our school. One of the most important tasks we complete as a department each year is a subject knowledge audit; we start with our curriculum and also include relevant GCSE specifications. This allows us to plan CPD to develop our own subject knowledge and learn from each other. This has been done in a range of forms, from an online form to a shared spreadsheet; the *how* is probably least important but it needs to be revisited, especially if staff don't teach all year groups every year or curriculum changes have been made during periods of absence.

As part of our school we have department 'curriculum planning time' and a department meeting every week. I have one of my non-specialists at our meeting but the other three are unfortunately not available. Although I feel lucky to have this time with some of the extended department, it often means that planning, assessment and moderation time is done outside of 'scheduled hours', which can hugely impact staff workload.

'People perform best when they feel included, safe, and have a sense of belonging' (Crome, 2023). Crome explains that there are many factors that contribute to successful teams but 'supportive teams are the bedrock of our success in schools'. This is something I have consistently tried to ensure happens in the departments I have led. In order to support non-specialists I have followed the 'model for thriving teams', which sets out the factors that make up a high-performing team. Although these factors also probably come under 'good departmental leadership', I think establishing clear structures is essential to supporting non-specialists.

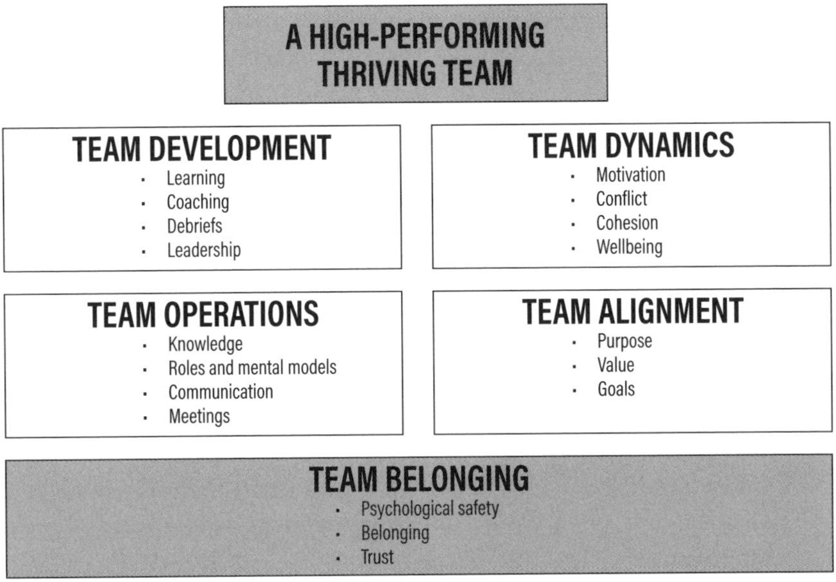

A HIGH-PERFORMING THRIVING TEAM

TEAM DEVELOPMENT
- Learning
- Coaching
- Debriefs
- Leadership

TEAM DYNAMICS
- Motivation
- Conflict
- Cohesion
- Wellbeing

TEAM OPERATIONS
- Knowledge
- Roles and mental models
- Communication
- Meetings

TEAM ALIGNMENT
- Purpose
- Value
- Goals

TEAM BELONGING
- Psychological safety
- Belonging
- Trust

Figure 23.1 Factors that make a high-performing team
Source: Crome, S. (2023) *The Power of Teams*

PROVIDING PROFESSIONAL DEVELOPMENT

As part of our curriculum planning time, we have a wide range of opportunities to provide CPD for all teachers within our department. We start by sharing resources, but the support does not stop there, and although we will get some shared development time with our non-specialists, it is unlikely to be regular. In order to support this and try to reduce time demands, I created a menu of CPD which is available both in person and also via video on our shared drive. This is clearly signposted in both department action plans, on meeting minutes and via our shared area. We keep this relevant and have a shared approach; we use our subject audit (mentioned previously) to share knowledge, but staff in my department also take 'control' of a year group and during department time they lead on sharing resources and CPD to the rest of the team.

Our CPD menu has a range of key topics, which I feel will help to support all teachers teaching geography in my school. The list below is some of the topics we offered this academic year.

- **Our curriculum** – I share the vision of our department, as well as the curriculum map for the full five-year curriculum, with links to further study at 16+. I also outline study at key stage 2 so that non-specialists have a clear understanding of what has been taught previously.
- **Key concepts** – Within our curriculum we have eight key concepts that underpin our topic areas and help students to make links between topics. The CPD session introduces these concepts and explains how they build throughout our curriculum and increase in both difficulty and depth as students progress through.
- **Geographical skills** – There are a wide range of skills taught in all geography classrooms, some of these may never have been taught by the non-specialist before and therefore confidence in understanding *how* to do the skills and teach them is important. In order for students to be able to 'think like a geographer' we need to ensure all staff fully understand how to do this too.
- **Adaptive teaching** – As part of our curriculum development and working in a school with mixed-ability classes, adaptive teaching is key to ensuring students of all abilities are challenged. Ensuring all teachers know how to encourage students to take their thoughts further or develop their answers is essential to ensuring our students make progress.

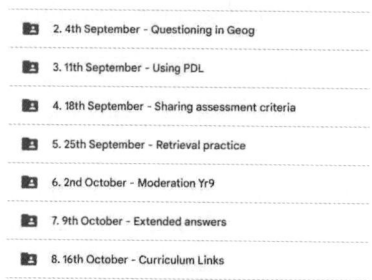

Figure 23.2 An example of our CPD area with clearly signposted topics

SHARING RESOURCES

When teaching staff join my department, they receive a range of different resources to support their teaching, which I outline below. The resources really are only a small part of the package staff receive in my department, however without any resources, teaching subjects outside of specialism is incredibly hard. Therefore, the resources themselves are probably one of the most efficient ways of developing staff. If the resources and planning is already taken care of, teachers can instead use the time to build their subject knowledge.

1. CURRICULUM OVERVIEW

We have a detailed and comprehensive curriculum overview, which is also linked to our medium-term plan. Our curriculum overview shows not only our key questions, but also where each topic meets the national curriculum, where our key concepts and themes are visited across topics, as well as any key knowledge and key skills we want students to take away with them.

Our medium-term plan is hyperlinked to make it easy to find lessons and resources. In our medium-term plan we also include core knowledge and core skills so that all staff (but especially non-specialists) know what knowledge and skills our students should leave each lesson with.

Lesson	W/C	Topic	Title	Core Knowledge	Core Skills	Assessment
1	04/09/23		What Geography happened over the summer?	I can	I know	Learning Journey
2			1. What do we mean by hazard risk?	why hazard risk varies around the world	analysing bar charts	Baseline
3	11/09/23		2. What is plate tectonic theory?	why plate tectonic theory exists	reading maps/atlas skills	
4			3. What happens at plate boundaries?	the processes that happen at four plate boundaries	explain physical processes with a diagram	
5	18/09/2023		4. What were the effects of the Indonesian earthquake?	social, economic and environmental impacts of earthquakes	categorising social, economic and environmental	
6			5. Why was the Indian Ocean Tsunami so catastrophic?	how factors such as location/time of day can affect a hazard	compare impacts of hazards	
7	25/09/23		6. Midpoint - Why do effects of tsunamis vary?	example of a tsunami	compare impacts of hazards	Midpoint
8			7. Midpoint feedback			Midpoint feedback
9	02/10/2023		8. How do volcanoes differ?	the difference between shield and composite volcanoes	annotating diagrams	

Figure 23.3 Extract from our school's medium-term plan

Another area we know non-specialists find more challenging is making links across the curriculum. For example, when I teach about population distribution, we revisit urbanisation which was previously taught. Without support this is much harder for non-specialists to do. Therefore, on our curriculum overview we have also made explicit links to other topics 'when else', so that all staff can see how the knowledge is building/linking to both other topics and other subjects.

Geography Curriculum 2023-24

				Year 7					
		How does weather vary around the world?	How are cities changing?	What are the characteristics of hot deserts?	What are the opportunities and challenges facing the Middle East?	What happens when the land meets the sea?	How do people use cold environments?	Should people be allowed to climb Mount Everest?	How can we manage TR more sustainably?
Content		Measuring weather, microclimate, types of rainfall, reasons for varying climates – altitude, latitude and proximity to sea, UK weather hazards, extreme weather in the UK, Tropical Storms	Urbanisation, rural-urban migration, megacities, challenges and opportunities of rapid urbanisation. Urban heat island effect, sustainable cities	World biomes, hot desert formation, characteristics of hot deserts, desertification, managing desertification	Climate in the Middle East, water shortage, population, Dubai – tourism, conflict	Waves, coastal processes, erosional landforms, depositional landforms, management of the coast, climate change	Characteristics of cold environments, life in Arctic, adaptations of plants and animals, natural resources, damage to the Arctic, Wilderness areas, climate change	Characteristics of alpine environments, Mount Everest tourism, earthquakes.	Characteristics of Tropical Rainforests, importance, challenges facing TRF, management of TRF
When else	Yr7	HT2 – climate of Hot Deserts, HT3 – climate of Middle East, HT5 – climate of Arctic, Alpine environments HT6 – conditions of alpine environments Yr8 HT1 – causes of uneven development HT2 – ice ages HT4 – India Monsoons HT5 – Russia physical Geography – climate Yr9 HT2 – spread of malaria, Haiti extreme weather HT5 – Climate change HT6 – flooding in Bangladesh	Yr7 HT3 – urbanisation in Middle East HT6 – base camp on Everest Yr8 HT2 – urbanisation on floodplains HT3 – population distribution, migration HT4 – cities in India HT6 – CBD Yr9 HT1 – hazard risk factors HT2 – spread of Cholera HT3 – safe places HT4 – sea level rise affecting Maldives HT5 – flood risk	Yr7 HT1 – weather and climate HT3 – Middle East Yr8 HT3 – migration Yr9 HT2 – Malaria Other subjects: Science – April ecosystems – characteristics/adaptation of animals.	Yr7 HT1 – weather and climate HT2 – How are cities changing HT3 – Middle East Yr8 HT1 – development opportunities HT3 – population distribution, migration	Yr8 HT2 – Ice on the land, processes of glacial movement HT2 – water processes, river changes Yr9 HT5 – Rivers and flooding Other subjects: Science – October- weathering and rocks		Other subjects: Science – April ecosystems – characteristics/adaptation of animals.	
Themes		systems, risk and resilience.	Sustainability, development, interdependence, inequality, globalisation, risk and resilience.	Sustainability, systems, interdependence, risk and resilience.	Sustainability, development, interdependence, inequality, risk and resilience.	Sustainability, systems, development, risk and resilience.	Sustainability, development, interdependence, risk and resilience.	development, interdependence, globalisation, risk and resilience.	Sustainability, systems, development, interdependence, inequality, globalisation, risk and resilience.

Figure 23.4 Extract from our curriculum overview

2. TOPIC GUIDE

Another way we have tried to support non-specialists was through the development of a topic guide. This outlines some of the key knowledge and thinking students need to know, but also some background information and knowledge that might help support all staff, but especially non-specialists.

Within the guide I include an overview and outline of our content. We also focus on the 'why' – I think while it is fairly straightforward for subject specialists to understand and make links across topics, it is fair to say that this may be harder for non-specialists or teachers new to the profession. It has also been useful to support our trainee teachers when they arrive at the school and start to teach some of our curriculum.

We use a wide range of texts in our curriculum and all of these are summarised in the appropriate topic guides. Some teachers with a particular interest in topics may read the whole text but this is not an expectation and therefore providing the summary allows teachers to share further context when reading extracts. To support understanding, we also provide key word lists and definitions; this encourages teachers to use the expected geographical language in lessons. We also include further links to find out more.

Further information or reading:

- Our World in Data – 'Material footprint per capita'
  https://ourworldindata.org/grapher/material-footprint-per-capita?country=~OWID_WRL
- UN SDG 12 'Responsible consumption and production'
  https://unstats.un.org/sdgs/report/2019/goal-12
- Study 34 has a straightforward explanation of sustainable fashion
  https://www.study34.co.uk/pages/what-is-sustainable-fashion
- True Cost (video) on environmental impact
  https://truecostmovie.com/learn-more/environmental-impact
- River Blue (movie) http://riverbluethemovie.eco
- Fashionopolis by Dana Thomas – JMK has a copy if you want to read.
- Stacey Dooley investigates - Fashion's Dirty Secrets – BBC documentary – JMK has a copy

Figure 23.5 Extract from our topic guide

3. SHARING MISCONCEPTIONS

'Support non-specialist teachers in how best to explain complex geographical ideas and how to identify and address misconceptions.' (Ofsted, 2023)

One of the key parts of our topic guide is the sharing of prior knowledge and common misconceptions. At the start of every topic, our students complete a baseline, and we share some of the key takeaways from these baselines across the department. This allows teachers to ensure that any misconceptions are addressed, and these misconceptions don't impact any new learning or the development of prior learning. This is particularly important for non-specialists who may not be aware of what misconceptions can occur in the classrooms, or how to tackle them.

Prior knowledge	Common misconceptions (from baseline)
Key Stage 3 • students have probably looked at or heard about hazards but not investigated causes. • Students may have discussed hazard risk in relation to other hazards e.g. weather hazards.	• tsunamis are weather hazards not tectonic hazards • hurricanes are called tornadoes • all volcanoes are the same • hazards only cause damage in LICs
Key Stage 2 • students should have learned about human and physical characteristics of North and South America on the eastern edge of the Pacific Rim. • students should also know about volcanoes and earthquakes but probably won't know their formation.	

Figure 23.6 Extract from topic guide showing prior knowledge and common misconceptions

4. OBSERVATIONS

As soon as I am aware of the timetable and of the non-specialists joining our department, I try to ensure that they are given time to observe different teachers across the department teaching geography. Where possible, I try to ensure these are the same classes or topics that they will be teaching. This doesn't ever stop, and my department have an open-door policy for all staff, but non-specialists are especially welcome in our department and we ensure we are always available for discussions wherever we can.

5. ASSESSMENT SUPPORT

We expect all staff in our department to mark assessments and involve all staff in the moderation process. Our medium-term plans clearly signpost assessments and, within this folder, we include a range of support for all teachers. We include the assessment itself, a mark scheme and some model answers that are annotated. We also include our feedback lesson, although this may change following moderation. Once students sit their assessment, we sit down as a team and check through our mark scheme, making any adjustments where we feel appropriate. We then mark assessments using the edited mark scheme and have a moderation meeting where we ensure we are all marking using the same criteria. The example assessments have really built confidence in marking and have given students and staff a benchmark to aim towards.

CONCLUSION

In conclusion, the resources given to non-specialists are probably the most important and time saving support. However, what is obvious to a specialist teacher may not be obvious to others. Simple adaptations such as including answers for every task will help to build knowledge and confidence for staff. It is essential that time is given both to non-specialists directly, but also time as a department team to ensure a cohesive vision for the department.

Five reflection questions

1. How can you ensure that you are supporting non-specialists who join your department?
2. Could you develop a knowledge and skills audit which could identify strengths and weaknesses of all teachers but also specifically non-specialists?
3. How could you ensure the department vision is shared by all staff, even those who may not be present at meetings/CPD where this is discussed?
4. What adaptations might be needed to current resources to support non-specialists further?
5. How could you support non-specialists in advance of them teaching your subject?

> **DIGGING DEEPER - THREE RESOURCES TO DELVE FURTHER**
> - Darlington, E. (2017) What is a non-specialist teacher? Cambridge Assessment Research Report. Cambridge: Cambridge Assessment.
> - Shreeve, J. (2018) Addressing the shortage of specialist geography teachers. *Teaching Geography*, 43(3), pp.98–100.
> - Roberts, M. (2017) Geographical education is powerful if .... *Teaching Geography*, 42(1), pp.6–9.

REFERENCES

Allen R., Mian E. & Sims S. (2016) Social inequalities in access to teachers. Social Market Foundation Commission on Inequality in Education: Briefing 2 Social inequalities in access to teachers.

Crome, S. (2023) *The Power of Teams: How To Create and Lead Thriving School Teams*. London: John Catt Educational Ltd.

Ofsted (2023) Getting our bearings – Geography subject report. https://www.gov.uk/government/publications/subject-report-series-geography/getting-our-bearings-geography-subject-report#discussion-of-findings (Accessed: 6 November 2023)

DfE (2016) 'Specialist and nonspecialist' teaching in England: Extent and impact on pupil outcomes. https://assets.publishing.service.gov.uk/media/5a75a6a140f0b67f59fce8a1/SubjectSpecialism_Report.pdf (Accessed: 28/05/2025)

DfE (2023) Calculation of 2023 to 2024 PGITT targets. https://www.gov.uk/government/statistics/postgraduate-initial-teacher-training-targets-2023-to-2024 (Accessed: 14 November 2023)

24. ONE APPROACH TO KEY STAGE 2 GEOGRAPHY - MOAT FARM JUNIOR

JON MCCORMICK

Moat Farm is a junior school located in the West Midlands. We see the power of geography in broadening the horizons and ambition of our children, helping them become more open minded, 'worldly' individuals and inspiring them to take more interest in the world.

There is so much that can be said about geography here and that is not including extra enriching activities, such as what we do for Earth Day or 'Moat Farm University', where selected children come out of class to participate in a geography project and amazing links with overseas schools. Here, we will concentrate on the key stage 2 curriculum: how we cover the objectives in what we chose to teach and how we chose to teach them.

PROGRESSION

When looking at the use of atlases, it was soon realised there was no progression. For example, Year 4 would start their unit on Europe with a first lesson of using an atlas to label the countries onto a blank map. The problem was Year 5 were doing the same thing with North America and Year 6 with South America! There was no progression in the type of maps being used and the vocabulary involved. Where are the geography skills in copying from an atlas onto a blank map? Another problem was the age of the atlases! Buying new atlases of the same type for each year group made this progression easier.

As you can see, each year group now has their own vocabulary to teach in lesson 1 of the unit signified, and then in lesson 2, this vocabulary is applied to maps from the atlas that progress through the year groups.

Year	Topic	Vocab to teach in lesson 1	Type of map to use in lesson 2 with the vocabulary
3	The World	Coastline, Border, Capital city (and symbol on Collins atlas map), north, east, south, west, Equator, Arctic/Antarctic circle, Continent, Country, Island, central.	Only 'simple' continent maps

Year	Topic	Vocab to teach in lesson 1	Type of map to use in lesson 2 with the vocabulary
4	Europe	Coastline, Border, Capital City, main city/town, (see key), north, east, south, west, Arctic circle, Continent, Country, Island, Landlocked, airport.	'Focus' maps of Europe

Year	Topic	Vocab to teach in lesson 1	Type of map to use in lesson 2 with the vocabulary
5	North America	Coastline, Border, Capital City, main city/town, (see key), north, east, south, west, northeast, northwest, southeast, southwest, Arctic circle, Island, Landlocked, Tropic of Cancer/Capricorn, American state, Canadian province, Independent, Altitude, Above sea level, Confluence, Estury, Mountain/highest point, Lake, Polar ice cap.	'focus' and 'physical'

Year	Topic	Vocab to teach in lesson 1	Type of map to use in lesson 2 with the vocabulary
6	South America	Partial coastline, Border, landlocked, Capital city, main city/town (see key), Coastal, inland, north, east, south, west, northeast, northwest, southeast, southwest, Tropic of Cancer/Capricorn, Equator, Independent Transport network, Altitude, Above sea level, Confluence, Estuary, Mountain/highest point, Lake, densely populated, sparsely populated.	'focus', 'physical' and 'population density' maps

Figure 24.1 Vocabulary to be introduced during the first lessons of KS2 geography in Years 3 to 6

INFERENCE

Geography is a great subject for you becoming a facilitator, providing the children with the information to investigate and them making their own inferences. For example, in Year 3 we get children making their own inferences on how temperature changes with latitude. First, children are given city names at different latitudes and their average temperatures. They then put these labels into the correct position onto a map and then use this to draw their own conclusions about latitude and temperatures. Mount Kilimanjaro is perfect for a plenary here, introducing other factors that influence temperature: what would children now expect the temperature to be like given where it is? A picture of a snow-covered Kilimanjaro is then shown to introduce altitude affecting temperature.

An important step in children understanding how imports/exports work, comes through inference. For example, in Year 3 children are given UK monthly averages for rainfall, sunshine hours, temperature and then growing/harvesting seasons for various fruit and vegetables. They can then make their own conclusions about what conditions are needed to grow this produce. Children are given average monthly temperatures for different temperate locations in the northern and southern hemispheres and make their own inferences about how the seasons differ between the hemispheres. We can then ask our learners, 'What do we do when we cannot grow strawberries here?' They are then in a position to make inferences about how we can potentially buy them from other countries when we cannot produce them here.

This is taken forward by the children collecting packaging and using this to calculate 'food miles'; this leads to work on the use of preservatives and organic vs non-organic food. It is important to ask, is there food that has been imported that could have been grown locally? (We have recently installed poly tunnels in our small school farm and hope this can be used to show the children how this is possible.)

A great way to introduce rivers in Year 4 is to give children physical maps and get them to follow the course of rivers to make inferences about where the sources are. They are then provided with precipitation maps to work out which UK areas receive the most rainfall. They can then draw conclusions about the connection between height of land, rainfall and the source of a river.

When studying Europe, our Year 4 learners are aware of wealth patterns within one continent through being given an 'average salary' map and then making their inferences, revealing the differences between east and west.

With topics such as earthquakes, we can investigate the sort of questions that children often ask. Year 5 look at earthquakes through the San Andreas Fault. They look at this area on Google Earth and note all the population centres nearby. Children will wonder, 'Why do people live here then?' We provide them with information about recent earthquakes in the region and how severity relates to the Richter Scale and then they answer questions such as:

1. How many significant earthquakes have there been over the last 34 years?
2. How many of the earthquakes were 'moderate', how many were 'strong' and how many were 'major'?
3. Between the oldest and the most recent 'major' earthquake, how many years were there?
4. What is the gap in years between the 'major' earthquakes? Would you say major earthquakes are frequent?
5. Do 'major' earthquakes always cause deaths?
6. How many earthquakes actually led to deaths over the 34 years?
7. Over the 34 years, how many people have actually died in earthquakes in total?
8. Over the 34 years, how many earthquakes have caused a significant amount of deaths?
9. What is more common in an earthquake – injuries or deaths?
10. Are earthquakes expensive? Do they cause a lot of damage? Who would pay to repair this damage?

Children should now be in a position to use their responses and inference skills to answer, 'Why do people continue to live in earthquake zones?'

A perfect plenary is to discuss earthquakes of the same magnitude but how they affect rich and poor countries in different ways, e.g. an earthquake hitting Haiti, killing and injuring many and one with a slightly higher magnitude hitting Japan but with no injuries or fatalities. Why?

Year 5 children investigate how one continent can have different climate zones. This lesson used to be a case of giving them the average monthly temperature, precipitation, sunshine hours, etc. for each place and them answering a series of questions much like a typical reading comprehension. The learning experience is improved through them flexing inference skills. Children are now given the same data on the different places but also given a physical map and biome map of North America with the places marked. They are asked to investigate why the differences in the data may be, using the maps as evidence, e.g. 'Las Vegas has much less rain than Winnipeg as it is in a desert biome' and 'I can see Miami is much closer to the tropics than Winnipeg so that is why it is warmer'.

After studying deforestation of the Amazon in Year 6, children use digital mapping (in this case Google Earth) to find evidence of deforestation and make inferences as to what they think the purpose of the deforestation may be, e.g. large-scale farming, urbanisation, industry, etc.

We discuss South American over-reliance on hydro-electric power in Year 6 and the problems this causes. Children are then asked to use their inference skills to work out the ideal locations for wind farms in this continent from maps: physical (does the site have the correct altitude?); political (is it close enough to infrastructure?); population density (is it too near a densely populated area?); biomes (which biome will be the least affected by a wind farm?); and tectonic plates (is it far enough away from one?).

COMPARISONS

In the curriculum there are comparison studies of regions. It is important that places are chosen that have an interesting contrast and that there is progression between what children in different years are comparing. Year 4 compare the Lake District with the Borgas region of Bulgaria for which there are great contrasts, e.g. the type of tourists that visit; where tourists come from; tourist activities. All this, as well as the obvious physical differences. Year 6 build on this when they compare the West Midlands region with South-east Brazil. They compare physical features but then compare standards of living between the two regions: the cost of living; unemployment rates; life expectancy; literacy rates; average salaries; air quality index; and infant mortality. All the time we are asking, 'What does this tell us about this region / the differences between the two regions?' The plenary is eye-opening when we reveal data for the world's poorest/richest countries.

When looking at South American exports, children often wonder why, despite the variety here, they have much lower average salaries, etc. compared to the UK. This is where Year 4 learning into types of industry – primary, secondary, tertiary (they carry out field work, mapping where examples are on our premises) comes in. What do children notice about these exports when compared to the exports of the UK and the type of industry they belong to?

ENVIRONMENTAL ISSUES

Our Year 3s spend time in our dinner hall: the wasted food, bound for landfill, is weighed/costed and used to work out how much food/money is wasted per day, week, month, year. Key questions can then be asked: if all primary schools in the UK are wasting the same amount, what would this amount to? Why does this seem particularly unfair given what we know about the world? It is great to talk about how other countries are addressing the problem of food waste. As a school, we are looking into this problem and are considering the purchase of 'Hot Bins' that can compost all the wasted food; the compost can then be used to grow vegetables (we have a small school farm) that can be sold cheaply to parents and carers.

Year 4 compare recycling within the UK and Europe. Comparing what can be recycled in the UK vs places like Sweden, Norway and Germany makes interesting learning as does comparing different UK boroughs. As part of fieldwork, children monitor at home what has been thrown away into non-recycling and then we study what could have been recycled from this in, for example, Germany. It is always very interesting to discuss the cost of recycling being an issue in some countries or asking the question, 'Do all countries need the same levels of recycling?' Do children understand how some countries may produce less items that need recycling?

In Year 6, we look at the Brazilian car industry and use this as a catalyst for working out how far we are away from targets for electric/hybrid vehicles. Children carry out fieldwork to ask the whole school (pupils and adults) what type of car their family drives. We then use this to work out (imagining the school is a microcosm of the world) how far we are away from targets. Any plenary here needs a discussion on what can be done to speed up the process.

It is important that children understand the 'knock on' effects of environmental issues and not just the immediate effects. Year 6 spend time 'mind mapping' the consequences of deforestation, what each consequence will lead to and what that will lead to, etc. It is always important for young learners to appreciate the 'positives' of such issues in, for example, what minerals mined provide the world with and how these are used in everyday items; how these are providing jobs, etc. This can lead to questions such as, 'How can we mine in a responsible way?'

FIELDWORK AND SCALE

It is important to have an end goal to fieldwork. When Year 3 study contours, they make a contour map of our field (fortunately we have a large field with slopes). They can then use this to design a cross-country running route to include challenging as well as not-so-challenging sections.

Year 4 visit a nearby town on the River Severn with the enquiry of how the river has changed the town's appearance. They study how the river, with the visitors it brings, has influenced the type of shops, attractions along the side of the river, the housing that provides river views, construction of defences as well as the river features that they have studied in class.

Year 5 do an immigration survey, asking about children, parents and grandparents who were born outside of the UK. Their enquiry is whether there is a difference between the countries where generations were born. They may find the grandparents' generation is mostly from the Commonwealth whereas for children and parents there may be more Eastern Europe and African countries of birth.

The scale of the Amazon Rainforest and its deforestation is difficult for young learners to grasp. In Year 6, to help children appreciate this, they simply measure the area of the playground and use what they find to answer questions like these: How many of our playgrounds can fit into the Amazon Rainforest? How many of our playground's worth of Amazon Rainforest is disappearing every day? If things continue at the same rate, how long will it be before it disappears?

When studying rivers in Year 4, do children appreciate the scale of a river such as the Nile? Children can measure the length of the playground and then time roughly how long it takes to walk across it. How many playgrounds roughly fit into the length of the Nile? Roughly how long would it take to walk the Nile?

DIGITAL MAPPING

Google Earth and Google Maps are excellent resources for primary school children, especially when complimented by 'Digimap' which, for what it is, is a very reasonably priced resource. It is important that children are given the chance to use digital mapping to explore what they have learned and to make observations. Without this, learning can seem abstract. It has already been mentioned how Year 6 children are given the opportunity to use Google Earth to observe real deforestation and make their inferences about what the purpose of it is.

Year 3 children, after having studied types of settlement/features and then mapping the local area, use digital mapping in Google Earth and Maps to compare our locality with Tokyo to see for themselves the differences and similarities between different types of urban area.

After having studied biomes of North America, Year 5 children are given the opportunity to 'go to' examples of each biome on Google Earth and make observations on what each biome actually looks like.

In Year 3, children use Google Maps to investigate what infrastructure would be affected within the blast radius of Mount Fuji. All their observations lead perfectly to the question that young learners ask: 'Why do people chose to build settlements near volcanoes?'

Similarly, Year 4 use the Digimap drawing function to highlight an area either side of UK rivers and then use this to investigate what could potentially be affected in the event of a flood. This is a nice link to flood defences which are explored in more detail in Year 5.

Year 5 use Digimap to look at UK land use and how it has changed. The teacher introduces this by using Digimap to access present and historical photos of UK towns. Can the children use their locational knowledge / map skills to make inferences as to why the city has seen such growth? For example, Milton Keynes is near London; Aberdeen is on the North Sea; would these have influenced growth? They can then use the Digimap drawing function to 'shade' a present map of a town but then overlay older maps which keep the same 'shade' on, meaning children can flip between the maps to see exactly how the town has grown.

When studying coastal defences, Year 5 children use present and historical maps to highlight the 'high water mark' on both maps and then measure the difference between them, i.e. how much land has been lost. They may see patterns, e.g. is it worse on the east or west coast?

QUESTIONING

It is important that children's thinking is challenged with good questioning, verbally or in marking. After having looked at the reasons/consequences of Amazon deforestation, Year 6 children are asked about developers and their proposals for more hydroelectric power stations in the region. Why does it need so many? What does this say about South America? Can the UK be so reliant on hydroelectric power? Why not? From an aerial image of a hydroelectric plant, can you see what other deforestation has had to take place because of the construction?

After having compared the West Midlands region with South-east Brazil, children are asked for example, 'What do the differences in average salaries and unemployment rates tell us about the two regions?'

In Year 4, when studying Europe, children look at immigration patterns into the continent; they are questioned, for example, 'Why do some European countries attract more immigrants than others?'

Five reflection questions

1. Is there clear progression in skills between year groups (not just the same skills but for a different topic)?
2. Is there an opportunity for children to take the lead in their learning through flexing inference skills?
3. Are comparisons and environmental issued carefully chosen?
4. Is fieldwork and digital mapping purposeful?
5. Do children have an opportunity to appreciate scale?

DIGGING DEEPER - THREE RESOURCES TO DELVE FURTHER

- *Primary Geography Journal*, The Geographical Association - see https://portal.geography.org.uk/journal/index/pg?utm_source=GA+Website&utm_medium=spring+journals&utm_campaign=Sep-Feb-2425-Web
- *Understanding and Teaching Primary Geography*, Catling and Willy (2018), London: Sage Publications Ltd.
- Developing Primary Geography, https://www.rgs.org/schools/resources-for-schools/guidance-and-support-in-developing-high-quality-primary-geography

25. THINKING SYNOPTICALLY IN THE A-LEVEL GEOGRAPHY CLASSROOM
PAUL LOGUE
@PLOGUEY

'Developing conceptual understanding by framing learning with "big ideas" helps students to think in more abstract ways and make connections between everyday and theoretical concepts.' (Rawlings Smith, 2017, p.265)

A-level geography is, without a doubt, the main factor that continues my burning passion for the teaching profession. I love thinking about every aspect of the A-level: topic and resource planning, fieldwork and international visits. Even marking is never too painful when it's one of the Year 12 or 13 classes' work.

For the past two academic years (September 2021 to July 2023) I have made exploring synoptic links a cornerstone of my subject improvement plan, what I see as the place piece of a coherent, functioning A-level geography curriculum. The bulk of this chapter exists thanks to the wonderful resource to the community that is the UCL Fawcett Fellowship, a marvellous programme that allowed me to explore pedagogical research out there to synoptic links.

WHAT IS SYNOPTIC THINKING?

Joining the dots. Weaving the threads together. Making links. Summing it all up. Seeing the bigger picture. All of these ideas hold the active ingredients that contribute to successful synoptic thinking. Synoptic thinking is the culmination of an A-level student connecting together the knowledge, understanding and skills that underpins the curriculum. It is unsurprising then that A-level examinations often use a combination of Assessment Objectives AO1, AO2 and AO3 when examining students on synoptic thinking. While the appearance of synoptic thinking can vary from exam board to exam board, it commonly takes the form of:

- Decision making exercises (DMEs)
- examining issues surrounding a region or sub-continent
- exploring a number of interconnected geographical issues and overarching themes.

Broadly speaking (as there will be some variation in tariffs and focus depending on the exam board), AOs can be summarised as:

Objective	Requirements	AS	A-level
AO1	Demonstrate knowledge and understanding of places, environments, concepts, processes, interactions and change, at a variety of scales	30 to 40%	30 to 40%
AO2	Apply knowledge and understanding in different contexts to interpret, analyse, and evaluate geographical information and issues	30 to 40%	30 to 40%
AO3	Use a variety of relevant quantitative, qualitative and fieldwork skills to: investigate geographical questions and issues; interpret, analyse and evaluate data and evidence; construct arguments and draw conclusions	20 to 30%	20 to 30%

Figure 25.1 A summary of the main assessment objectives (AOs)
Source: Ofqual (2017). Published under Open Government License v. 3.0.

WHERE DO WE FIND IT IN GEOGRAPHY?

In his book, *Powerful Geography,* Mark Enser (2020) laid out a very clear case that 'anything isn't geography' and the slippery slope that we as teachers could and have in the past put ourselves in in order to make the connections geography emanates to real world topics and dilemmas.

Creating false dichotomies of thinking and dot connecting can be a troublesome factor for teachers teaching a topic. Instead of geographical thinking along the lines of 'A could connect to B', students may make a series of logical leaps and get to 'A is definitely connected to B', and content is shoehorned together in a very unconvincing manner. For example, the number of students who answer my probing questions with 'climate change' has exploded in the past number of years. In many cases climate change would of course be a contributing factor, but there is failure to recognise the root causes.

So, where *can* we find it in geography?

Simon Oakes (2020) set out a very clear summary of where synopticity can present itself in the curriculum. Establishing four core approaches to synoptic links (see Figure 25.2), Simon proposes that these approaches allow for 'the big picture' to emerge more clearly and ensure that students are 'thinking like a geographer'. Like most of Simon's ideas, this has been something that I now pull on frequently for exploring synoptic links in the A-level.

Approach	What geographers do
Topic links	Make links ('joining the dots') between topics, theories, processes or case studies. For example, understanding how two places are interdependent or connected with one another
Case-study connections	Explore how a particular local place is influenced by a range of processes or issues associated with different geographical topics, such as globalisation and tectonic processes
Concept connections	Apply important 'overarching' geographical concepts, such as positive feedback, across different geography topics
Wicked problems	Reflect critically on the complexity of geographic decision-making, especially in relation to wicked problems such as climate change and plastic pollution (these are issues that are hard to tackle because of the diverse and numerous factors at play)

Table 25.1 Where synopticity can present itself in geography
Source: From Oakes (2020) 'Thinking synoptically'

THE CORE CONCEPTS AND SYNOPTIC THEMES OF THE A-LEVEL CURRICULUM

These are over-arching themes designed to help students make links between different geographical themes, ideas and specialist geographical concepts on the curriculum. The synoptic themes incorporate core concepts such as: causality, systems, feedback, inequality, identity, globalisation, interdependence, mitigation and adaption, sustainability, risk, resilience and thresholds.

I have seen some wonderful examples of curriculum leaders using these concepts (see Digging Deeper at the end of the chapter) to turbocharge synoptic opportunities at key stage 3.

For me, the core concepts have had a large effect on me and how I organised my department and our subject knowledge when I made the decision to restructure our curriculum when I joined my current place of employment, a sixth form college, as head of geography.

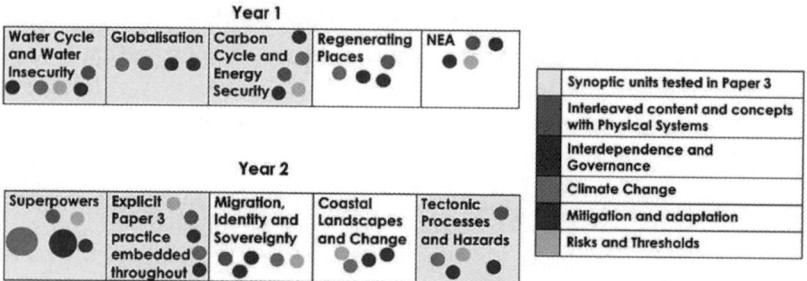

Figure 25.2 An expression of the main concepts and themes that drive my curriculum (for Edexcel A-level specification)

A colour version of this diagram can be found by scanning the QR code below:

Additional to the core concepts, the Edexcel specification uses three synoptic themes to link directly areas of study: the **Players** involved in geographical issues and decisions; **Attitudes and Actions**, how attitudes to geographic issues affect actions in terms of policies and strategies; and finally **Futures and Uncertainties**, decision-making about geographical issues that will affect people in the future.

Using the language of the specification is a core idea throughout our A-level course. We incorporate the use of PAF in our lessons, case study templates and homework/independent learning so students are constantly immersing themselves in synoptic themes.

St Dominic's Sixth Form College Geography Department

Independent research task

Focus: _____

What are the main issues identified?	
Who are the main players identified? What is their role?	
What are the main actions undertaken? How successful are they?	
What are the connections to the A-level specification?	

Figure 25.3 An example of our independent learning research templates incorporating PAF (players, attitudes and actions, future uncertainties) and encouraging synoptic links

PLANNING YOUR CURRICULUM TO EMPHASISE SYNOPTIC THINKING

Below are a number of big (but also easy!) changes that I have made to the geography curriculum at key stage 5 over the past number of years to make it synoptic that you could use to develop your own curriculum to emphasise synoptic thinking.

RETHINKING THE WAY AND ORDER YOU THINK YOUR CURRICULUM

Tectonics. Then Globalisation. Coasts next. Changing Places after that. Sound familiar? This order appears to be the norm in many schools and colleges across England and Wales. The KS5 geography curriculum is traditionally structured and dominated by the topics that are examined in AS exams and A2 exams, where topics such as tectonics, coasts, globalisation and regenerating places are examined in Year 1 and the remainder of the curriculum is taught in Year 2. It is interesting to think that many departments across the country still follow this model despite the sharp decline of AS level entries over the course of the past decade. AS entries have steeply declined from 55,801 in the UK (excluding Scotland) to 5,384 in 2022 (GA, 2022 using JCQ statistics).

Additionally, I have always found this to be a very repetitive geography curriculum for students as many topics are not 'new' to them and merely build on previous learning from those topics studied in KS3 and KS4. As we do not enter students for AS exams, I have unchained the curriculum from the 'traditional' order.

I also changed the curriculum order, so our two-year A-level looks like the this:

Year 1 Curriculum				
The Water Cycle and Water Insecurity	Globalisation	The Carbon Cycle and Energy Security	Regenerating Places	NEA and Introduction to Superpowers

Year 2 Curriculum			
Superpowers	Migration, Identity and Sovereignty	Coastal Landscapes and Change	Tectonics Processes and Hazards
Paper 3 skills throughout the year			

Table 25.2 An overview of our A-level curriculum

Our curriculum aims to break the mould and establish a fully synoptic curriculum that engages students in the core principles and themes / core concepts that encompass the A-level curriculum and geography as an academic discipline. Year 12 students now are taught a reformed curriculum that begins by them examining the Earth's life support systems and the human and physical factors that are accelerating to their physical breakdown. In this curriculum, I place the two physical systems units at the heart of what we do. Additionally, I believe that these two topics also allow students to cut their teeth on the skills and writing opportunities available on a much more efficient and synoptic level. Between the topics, students are introduced to some of the core synoptic themes of the A-level through globalisation to build the connections between the subjects.

This curriculum is designed for a single teacher. However, I believe it is also perfect for the two-teacher class sharing formula, particularly in Year 1. It allows for beautiful exploration of the synoptic links and interleaved concepts and case studies.

TEACHING THE CURRICULUM THROUGH THEMES AND ENQUIRY QUESTIONS

A few years ago, I blogged about my frustration with, and experimentation of, how I taught the Regenerating Places topic. I committed what some would account to as a cardinal sin by changing the order in which we taught and sequenced the unit in order for the murkier parts of the specification to seem more relevant and tie together the repetitious parts of the specification. Since then, all of our A-level topics have been grouped under Key Questions (KQs) that drive lines of enquiry and themed exploration. I have blogged about how I use this for all the physical systems units and Regenerating Places in my 'How I teach' series on my blog (https://geogpaul.wordpress.com/2020/09/06/how-i-teach-regenerating-places/, see also chapter 2).

KQ1: What are the issues associated with energy demand?
6.4abc Consumption and access to resources
6.5abc Reliance on fossil fuels- mismatches, energy pathways, unconventional fossil fuel energy

KQ2: How does the carbon cycle operate?
6.1abc Biogeochemical and geological cycle
6.2abc Biological sequestration
6.3abc Balance and human interaction

KQ3: How are the water and carbon cycles linked?
6.7abc Human activity threats
6.8abc Impacts on wellbeing from degradation of water and carbon cycles

KQ4: What are the alternatives to fossil fuels?
6.6abc Renewable and recyclable energy, biofuels, radical technologies

KQ5: What are the options for the future?
6.9abc Future emissions, feedback mechanisms, adaptation, mitigation

Figure 25.4 The key questions that drive the synoptic thinking in our Carbon Cycle and Energy Security Scheme of Work for Edexcel A-level specification

THE IMPORTANCE OF INTERLEAVING

Interleaving involves mixing and alternating different topics, concepts or skills within a single study or multiple study sessions. Instead of focusing on one topic exclusively before moving on to the next, interleaving encourages students to switch between related or unrelated subjects (Enser, 2020). This approach is often contrasted with the more traditional 'blocked' or 'massed' practice, where students study one topic intensively for an extended period before moving on to the next.

Interleaving is so important for teaching the A-level geography curriculum for many reasons, including:

- **Enhanced retention**: Interleaving helps students retain information better than blocked practice. When topics are interleaved, students are forced to retrieve information from long-term memory and switch between different concepts, which strengthens their memory recall and retrieval processes. This improves long-term retention of the material.
- **Improved transfer of knowledge**: Interleaving helps students see the connections between different geographical topics and promotes the transfer of knowledge from one context to another. This is crucial for A-level geography, where students are expected to apply their knowledge to various case studies and scenarios.
- **Cognitive flexibility**: Geography students must develop cognitive flexibility to adapt to different geographical contexts and challenges. Interleaving encourages this flexibility by requiring them to switch between topics, adapt to different questions, and make connections between seemingly unrelated concepts.
- **Reduced interference**: Interleaving can help reduce interference, where similar concepts or topics compete for attention and cause confusion. By spacing out the learning of related topics, students are less likely to mix up similar information, resulting in clearer understanding.
- **Joining the dots between topics**: A simple, yet effective approach is getting students to mind map connections between topics. We often close units by getting students to sit in groups with copies of the specification and asking them to make the connections between topics, exploring the core concepts and PAF synoptic themes.

WHAT IS GEOGRAPHY TEACHING, NOW?

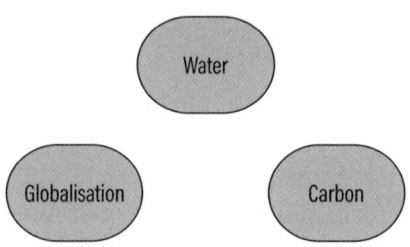

Figure 25.5 Joining the dots between topics

MAPPING THE SYNOPTICITY IN YOUR CURRICULUM

We have become better each year at signposting the overlapping content across the specification and getting students to independently complete or review content.

Lesson	What's covered?	Spec Points	Examples/ case study	Homework/ Consolidation	Lesson	What's covered?	Spec Points	Examples/ case study	Homework/ Consolidation
1	Introduction to regeneration			Lesson 3 is on the Rust Belt. Revise this lesson from the Globalisation unit and be ready to be an expert for this lesson.	4	UK Economic theory & migration	8B.2 bc	Deregulation Open borders Regional movements in the UK	Complete this lesson independently by reading through slides and making notes
					5	Changes and tensions because of cultural and ethnic change	8B.3 Ab	US-Mexico border flows EU migration	Strongly advise you to revise this aspect of Globalisation spec point 3.8a which covers this. TEXTBOOK p.233-235
10	Migration between former colonies and imperial care	8B.5c		Complete this lesson by independently by reading through the slides and making notes	14	Role of IMF, WB, WTO	8B.8 Abc	Jamaica structural adjustment NAFTA	Complete this lesson by independently reading through slides and making notes

Figure 25.6 Examples of overlapping specification globalisation content from Regenerating Places and Migration, Identity and Sovereignty in the Edexcel A-level specification

EXPLORING SYNOPTIC LINKS THROUGHS CASE STUDIES

Case studies are one of the strongest and thickest threads holding the geography curriculum together. A broad case study curriculum is excellent, but using case studies again and again to explore new deeper layers can be a golden thread. Some examples of case studies that are interleaved throughout the curriculum (all topics are from Edexcel) are given in Table 25.3 (see also chapter 16).

Example 1: Russia/Ukraine

Focus	Topic
Superpower tensions – exerting spheres of influence	Superpowers
Global energy insecurity – players and insecurity issues	Carbon Cycle and Energy Security
Ethnicity as a factor leading to contested borders	Migration, Identity and Sovereignty

Example 2: The Rust Belt, USA

Focus	Topic
Effects of the global shift on a developed nation	Globalisation
A failed place	Regenerating Places
Future uncertainty for superpowers / the need for economic restructuring	Superpowers
Anti-globalisation movements / Trump's America First ideology wins Rust Belt states in 2016 election	Migration, Identity and Sovereignty

Table 25.3 Examples of interleaved case studies

This is undoubtedly easier to manage if you teach the whole cohort on your own like I do. In departments where there are two or more teachers sharing a class (often with one teaching the physical units and one teaching the human), I recommend that you devote a series of curriculum/department times to explore the topics and case studies you teach and how you can link the topics together most effectively.

EXPLICIT PRACTICE / USING THE CORE CONCEPTS EXPLICITLY

Finally, we use explicit practice to map out where we find synopticity in our curriculum. We use the template shown in Figure 25.8 for students to identify the themes and connections they have explored when we read essays, articles and resource booklets from past papers.

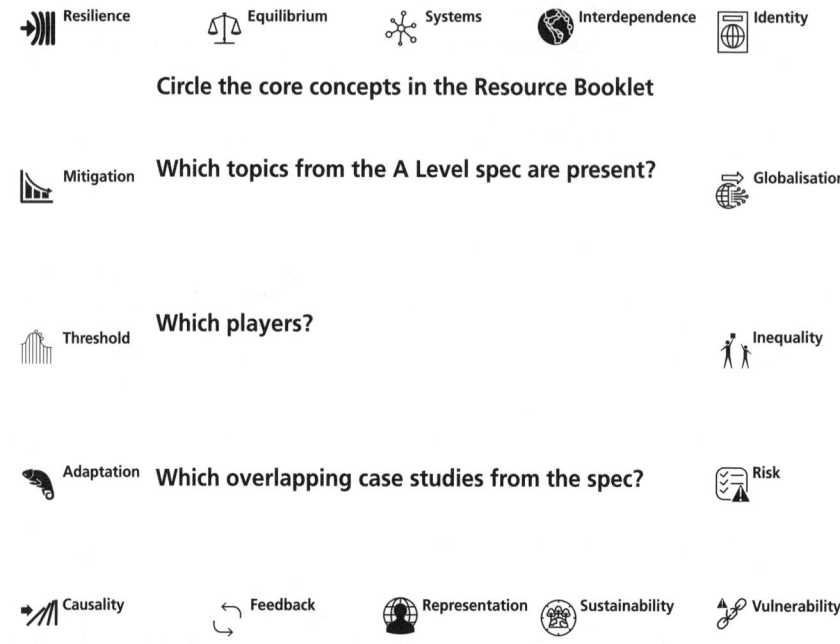

Figure 25.7 Template for identifying synopticity

Five reflection questions

1. How do you plan your curriculum so it reflects synoptic thinking?
2. How do you prepare your students for synoptic thinking through KS3 and KS4?
3. How do you embed synoptic thinking through interleaving in your curriculum?
4. How clear is synoptic thinking mapped across your curriculum and articulated to students?
5. How widely do you use core concepts in your lesson planning and teaching?

> **DIGGING DEEPER - THREE RESOURCES TO DELVE FURTHER**
> - A-level geography specialised concepts - knowledge organisers: https://resources.eduqas.co.uk/Pages/ResourceSingle.aspx?rlid=1990 (Accessed: 28/05/2025)
> - Using core concepts to inform your KS3 curriculum: https://www.youtube.com/watch?v=2v6YbP7rqUUandab_channel=SenecaCPD (Accessed: 28/05/2025)
> - RGS Choose Geography website, https://www.rgs.org/choose-geography

REFERENCES

Enser, M. (2020) Interweaving geography: retrieval, spacing and interleaving in the geography curriculum. *Teaching Geography*, 45(1), pp.15–17.

Enser, M. (2021) *Powerful Geography*. Carmarthen: Crown House Publishing, pp.69–72.

Geographical Association (2022) A level geography results analysis 2022. https://geography.org.uk/wp-content/uploads/2023/05/GeographyAlevelresultsanalysis2022.pdf (Accessed: 28/05/2025)

GeogPaul. (2022) Blog. Carbon cycle and energy security resources. https://geogpaul.wordpress.com/2022/01/17/carbon-cycle-and-energy-security-resources/

Geography.org.uk. (2022). *Geography A-level Results Analysis 2022*. Available at: https://geography.org.uk/wp-content/uploads/2023/05/GeographyAlevelresultsanalysis2022.pdf (Accessed: 28/05/2025)

Oakes, S. (2020) 'Thinking Synoptically', *Geography Review*, February 2020, London: Hodder Education. Available at: https://www.hoddereducation.co.uk/magazines/magazines-extras/geography-review-extras (Accessed: 28/05/2025)

Ofqual. (2017) *GCSE, AS and A level Assessment Objectives: Geography*. https://www.gov.uk/government/publications/assessment-objectives-ancient-languages-geography-and-mfl/gcse-as-and-a-level-assessment-objectives#geography-1

Pearson Edexcel. (2019) *GCE A level Geography Specification*. (5th ed.) https://qualifications.pearson.com/content/dam/pdf/A%20Level/Geography/2016/specification-and-sample-assessments/Pearson-Edexcel-GCE-A-level-Geography-specification-issue-5-FINAL.pdf

Rawlings Smith, E. (2017) 'Post-16 Geography'. In Mark Jones (ed.). *The Handbook of Secondary Geography*. Sheffield: Geographical Association, pp.260–277.

Weinstein, Y. & Sumeracki, M. (2018) *Understanding How We Learn: A Visual Guide*. Abingdon: Routledge.

26. TEACHING CAREERS IN GEOGRAPHY

ROUNA ALI
@ROUNAALI

'Teenagers taught about the world of work are more motivated to get higher GCSE results.' (Coughlan, 2019)

CAREERS IN CONTEXT: THE BIGGER PICTURE

The quote above is taken from a BBC News article with the headline 'Careers Lessons push up GCSE grades', but it is actually based on a study by Percy and Kashefpakdel (2018) regarding how school students' contacts with outside speakers and influencers such as employers and training providers, can lead to increased drive, ambition and higher than predicted exam grades for those students. A report in 2020 by researchers at UCL also came to similar conclusions: 'We show that driven and ambitious young people outperform their peers by 0.36 standard deviations (or about half a GCSE grade) on their total GCSE points.' (Jerrim, Wyness and Shure, 2020).

From 2018, the Department for Education has placed legal duties on all 'schools and academies to provide opportunities for a range of education and training providers to access all Year 8 to 13 pupils to inform them about approved technical education qualifications and apprenticeships. Through the Skills and Post-16 Act 2022, the government has strengthened this legislation by introducing a minimum number of six provider encounters that every school must provide...' (DfE, 2023). This means that every student in every year group from Years 8 to 13 must have at least one encounter with an employer or apprenticeship/training provider or technical education provider at some point during the academic year. The importance of this repeated outsider contact is twofold: it will potentially motivate students to achieve higher grades but also give them more of an in-depth perspective on different post-16 options, helping them make informed decisions.

The BBC journalist quoted at the start of this chapter used a generic term 'careers lessons' (or, in this case, contact with employers) for what is actually known as a much wider programme of careers education, information, advice and guidance (CEIAG) that takes place in schools and colleges. Figure 26.1 shows an example of what CEIAG can look like in a school or college. I am both a geography teacher and professionally qualified level 6 careers guidance adviser with 17 years' professional experience, specialising in working with children and young people. During my PGCE geography teacher training in 2012 there was absolutely no mention of how to embed careers education within the geography curriculum (the 'CE' part of CEIAG which involves all careers leaders and subject teachers, including geography teachers, as shall be made clearer as this chapter progresses). On the flip side, when I trained to become a careers adviser in 2006, two-thirds of the programmes' focus was on delivering 1:1 careers guidance interviews with children and young people (the 'IAG' part of CEIAG which usually involves the work of an independent, external and impartial careers guidance adviser who comes into a school to interview students on a 1:1 basis and is not employed by the school). There was very little emphasis on careers education. It is therefore to my advantage that I can now bring these two sides of my professional identity together and share my combined knowledge, skills and experience with fellow geography teachers as the main purpose of this chapter.

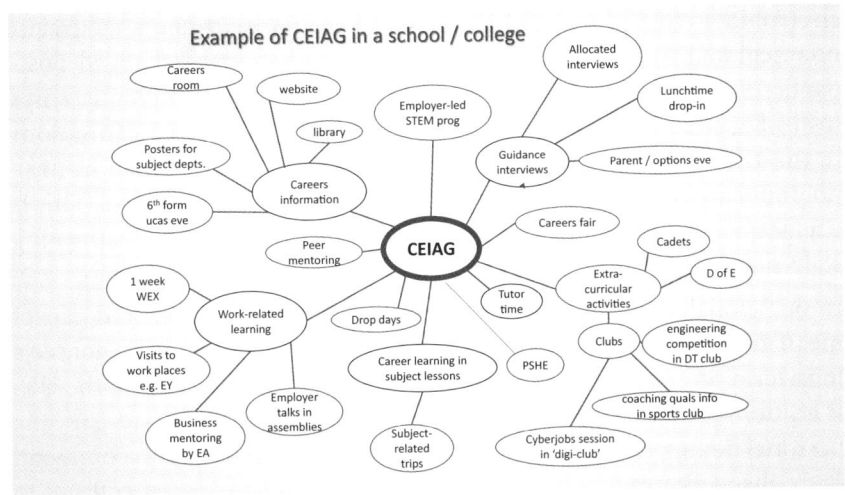

Figure 26.1 An example of what CEIAG can look like in a school or college
Source: Bob Neame (2022), with permission

Geography teachers, as with all teachers, are significant adults in a young person's life, and we want them to achieve high grades, but we also want to deliver to them the highest quality geography education and fieldwork that we can, and inspire them to become avid lifelong learners, responsible yet active global citizens and future protectors of our planet. These are fundamental reasons why careers work in a school should not be seen as a small 'add-on' or footnote to the curriculum.

As a geography teacher, you had to come to the decision to teach and had to plan and organise your own career. You will have developed a myriad of career development and management skills along the way, such as investigating different careers, arranging relevant work experience opportunities, researching the labour market, networking and creating contacts, applying to university, mastering the CV, negotiating the job application and interview process and so much more. Today's students also need to consider: how artificial intelligence and rapid advances in technology are making inroads into different jobs, careers and professions; jobs involving sustainability of the Earth and solving real world problems such as mitigating the impacts of climate change and the increasing need for mental health professionals in the future; getting up to speed with how employers' recruitment practices are changing with electronic screening of CVs, use of social media, zero-hours contracts; how regional, national and global labour markets have changed post-pandemic; the gradual societal shift towards degree-level apprenticeships instead of going to university. Also, as most companies and organisations in the UK are now expected to work towards becoming carbon neutral, they are employing climate change project officers and managers. This is a huge growth area in the labour market of this country and is happening right now – geographers need to be at the forefront!

In our globalised and diverse world some students also ask and expect me to give IAG on how to live and work abroad because they do not see a long-term prosperous and fulfilling future for themselves here in the UK. It is also worth noting that Ofsted 'includes careers guidance as part of a personal development judgement. Ofsted is legally required to comment, in an inspection report, on the careers guidance provided at colleges to 16- to- 18-year-olds and students aged up to 25 with an education, health and care plan' (DfE, 2023).

THE GATSBY BENCHMARKS

There are eight Gatsby benchmarks that schools have to meet with regards to their CEIAG (DfE, 2023):

> **Benchmark 1:** A stable careers programme.
>
> **Benchmark 2:** Learning from career and labour market information.
>
> **Benchmark 3:** Addressing the needs of each pupil. Young people have different career guidance needs at different stages.
>
> **Benchmark 4:** Linking curriculum learning to careers. All subject staff should link curriculum with careers, even on courses that are not specifically occupation-led. This is where the geography teacher has got an extremely important and relevant role to play. (We will look at this in more depth below.)
>
> **Benchmark 5:** Encounters with employers and employees.
>
> **Benchmark 6:** Experiences of workplaces.
>
> **Benchmark 7:** Encounters with further and higher education.
>
> **Benchmark 8:** Personal guidance. Every student should have opportunities for guidance interviews with a career adviser.
>
> (For more details see DfE, 2023 and The Gatsby Charitable Foundation, 2023).

WHERE DO I FIT INTO ALL OF THIS AS A GEOGRAPHY TEACHER? – BENCHMARK 4

Geography teachers may be asked to deliver lessons on how geography subject knowledge and geographical skills such as GIS are connected to a range of jobs and careers for Careers Week, but schools also need teachers to embed careers in schemes of learning. I have used resources from Parkinson (2020), the Geographical Association and the Royal Geographical Society to create geography and careers lessons which will inspire some students to take up geography at GCSE and A-level and hopefully empower them to think about their long-term career development. An approach which I have found successful is included below.

1. The learning objective for these lessons is to learn about, explore and further investigate the range of jobs and careers that are connected to the subject of geography as a whole. It is also to help students develop their self-awareness and think about whether they would be a good fit for these jobs and careers that are more explicitly linked to the subject and skills of geography.
2. How many job titles can you think of starting with the letters G.E.O.G.R.A.P.H.Y? (For example, geologist, energy engineer, oceanographer, graphic designer, refugee adviser, air cabin crew, pilot, housing officer, youth and community worker.)
3. Next it is time to ignite the 'fire in their bellies'! We can take themes and concepts from our subject knowledge in conjunction with the various geographical skills that students develop and link these to certain jobs and careers. For example:
   - **Are you fascinated by maps?** Geographers use maps to understand the patterns in places and landscapes. You will learn map work and GIS from paper maps to computer-generated maps and their analysis. You might become a GIS specialist, cartographer, utilities manager or remote sensing analyst.
   - **Do you like to people watch?** Geographers are curious about why places change and why cities are made up of very different areas. You will learn about population and urban issues and challenges – about where people live and how places and

populations are changing. You might become an urban town planner, social worker, market researcher, housing officer, paralegal or an estate agent.

- **Do you want to know why people do the work they do?** Geographers can explain the reasons for different types of work. You will learn about the changing economic world. You might become an economist, transport manager, location analyst, an online retailer or a regional developer.
- **Do world events interest you?** Geographers know that we live in an increasingly interconnected world that is full of widening social inequalities. You will study development and global issues. You will learn why places and societies are different and why higher income and lower income countries face different challenges. You might become a development aid worker, a diplomatic service officer, a refugee adviser or a charity fundraiser.
- **Do you enjoy watching the weather?** Geographers analyse developing weather systems and can predict natural hazards. You will study climate and weather hazards including changing global climates, extreme weather events and their interactions with people. You might become an environmental lawyer, weather scientist or meteorologist, a climate change policy or projects officer for a local council, a TV weather presenter, disaster manager, flood prevention officer, an actuarial scientist or risk assessor or a water treatment worker.
- **Do you enjoy being outdoors in the landscape?** Geographers can explain the features of our landscapes – from mountains to molehills. You will learn how plate tectonics, water, ice and wind shape the surface of the land and how people can influence and adapt to these processes. You might become a hydrologist, coastal manager, coastguard, geologist, civil engineer or soil conservationist.
- **Do you like discovering new places?** Geographers are interested in different cultures, peoples and places. You will delve deeper into many case studies from around the world. You might

become air cabin crew, an online travel agent, TEFL teacher, tourism officer or tour manager, social or media researcher.
- **Do you care about the future of our planet?** Geographers appreciate the importance of sustainability and work to solve real world problems. What are the different types of threats that are going to have an impact on our world in future? You will study ecosystems (such as the tropical rainforest) and the environment. You will learn how the environment works, the increasing human pressures placed upon it and the need for sustainability. You might become an energy or environmental engineer, estate manager, a forestry ranger, an environmental or sustainability consultant, a climate change manager, a pollution analyst or a nature conservation officer.

4. The next step is to encourage students to start investigating one or two career ideas using the National Careers Service website. This is an online database where you can read and learn about hundreds of jobs and how you can get into them. (See National Careers Service Explore Careers, 2023.)
5. The main task is for students to answer as many questions in as much detail as they can for the careers they are interested in. Here is a suggested set of questions:
   - What qualifications will I need for this job?
   - Do I need to go to university?
   - What GCSE grades do I need to get started on this particular career path?
   - What knowledge and skills developed through geography will help me with this career?
   - What other subjects apart from geography will I need?
   - Am I good at and will I enjoy these subjects?
   - How much money can I expect to earn? What is the starting salary for this job role? (Average salary in the UK in 2023 was £29,588.)
   - What do they actually do in these jobs on a day-to-day basis? (Tasks and activities.)

- Would I enjoy working in this kind of environment (e.g. laboratory, library, train station, airport, school, clinic, prison)?
- What are the most important skills for this job?
- What kind of personal qualities would help me with this type of work?
- Will I need to learn other languages or move to another country? (Development aid worker is a job that is often used in geography and careers lessons but it often requires the postholder to move from country to country. How many of your young people would be able to do this in their futures? It is useful for students to consider these kinds of factors that will impinge on their career decision making.)
- What would be the main advantages and disadvantages for me if I had this job?
- What would be the main positive and negative impacts on society and the environment if I carried out this type of work?
- Are there vacancies for these types of jobs available right now in my local area?
- How diverse are these occupational groups?
- What potential barriers and obstacles could get in the way of reaching your career goals?
- What other challenges could you face in the workplace after you have obtained this job and are trying to further develop and progress your career?
- What do I need to do now in order to eventually reach this particular career goal?

6. Getting students to feedback to you their findings should hopefully dispel and challenge any myths or stereotypes that might surround certain job ideas and give them a balanced, objective and realistic view of a specific job instead of indulging in naïve fantasies and fake news fuelled by social media platforms.
7. In theory, after points 1 to 6 above have been carried out, this last step ideally would be to get students to start creating their own career action plans. If there is time, the students could even deliver

a presentation at the end that describes and explains the career journey that they have planned out for themselves in order to become a civil engineer, for example.

Benchmark 4 is about going much further than one or two lessons during Careers Week. It is about linking curriculum learning with careers, and how geography departments are addressing this (if at all) will vary greatly from school to school. The Royal Geographical Society has pragmatically made explicit 'Ten reasons why we need to embed careers education and guidance into the geography curriculum' (see Royal Geographical Society, Careers Education in the Geography Curriculum, 2023).

Careers is a huge topic, so where do we start? Even though CEIAG in a school or college setting is a large umbrella term that can encompass a whole range of tasks, lessons, activities and experiences, the three main aims of CEIAG (i.e. students' self-awareness and development, career exploration, and career management), should help all geography teachers to focus their efforts on specific themes when it comes to embedding careers in geography. Ask:

- Are there good examples to refer to of geography departments that have prioritised careers within their geography curriculums over a number of years?
- What was the impact on their students in terms of their exam grades and their short-term and long-term career development?
- How can positioning careers as a central, compulsory and constant presence in schools help our students in a globalised and increasingly complex world?

We also need to consider different experiences our students may face in the workforce. The Equality and Human Rights Commission has recently reported on how those with protected characteristics are more negatively impacted by three long-term British labour market trends: 'the increase in flexible ways of working (whether by time or place), the growth of self-employment and the gig economy, and the increasing use of automation and artificial intelligence (AI)' (Avanzo Windett et al, 2022).

CAREERS IN GEOGRAPHY

The main professional body for career guidance and career development practitioners in the UK is the Career Development Institute (CDI) and their framework is an extension of the three main aims of CEIAG. On page 13 of their *Framework Handbook* for schools, they give a table of ideas and suggestions on how to incorporate careers in your subject. For example, make explicit the skills that will be developed throughout the schemes of work and their relevance to everyday life and careers. Putting up careers-related displays in the classroom and corridors. Getting school alumni and guest speakers or employers into geography lessons and know when to refer and signpost students to specialised advice and guidance (CDI, 2021). Geography schemes of learning can often include several job idea links. For example, a Year 7 scheme of learning on 'Why are rivers important?' could include a lesson that can be connected to the work of a hydrogeologist and hydrologist. At the time of writing, the Indeed jobs website had many live vacancies throughout the UK for junior hydrologists, graduate hydrogeologists, flood hydrologists and senior water scientists to name but a few, making this a realistic, possible and positive career option for our students.

One of the most obvious job links to tectonic hazards in geography lessons is the job of a volcanologist. While it is good to have high career aspirations, this also needs to be balanced against Gatsby benchmarks 2 and 3 to do with labour market information and addressing the needs of each student respectively. As most jobs for volcanologists are in the USA with NASA these may be unrealistic for our students. Are we setting them up to fail by suggesting such careers to them? Discussing this kind of career may be too idealistic and theoretical for your students, but could help to bring more equality, diversity and inclusion to such careers. Would this career idea address your students' needs or could other job links be found in your local / regional area that link to tectonic activity?

As geographers, we might also be secretly and silently assessing labour markets on different scales: the local, regional, national and international labour market economies. Depending on who inhabits your classroom, how accessible are these careers for those from working class backgrounds, women, young people, people of Asian, African, Chinese, Latin American, Eastern European, Roma or Gypsy descent?

Are there careers that actively favour and do not discriminate against people with disabilities or who are from LGBTQIA+ communities?

Rather than volcanologist, it may be more meaningful to look at the job roles of engineering geologist and geophysicist as employers are actively recruiting for many such positions here in the UK right now. In some schools there are end of unit summaries which involve students revising and reflecting upon the knowledge and skills that they have gained throughout the unit and making a link with careers ideas. How impactful is this approach? How could it be improved?

For KS3, could some of your geography schemes of learning be developed with your school's careers guidance adviser and/or a local employer? It might not be practical or appropriate to embed careers into every single geography KS3, KS4 and KS5 scheme of work, however embedding it into alternate schemes at each key stage could be enough to 'ignite that fire in their bellies' which would encourage and motivate students to start taking action and start their lifelong career planning and development. In the past I have taught lessons on the tropical rainforest (TRF) to Year 8 students, explaining to them how they can campaign and work towards protecting the TRF in future in terms of jobs and career ideas linked to the TRF. Some of the students almost got out of their seats and wanted to start protests! Knowledge is power – power that I have always strongly believed in and thoroughly enjoy passing on to our young people.

A KS3 geography curriculum with career links would look at one or two jobs or career ideas that are linked to that specific unit of work in depth so that the topic of careers in geography is given quality time and space, using the framework suggested above. Don't forget to consider apprenticeships – many students will be keen to hear more.

APPRENTICESHIPS

When considering an area of work, find out what apprenticeships are available in this area of work in order to begin this career pathway. Look at labour market information. Do a quick search on the Government's apprenticeship website (see Find an apprenticeship – www.gov.uk). If you find an apprenticeship, study the live vacancy carefully and in depth alongside your students on the whiteboard. They will be fascinated by this, trust me! Show them how long the apprenticeship is for (18 months,

24 months?), the salary, GCSE and other entry-grade requirements, working hours, location, job description, the qualification earned and how to apply. Make it clear to students that an apprenticeship is very similar to job searching and that they need to keep applying for them if they want one! Some students wrongly assume that their teachers and advisers at school will just provide them with an apprenticeship at the end of Year 11 if they want one.

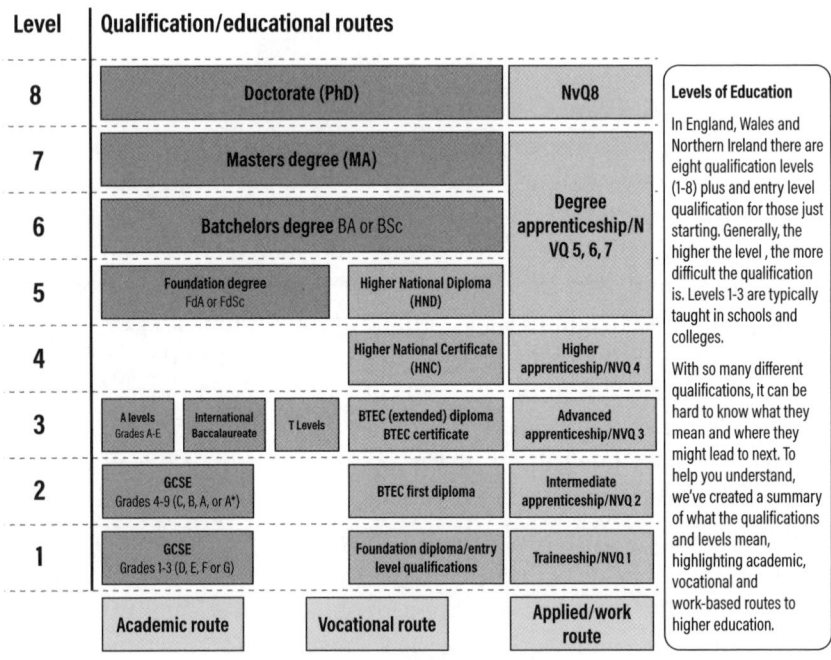

Figure 26.2 Routes through Education showing academic levels 1–8 on the lefthand side of the chart
Source: The Parents Guide to Post-18 Options 2023–2024 (2023)

GEOGRAPHY TEACHERS, YOU ALSO HAVE A ROLE TO PLAY HERE! - BENCHMARKS 2, 5, 6 AND 7

In your own professional network or school community, do you have contacts or are you aware of any local opportunities that your mini geographers could benefit from? These might be former students who have now excelled in the world of work, and geography is the subject that helped them to reach their goals and ambitions. How can you get them involved in your new careers in geography campaign or in your school's overall careers programme? Could they deliver some of the course content? This could be done virtually online, through a recording or in person. What about them offering work experience or getting involved in Careers Week?

Geography teachers will have been to university. Are you still in contact with your *alma mater*? What about your former lecturers, dissertation supervisors, tutors and the people you studied alongside? What occupations are they working in currently? How did geography help them get to where they wanted to be? As geography graduates, could they speak to your A-level students, for example, to explain to them in detail what it is like to study geography at university level (Benchmark 7)?

Careers in the Geography Curriculum			
	Year 7	Year 8	Year 9
Autumn Term	What are map skills? Geospatial Technician Digitisation Officer	How do we use natural resources? Energy Engineer Environmental Consultant	What are sustainable ecosystems? Sustainability Co-Ordinator / Consultant Recycling Engagement Officer
Spring Term	What is international development? Children's Social Worker Charity Fundraiser	How is the world's population changing? Journalist (Newspaper, Magazine, Broadcast) Social Science Researcher (e.g. Criminologist)	What is an economy? Economist Data Analyst
Summer Term	Why are rivers important? Hydrologist / Hydrogeologist Nature Conservation Officer	What happens where the land meets the sea? Marine Biologist Environmental Education Officer	How is the world's climate changing? Climate Change Policy or Project Officer Planning and Environmental Lawyer

Table 26.1 KS3 geography curriculum with careers mapped

WHAT ABOUT MY OWN CPD AND CAREER DEVELOPMENT?

By reading this chapter, there is a chance that I might have ignited the fire in the bellies of some geography teachers! Would you like to become a careers leader for your school or train to become a careers guidance adviser? If the answer is 'Yes' to any one of these questions, then please look at the new CDI website (to be launched in due course at the time of writing). The CDI offers many important, up-to-date and relevant training events, webinars, newsletters and networking opportunities for both members and non-members of the CDI.

Five reflection questions

1. How could your geography curriculum address the three main aims of CEIAG?
2. How could you develop some of your geography schemes of learning in partnership with a Level 6 professionally qualified careers guidance adviser or local employer?
3. How could you balance high career aspirations with your local, regional and national labour market information when putting into practice your teaching of careers in geography?
4. What impact would linking the geography curriculum to careers have on your specific student population?
5. How can you best incorporate your own past and present personal and professional contacts and networks to enhance your school's and students' careers education programme?

DIGGING DEEPER – THREE RESOURCES TO DELVE FURTHER

- Andrews, D. and Hooley, T. (2022) *The Careers Leader Handbook: How to create an outstanding careers programme for your school or college*, 2nd Edition. Trotman Indigo Publishing. (This is for those interested in leading CEIAG across a whole school setting.)
- Parkinson, A. (2020) *Why Study Geography?* London: London Publishing Partnership. (Good for career exploration.)
- Tupper, H. and Ellis, S. (2020) *The Squiggly Career: Ditch the ladder, discover opportunity, design your career*. London: Penguin. (How to articulate career management and development skills both for ourselves and our students in the post-industrial world.)

REFERENCES

Andrews, D. & Hooley, T. (2022) *The Careers Leader Handbook: How To Create an Outstanding Careers Programme for Your School or College*, 2nd Edition. Bath: Trotman Indigo Publishing.

Avanzo Windett, S., Bivand, P., Atay, A., Aleynikova, E. & Ahmed, J. (2022) *The Future of Work: Protected Characteristics in A Changing Workplace.* Manchester: Equality and Human Rights Commission.

Barnes, I. & Kent, P. (2010) *Leading careers education information advice and guidance (CEIAG) in secondary schools.* National College for Leadership of Schools and Children's Services.

CDI. (2021) Career Development Framework Handbook KS3, KS4 and Post 16. CDI_107-Framework_Handbook-web_Updated.pdf (Accessed: 12/09/2023)

CDI. (2023) CDI Academy: What we offer. https://www.thecdi.net/cdi-academy (Accessed: 17/09/2023)

CDI. (2023) Code of Ethics. https://www.thecdi.net/Code-of-Ethics (Accessed: 17/09/2023)

CDI. (2012) The ACEG framework for careers and work-related education: a practical guide. https://www.thecdi.net/policy-campaigns-and-media/research-and-reports (Accessed: 23/09/2023).

CDI. (2023) Training and Events. https://www.thecdi.net/training-and-events (Accessed: 17/09/2023)

CDI. (2023) UK Register of Career Development Professionals. https://www.thecdi.net/professional-register (Accessed: 17/09/2023)

Coughlan, S. (2019) Careers lessons push up GCSE grades. BBC News, 17 May 2019. https://www.bbc.co.uk/news/education-48268267 (Accessed: 17/09/2023)

Cre8tive Resources. (2023) Careers explored learning journey map: Year 7 – Year 11. https://www.cre8tiveresources.com/wp-content/uploads/woocommerce_uploads/2023/05/Careers-Curriculum-Map-efti7x.pdf (Accessed: 17/09/2023)

DfE. (2023) Careers guidance and access for education and training providers: Statutory guidance for schools and guidance for further education colleges and sixth form colleges. https://assets.publishing.service.gov.uk/government/uploads/system/uploads/attachment_data/file/1127489/Careers_guidance_and_access_for_education_and_training_providers_.pdf (Accessed: 17/09/2023)

Geographical Association. (2023) Promoting geography. https://geography.org.uk/support-guidance-promoting-geography/ (Accessed: 29/08/2023)

Gov.uk. (2023) Find an apprenticeship. https://www.gov.uk/apply-apprenticeship (Accessed: 12/09/2023)

Indeed. (2023) https://uk.indeed.com/ (Accessed: 12/09/2023)

Jerrim, J., Wyness, G. & Shure, N. (2020) Driven to succeed? Teenagers' drive, ambition and performance on high-stakes examinations. Working Paper No. 20-13. UCL Centre for Education Policy and Equalising Opportunities (CEPEO).

National Careers Service. (2023) Explore careers. https://nationalcareers.service.gov.uk/explore-careers (Accessed: 29/08/2023)

Neame, B. (2022) *CEIAG* 'Refresher: Recent developments, frameworks, expectations prevalent and recent theories current delivery landscape'. (Unpublished)

Parkinson, A. (2020) *Why Study Geography?* London: London Publishing Partnership.

Percy, C. & Kashefpakdel, E. (2018) 'Social advantage, access to employers and the role of schools in modern British education'. In Tristram Hooley, Ronald Sultana, Rie Thomsen (eds.) *Career Guidance for Emancipation*, pp. 148–65. Abingdon: Routledge.

Prospects. (2023) Job profiles. https://www.prospects.ac.uk/job-profiles (Accessed: 12/09/2023)

Reed. (2023) Gateway to Work. https://www.reed.com/tools/gateway-to-work (Accessed: 12/09/2023)

Royal Geographical Society with IBG. (2023) Careers Education in the Geography Curriculum. https://www.rgs.org/schools/resources-for-schools/careers-education-in-the-geography-curriculum (Accessed: 13/09/2024)

Royal Geographical Society with IBG. (2023) Choose Geography. https://www.rgs.org/choosegeography (Accessed: 29/08/2023)

Skills and Post-16 Education Act 2022. (2022) https://www.legislation.gov.uk/ukpga/2022/21/contents/enacted (Accessed: 28/08/2023)

The Gatsby Charitable Foundation. (2023) Good Career Guidance. https://www.gatsby.org.uk/education/focus-areas/good-career-guidance (Accessed: 23/09/2023)

The Parents Guide to Post-18 Options 2023-2024. (2023) https://www.theparentsguideto.co.uk/samples (Accessed: 23/09/2023)

Tupper, H. & Ellis, S. (2020) *The Squiggly Career: Ditch The Ladder, Discover Opportunity, Design Your Career.* London: Penguin.

UNIFROG (2023) Working in geography. Working_in_Geography_UK.pdf (Accessed: 12/09/2023)

Universities UK. (2023) Jobs of the future. jobs-of-the-future.pdf (Accessed: 12/09/2023)

27. THE WONDER OF PHYSICAL GEOGRAPHY

CATHERINE OWEN (WITH THANKS TO DUNCAN HAWLEY FOR FEEDBACK AND SUGGESTIONS)
@GEOGMUM

'To release "hidden" physical geography teachers need to engage with deep thinking about the subject and what this means in terms of powerful knowledge. In turn, this enables teachers to shape the curriculum so as to reveal and encourage exploration of awe and wonder that would otherwise remain hidden or mysterious to the student.' (Hawley, 2020, p.29)

PHYSICAL GEOGRAPHY CALMS ME

As I learn or teach about the natural world, the way different elements connect gives me a deep sense of satisfaction. With a background in geomorphology, I used to feel less comforted when studying ecosystems, but having taken time to learn more, I am now equally entranced by the beauty of nutrient cycling, adaptation of species and more. Studying physical geography also involves the exploration of physical–human interactions in different contexts, which can be upsetting as we learn how people have altered the natural environment but can also be hopeful.

Understanding these interactions may enhance our wellbeing as we realise that changes such as in the climate and due to hazard events can be explained, potentially making them less overwhelming (see chapter 38).

The systems approach often dominates physical geography teaching in schools, for example the DofE (2014) stated that for the water and carbon cycles topic 'study must ... take place within a systems framework emphasising the integrated nature of land, earth and atmosphere', putting this approach at the heart of A-level geography specifications in England from 2015. Showing physical systems using this stable, reassuring approach is beautiful in its own way, but nature is dynamic, so the systems in the textbooks don't always exist in the same way in reality. The systems are generalised and abstracted from reality, often creating an 'average' from a range of data sets. They depend on their context in a way the systems in textbooks don't. Hawley (2018) discusses how 'academic understanding has shifted from this empirical, rigid world to acknowledge simplicity doesn't exist and the real world is more "naughty", complex, approximate and our interpretations of it are socially constructed.'

When we embrace complexity, we can learn more. When carrying out a river study we are likely to find that elements don't fit the Bradshaw model, giving us the opportunity to consider other variables and develop a better understanding of a place. We are thinking critically, recognising that things are not always as they seem.

Acknowledging and learning about the 'naughty' nature of physical geography only increases my satisfaction. Engaging with the complexity of physical geography makes me happy as I develop my understanding, but also realise that I am only dipping my toes into the wonder of the world.

PHYSICAL GEOGRAPHY STIMULATES MY THINKING

While learning and teaching about physical geography connects me to the world and calms me, it also stimulates my thinking, both in terms of content and pedagogy. Having left university 30 years ago, it is important to keep abreast of developments. I have found my membership of the Geographical Association invaluable in doing this and particularly appreciate the work of the Physical Geography Special Interest Group (see https://teachingphysicalgeog.wordpress.com/). Social media has enabled me to follow specialists in a wide range of fields and also access resources produced by teachers who have expertly delved into particular topics, such as Alistair Hamill's work on plate tectonics (for example, see 'The Mantle Isn't Molten', https://www.youtube.com/watch?v=Ag7ryVQRmrU).

The dynamic nature of physical geography not only necessitates keeping up to date with developments in our understanding of phenomena and processes, but it is also important to keep reflecting on our pedagogy. When we teach physical geography through models and stories we simplify reality for our students, but we need to be aware that 'When a construct is oversimplified, outdated, offered as the sole explanatory model, or as through no understanding about the physical world existed previously, and without critical evaluation, it can obfuscate rather than clarify' (Hawley, 2018). We can encourage our students to 'think like a physical geographer', comparing evidence to a model, considering the context (place, feature, data set) and potentially exploring possible alternative explanations.

Planning how to teach physical geography concepts and processes requires careful thought about our students' prior learning and context as well as what we want them to learn. My school is located on the Somerset coast, on the edge of the Levels. This means that my students have a good day-to-day understanding of the coast, but little experience of natural rivers (we are on reclaimed land, criss-crossed with drainage ditches and carefully managed) and even less awareness of glacial landscapes. Our baseline assessments (see https://geogmum.wordpress.com/2019/08/31/first-lesson-with-yr-7/) suggest that our students encounter little physical geography at KS1 and KS2, so we plan carefully to ensure we build their knowledge step by step (see example in Figure 27.1), while not just repeating topics such as 'rivers and coasts' in different key stages.

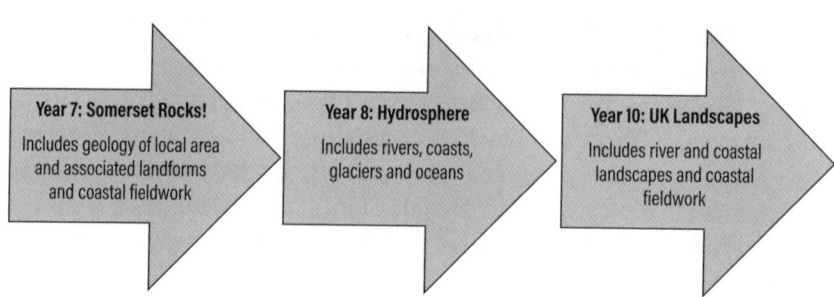

Figure 27.1 One example of developing students' understanding during KS3 and KS4

The Geographical Association (2023, p.7) has suggested a curriculum framework which includes two key concepts which are particularly relevant to teaching physical geography:

- **Earth Systems** *is key to a network of ideas about physical processes and cycles, dynamic biological, chemical and physical changes, exemplified in a range of landforms, landscapes and environments.*
- **Environment** *lies at the top of a network of ideas about interactions between physical and human geography, ecosystems, environmental change and impact, resources and sustainability, again followed up and revealed in a variety of contexts at micro to macro scales.*

These concepts are then explored in terms of five organising concepts – time, scale, diversity, interconnection and interpretation. Figure 27.2 shows examples of how this could be used in school geography. This approach may be useful for curriculum creators wishing to review physical geography teaching across their curriculum. However, this figure is in itself a model and we need to adjust it to fit our contexts. We also need to recognise that it is a simplification, for example 'recognising the Anthropocene' is included under 'scale' but is also an 'interconnection' concept and a 'time' concept.

As shown in Figure 27.1, we need to think carefully about fundamental concepts in physical geography, putting a firm foundation of knowledge in place before building on it. In Figure 27.1, students develop a basic understanding of geology and processes of weathering, erosion, transportation and deposition in their local area in Year 7, then build upon this in Year 8 to apply these concepts to rivers, coastlines and glaciers, putting them in a good position to study the UK landscapes GCSE topic in Year 10. While we are careful not to repeat topics, previous concepts and processes are reviewed before being used in new contexts, enabling students to consolidate their learning. Students can be supported in thinking like physical geographers by observing/considering:

- Materials (solid and fluid)
- patterns (form)
- flows
- distribution
- rates of operation (time)
- mechanisms (processes)
- dynamic changes.

Using these disciplinary concepts to underpin physical geography teaching across topics allows students to make connections between the substantive knowledge of different topics and allows them to compare the physical geography of different locations. Planning to use these concepts in more sophisticated ways allows for progression. In the words of Hawley, they provide the 'lens' through which physical geographers bring 'order' to the world.

Table 27.1 also shows students moving from the familiarity of their local area to areas beyond their own experience. Having personal experience of a location means students have a richer understanding of its context; they can connect their new physical geography knowledge to the place knowledge they already have.

As there is a lot of specialist vocabulary associated with physical geography, we provide students with a knowledge organiser for each topic (two sides of A4 paper) and make sure key terms are explicitly defined in lessons. If students don't use these terms when answering questions, we challenge them to say it again, but better. A useful podcast to listen to which considers why specialist vocabulary is needed and how it makes students' answers better can be found here: https://www.katestockings.com/geographycurriculum/geogpod-how-to-speak-like-a-geographer.

Table 27.2 A conceptual grid for physical geography

	Time continuity/change	Scale micro/macro	Diversity similarity/difference	Interconnection independent/contingent	Interpretation perspectives/representations
Earth Systems (Stable Earth/dynamic Earth)	Timescales for operation of Earth processes (millions of years to seconds) Stable Earth/dynamic Earth Systems equilibrium, system change and tipping points	Earth systems and processes at micro, or global scale Recognising the Anthropocene	Diversity in the Earth and natural world, adaptations and modifications Similarity in Earth processes over different timescales	Connections and interdependence in Earth processes and consequent developments Feedback processes affecting natural landform and phenomena	Changing perspectives on Earth processes and systems Impact of technologies
Environment (Environment at risk, environment as a responsibility)	Pace of environmental change • accelerating/declining • recognising the Anthropocene Implications of recognising the Anthropocene for human life and actions	Lived environments on a continuum from natural places to human spaces Changing environments and human impacts at all scales, from individual to global	Human environments and natural ecosystems modified by human actions Shared and varied responses and commitments to environmental responsibilities	Human dependence on a changing planet characterised by significant environmental change Understanding and sense of responsibility fostered by scientific and cultural activities	Technical management solution vs natural change Conceptions of nature as separate from or on a continuum with human life Approaches to environmental issues and differing interpretations of sustainable development

Source: Adapted from Geographical Association, 2023, reproduced with permission of the Licensor through PLSclear

Physical geography concepts and processes can also be communicated to students through diagrams, although we must remain wary of the dangers of simplifying reality. Luke Tayler (2024) discusses how a geography teacher looking at a photograph of a slumping cliff will know what is happening and also be able to extrapolate to consider a wider picture, making links between coastal erosion, geology, coastal processes, climate change, coastal management and more. We developed this framework for thinking through our own physical geography education and subsequent experiences of reading, watching others teach and more. Tayler acknowledges that we want our students to be able to do this, but for this to happen our students need to have 'building blocks of knowledge'. Teachers can provide these building blocks through explicit instruction, presenting information to students carefully, possibly aided by diagrams. When a teacher draws a diagram, they are providing information in a logical structure, with opportunities to pause to check understanding. When teachers draw diagrams live, giving a commentary about what they are doing and why, they are making their expert understanding of physical geography transparent to their students, helping them understand how the diagrams show interactions and interconnections between different elements in a system.

Hawley (2024) suggests that we need to go beyond Tayler's recommendations to teach students drawing techniques, emphasising how 'line-drawing in a diagram (or field sketch) produces analytical geography – delineating natural, recognisable boundaries, areas and flows that mean something geographically (an idea, once taught, that is easy for students to grasp)', encouraging students to go beyond mastering the diagrams we draw with them to being able to create their own diagrams using the techniques they have learned.

Drawing a diagram rather than presenting it as a completed piece also means students will focus on a particular part at a particular time. We can adjust what we emphasise in the diagram to make learning more effective. When physical geography is beyond students' experience, such as in my classes when studying glaciated landscapes, diagrams can help students to visualise processes, making the abstract more concrete. Completed diagrams on screens may look more professional, but perhaps it is time for more of us to get our board pens or visualisers out and create our own diagrams in real time!

There is a growing range of resources out there for teachers seeking to bring physical geography to life for students with limited experiences. I highly recommend Earth Learning Idea resources (https://www.earthlearningidea.com/), having had great success with activities such as party popper volcanoes, the toilet roll of time and the water cycle in a bag. The activities on this website have been designed by earth science teachers and include careful explanations of how to use them in the classroom, taking them far beyond the gimmicky approaches teachers may come across elsewhere. I have also found the videos from Time for Geography (https://timeforgeography.co.uk/) very useful. Virtual fieldtrips are also great for bringing unfamiliar environments into the classroom – my Year 7 students love this resource for exploring glaciated areas as they are able to immerse themselves in a landscape which is unfamiliar, seeing the physical geography (https://vrglaciers.wp.worc.ac.uk/wordpress/). We can explore landscapes using GIS – in my school we use Digimaps to explore how our local coastline has changed over time in Year 7, then build on this over time. However, whatever resource used, we need to think critically about the information it is presenting and how, being ready to step in to deal with any oversimplification, etc.

I am grateful to the Decolonising Geography Collective (https://decolonisegeography.com/) for challenging me to think about physical geography through a decolonial lens and how my teaching can be decolonised. Radcliffe (2022) suggests that physical geographers need to 'engage with a more diverse set of knowledges to understand its objects of study'. She gives the example of geographers studying the wetlands of Chicago acknowledging colonisation and the technocentric nature of wetland modelling and listening to indigenous voices to widen their understanding. The emerging concept of ethno-geomorphology acknowledges that there are many different ways of 'knowing' landscapes and recognises that systems approaches to physical geography and indigenous knowledge of landscapes and processes often overlap.

A simple step we can take is using indigenous names for landscape features, rather than colonised names; when I teach about Lake Victoria as part of our GCSE study of Uganda, I can explain that it is known as 'Nnalubaale' in Luganda, which means 'place of fish'. Another area which presents immediate opportunities for discussing colonisation is conservation around the world, where people may be displaced or have their lives limited by 'fortress conservation' measures which see indigenous people as a danger to a protected ecosystem, rather than part of it (Mbaria and Ogada, 2016). This is a political issue of power, taking it beyond the remit of the natural systems of physical geography, but such issues become important when GCSE specifications in England require students to explore issues such as managing tropical rainforests sustainably as part of their physical geography content.

PHYSICAL GEOGRAPHY TAKES ME PLACES

When I was at school near Bristol, I took part in physical geography fieldwork on the Quantock Hills and in the Lake District. At University of Plymouth I carried out fieldwork in Snowdonia in Wales, County Clare in Ireland and in Mallorca (we were due to explore the limestone of Yugoslavia, but war broke out). I am now privileged to lead physical geography fieldwork to Watchet in Somerset for my Year 7s (see https://geogmum.wordpress.com/2015/06/20/wonderful-watchet/) and Lyme Regis for my Year 10s. We continue to run these field visits against a background of it getting harder and harder to take students out of the classroom, as Steve Brace (2024) acknowledged in his letter to *The Guardian*:

> 'Over the last 20 years, Ofsted reports have shown that school fieldwork has been declining. And a survey of geography teachers in 2023 indicated that since Covid, up to 40% of secondary schools may have cut their provision of fieldwork. This trend affects smaller schools and those serving disadvantaged pupils the hardest. A combination of costs, Covid catch-up and other administrative hurdles are limiting the work of many geography teachers who want to offer their pupils high-quality fieldwork.'

It is vital to take our students on fieldwork, particularly that involving physical geography, where they are able to make connections with the natural world. Not only do they learn about geographical concepts and processes, but they also learn about how to interact with different landscapes. I love visiting different places; fieldwork helps our students develop a framework for understanding places and can spark interest in lifetime learning about the outdoors.

PHYSICAL GEOGRAPHY EMPOWERS ME

Hawley (2020) acknowledges that 'awe and wonder' are important in geography, grabbing students' attention and providing a 'wow factor', but stresses that 'Powerful knowledge has the capacity to move students beyond the emotional and obvious and achieve enduring understanding by providing new ways of thinking about the physical world'. He likens the process of helping students to access 'hidden' physical geography to an art expert helping an observer understand a painting, revealing its deeper significance. As teachers, we are in a powerful position to reveal the depths of a landform or landscape to our students.

This powerful knowledge will, in turn, enable our students to apply what they have learned in different contexts, empowering them to understand issues in the world around them. Corbridge and Hawley (2020) consider it is important to have a understanding of specific processes at work in a coastal area before assessing how useful different coastal defences may be, using the Fylde coastline as an example.

CONCLUSION

There are so many reasons to make sure that we continue to learn and teach about physical geography – for our own benefit as well as that of our students. A shout out on 'X' asking how learning and teaching physical geography makes people feel led to these responses:
- 'It's like putting all the pieces of a jigsaw together'
- 'It's being able to "read" the landscape and understanding that unlocks a love for physical landscapes and understanding why settlements are where they are.'
- '"For what we do not know we fear. For who we fear, we fight." With geography knowledge, we have less fear of the world and less desire to fight.'
- 'It's the logicality of it that I love – A leads to B leads to C and so on.'

- 'Balanced? In order to have understanding of the human, you need to know and understand the physical. Often at the end of a school year I love the connections that become so clear to the learners between all the individual topics ... and therefore I suppose "clarity".'
- 'Satisfied. There's a clear joy in making sense of the world, the landscapes, and there's a logic and clarity to a lot of it that allows access, process and exploration of data and complexity...!'
- 'Like a QUEEN. And by this I mean as a writer and speaker about nature and landscape, I feel privileged to know the intimate details of the complex science behind the aesthetics. It's my privilege to share that knowledge.'
- 'I love knowing more detail about the landscape around me and being able to "read" it. And that also feeds my soul – I always feel recharged after time in landscapes (especially mountains and the sea).'

(Thanks to @DoctorPreece, @jurassicg1rl, @GeographyLizzie, @bobdigby, @davidErogers, @Mgs_geog, @ClickAndLearnUS and @BeKind_KLB for their responses to my 'X' shout out!)

Five reflection questions

1. How does physical geography make you feel?
2. How do you engage with the 'naughty' and complex nature of physical geography?
3. How do you keep up to date with developments in physical geography knowledge and pedagogy?
4. How do you plan your physical geography curriculum so that it meets the needs of your students?
5. How do you bring places beyond your students' experiences to life through use of different resources, GIS and fieldwork?

> **DIGGING DEEPER – THREE RESOURCES TO DELVE FURTHER**
> - Hawley, D. (2020) Beyond awe and wonder: using powerful knowledge to release 'hidden' physical geography. *Teaching Geography*, Spring 2020. Sheffield: The Geographical Association.
> - Tayler, L. (2024) *Visualising Physical Geography*. Routledge.
> - Geography Education Online website at: https://www.geographyeducationonline.org/a-level/physical-geography (Accessed: 28/05/2025)

REFERENCES

Brace, S. (2024) Geography students are losing access to nature as fieldwork falls. *The Guardian*, 8 April 2024. https://www.theguardian.com/environment/2024/apr/08/geography-students-are-losing-access-to-nature-as-fieldwork-falls

Corbridge, A. & Hawley, D. (2020) 'Holding the line': a case study of the physical geography and coastal management of the Fylde coast. *Teaching Geography*, Summer 2020. Sheffield: Geographical Association.

DofE (2014) Guidance: GCE AS and A level geography. https://webarchive.nationalarchives.gov.uk/ukgwa/20210923094621/https://www.gov.uk/government/publications/gce-as-and-a-level-geography (Accessed: 28/05/2025)

Geographical Association. (2023) A framework for the school geography curriculum. https://geography.org.uk/wp-content/uploads/2023/07/GA-Curriculum-Framework-2022-WEB-final.pdf (Accessed: 28/05/2025)

Hawley, D. (2018) 'Physical Geography' in *Debates in Geography Education*, 2nd Edition. London: Routledge.

Hawley, D. (2020) Beyond awe and wonder: using powerful knowledge to release 'hidden' physical geography. *Teaching Geography*, Spring 2020. Sheffield: Geographical Association.

Hawley, D. (2024) Book review of 'Visualising Physical Geography'. *Teaching Geography*, Spring 2024. Sheffield: Geographical Association.

Mbaria, J. & Ogada, M. (2016) *The Big Conservation Lie*. Lens and Pens Publishing.

Radcliffe, S. (2022) *Decolonising Geography: An Introduction*. Cambridge: Polity Press.

Stocking, K. (2020) GeogPod: How to speak like a geographer. https://www.katestockings.com/geographycurriculum/geogpod-how-to-speak-like-a-geographer (Accessed: 28/05/2025)

Tayler, L. (2024) *Visualising Physical Geography*. London: Routledge.

28. THE DIAGRAM, THE VISUALISER AND THE VIGNETTE: HOW TO TEACH ABOUT THE MULTIPLIER EFFECT

ABDURRAHMAN PÉREZ-MCMILLAN
@MR_PEREZ5

'The human mind seems exquisitely tuned to understand and remember stories – so much so that psychologists sometimes refer to stories as "psychologically privileged", meaning that they are treated differently in memory than other types of material.' (Daniel T. Willingham, 2021)

WHEN, NOT IF

Whether you explicitly aim to do so or not, it is highly likely your current KS3–5 curriculum deals with the multiplier effect at least once, if not several times. This chapter, therefore, is not about encouraging you *to* teach it, but rather written to make you think about *how* you teach it. While not mentioned by name in the national curriculum, nor in the specifications of any of the major awarding bodies for GCSE and A-level, the multiplier effect (sometimes referred to as cumulative causation) is a concept which is central when teaching about urban, development or economic geography.

Put simply, the multiplier effect refers to the 'snowballing' of economic activity; the introduction of a new industry or the expansion of an existing economic activity in an area encouraging growth in other industrial sectors, and/or further investment in said area (followed by improvements in quality of life and/or infrastructure).

While you might currently be considering how this chapter will be helpful come the spring term of Year 8 (when your students are first taught the multiplier effect), we know that the multiplier effect will return at some point or another in our curriculum. It may not feature in your 'big ideas' or 'threshold concepts' as a department, but the Report of the ALCAB Panel on Geography (July 2014) found 'systems' to be one of 12 key concepts in A-level geography. What is more, covering the multiplier effect also 'ticks off' many of the remaining 11 (causality, difference, inequality, representation, interdependence, mitigation and adaptation, sustainability, risk, resilience and thresholds). While this chapter explores teaching of the multiplier effect as a concept, it is also an exploration of the use of diagrams, visualisers and vignettes in the classroom.

WHY A DIAGRAM?

When discussing the need for using diagrams in our teaching of geography, Luke Tayler (2023) notes: 'It is our ability to see and use the links between (individual processes, concepts, and systems) that makes us relative experts in this field, and it is this type of thinking that we want for our students'. He goes on to discuss 'how the teacher needs to be in complete control of the "information exchange"' involved in delivering new information to students. He argues that diagrams help us (teachers) do this in two ways:

1. Diagrams regulate the flow of information: Students can provide teachers with quick and frequent feedback if they actively participate in the explanation process. He adds: 'This allows the teacher to pause a drawing or to revisit any parts of the explanation.'

2. Direct students' attention: 'Drawing a diagram in real time (as opposed to just showing the completed version) allows the teacher to ensure the students are looking at the relevant part' (Tayler 2023).

Writing in 2021, Caviglioli and Goodwin had been equally positive on the value of diagrams:

'Word-diagrams — a more accurate and useful term than graphic organisers — are like projections of your schema. Their non-linear spatial arrangements of words, with explicit connections, reflect back to your current understanding. That, in turn, triggers in you an evaluation of the accuracy of the links before your eyes. In this cognitive loop, you react by automatically considering improvements. Only a few among us are able to achieve this level of thinking purely in the disembodied mind.'

Clearly, diagrams have a place when teaching about the multiplier effect, but the work done before, during and after the diagram is drawn is crucial.

Before: It goes without saying that subject knowledge is crucial; you will be imparting new knowledge and, what's more, the students will have a thousand and one questions, and you should be prepared to answer them. Some might be simple ('What is disposable income?'), but others might prove more challenging ('Why does an area's "image" improve after rising demand, and why does this mean other companies are attracted?'). Therefore, while I believe the diagram should be completed live along with students, you should probably practise it beforehand (as I am sure you have practised your global atmospheric circulation model diagram!), or at least have a rough idea of where you are going with every box to be drawn.

During: When learning is understood as 'the long-term retention of knowledge and the ability to transfer it to new contexts' (Didau 2019), then the purpose of this diagram is for students to:

a) grasp an understanding of the 'snowballing' of economic activity involved in the multiplier effect
b) understand how the multiplier could function in other, unfamiliar contexts.

THE DIAGRAM

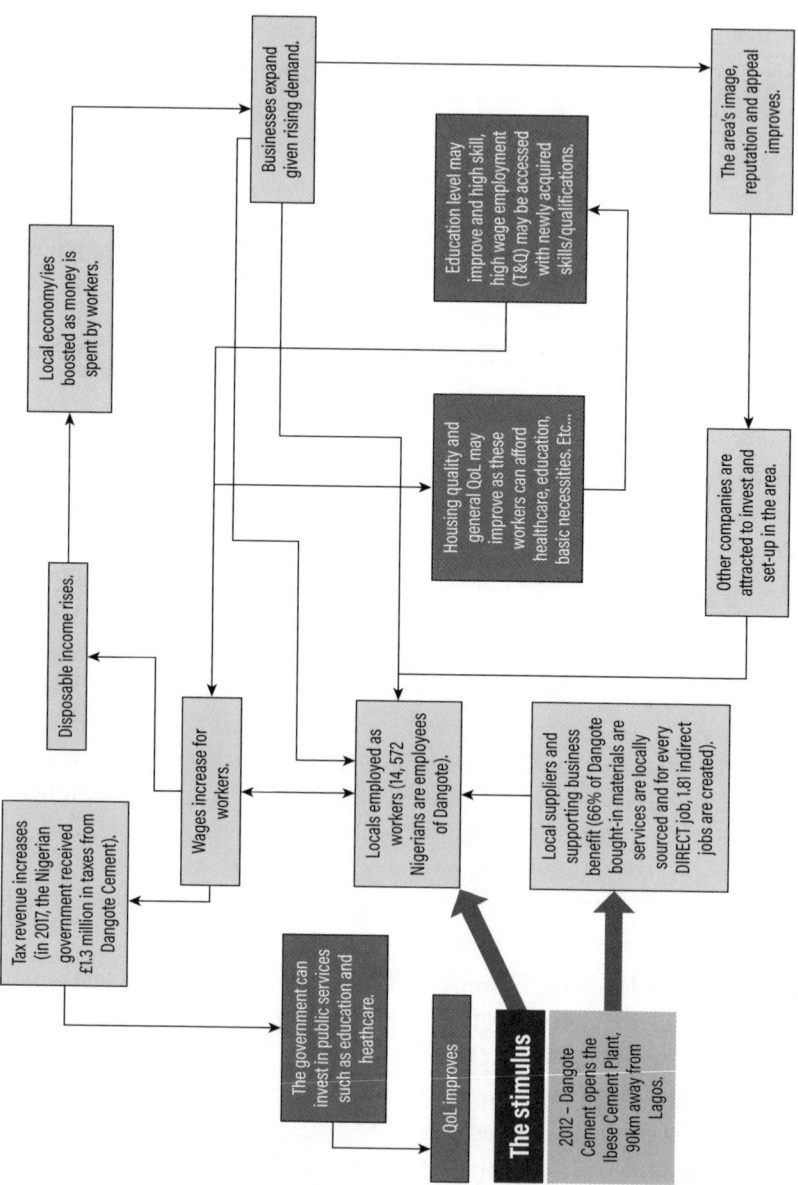

Figure 28.1 A diagram of the multiplier effect

Therefore, questioning during the process of drawing the diagram together as a class is key. Some of these questions no doubt will build on students' own suggestions, but some will have to be planned by you (i.e. 'Why might the quality of life of the workers in the cement plant improve as a result of being employees?')

After: Given that we want to understand how the multiplier could function in other, unfamiliar contexts, it is crucial that this lesson, this one episode of teaching, is not the only time students are exposed to it. Furthermore, it is wise to follow up the diagram and vignetter with independent written practice, where students have the chance to put into words exactly what they think the inner workings of the multiplier effect are. These would then be followed up with effective feedback to ensure any misconceptions are caught early.

THE VIGNETTE

We know from the writing of Daniel Willingham, which was later built upon by Tharby (2018) and then Chiles (2021), that stories are 'psychologically privileged' (Willingham 2021). When teaching about the multiplier effect, I think it is wise to use a carefully constructed vignette as a vehicle with which to explain the concept, while rooting it firmly in non-abstract terms. I find the term 'vignette' more closely describes how I use this with students: a brief evocative description, account or episode.

We remember stories and we are more likely to remember relevant and interesting ones – ones that carry a message. I chose these vignettes based on one or more of these criteria:

1. The location/setting's familiarity to me.
2. The location/setting's familiarity to the students.
3. The general awe and wonder factor behind the story.

Number 2 usually trumps them all, but I do consider Number 1 a lot if the concept I am trying to get across is tricky and Number 3 if someone who is a better story-teller than I am (like Dipo Faloyin) has already written a relevant story!

I will often tell students how, as a student at a nearby school, I noticed that there had been a Starbucks and Paul's near Holland Park station since 2008. Its success was clear for all to see and in 2019 a Pret a Manger popped up, followed by a Joe and the Juice in 2020. Most recently, these have all been joined by the up-and-comer Buns from Home's flagship bakery.

I realise there are many ways in which this vignette is far different from the situation in Ibese, some 4,000 miles away, I do. However, the key messages remain:
- The multiplier effect is the process by which economic activity in an area encourages growth in other sectors, and/or further investment in an area.
- It doesn't just happen in far-flung places.
- It doesn't happen overnight.
- Events can have many and concurrent 'domino effects'.

As with any good analogy, the key components need to be present, so I will use the vignette to 'tick off' many parts of the diagram, such as:
- locals being employed as baristas
- the area's image, reputation and appeal improving, as evidenced above
- other companies are attracted to invest and set-up in the area.

In that way, the vignette has been used as a vehicle with which to further explain the concept of the multiplier effect, not as a gimmick designed simply to superficially engage or interest students, or one which needlessly distracts from the aim of the lesson.

WHERE ELSE?

Using the diagram and vignette approach is clearly a strong pedagogical approach in other human geography topics. Given that KS3 curricula are so varied, the following examples are based on the specification of two major exam boards at GCSE and A-level, OCR and AQA. Where these would not be appropriate, a well-drawn word-diagram could get 'the message' across to students quite efficiently and in an engaging manner too:

AQA specification point	Vignette idea
An overview of the strategies used to reduce the development gap: microfinance loans	With due care to avoid the 'the danger of a single story', this story could be used to create a narrative around how microfinance loans can help reduce the development gap: https://opportunity.org.au/in-action/stories/lydia
A case study of a major city in an LIC or NEE	If you cover Lagos, a popular case study, a vignette based on Dipo Faloyin's wonderful chapter on the city would make a hugely informative and engaging lesson segment

OCR specification point	Vignette idea
Investigate the contemporary challenges that affect life in the AC city, such as housing availability, transport provision, access to services and inequality	Any of the harrowing stories covered by Kwajo Tweneboa, the social housing activist and content creator, could be used to paint a stark picture of the housing inequality which exists across many UK cities
An understanding of the causes, effects, spatial distribution and responses to an ageing population	Any of the accounts contained within this would provide a fascinating 'insider' perspective into the UK's ageing population: https://www.ons.gov.uk/peoplepopulationandcommunity/birthsdeathsandmarriages/ageing/articles/voicesofourageingpopulation/livinglongerlives

Five reflection questions

1. Where in your curriculum are students first taught about the multiplier effect?
2. Which activities are used and how effective have they proven to be?
3. How could visual aids, such as a hand-drawn diagram under a visualiser, help students make sense of the multiplier effect?
4. How could vignettes – even if they are hypothetical scenarios – be used to reshape or reframe the way in which the multiplier effect is taught in your school?
5. How could diagrams and/or vignettes be used in teaching other human geography topics?

> **DIGGING DEEPER - THREE RESOURCES TO DELVE FURTHER**
> - Tharby. A. (2018) *How to Explain Absolutely Anything to Absolutely Anyone: The Art and Science of Teacher Explanation:* Carmarthen, Crown House Publishing.
> - Chiles. M. (20201) The Sweet Spot: Explaining and modelling with precision. London: John Catt Educational Ltd.
> - Caviglioli. O & Goodwin. D. (2021) Organise Ideas: Thinking by Hand, Extending the Mind. London: John Catt Educational Ltd.

REFERENCES

Didau, D. (n.d) Blog. What's wrong with Ofsted's definition of learning? https://learningspy.co.uk/featured/whats-wrong-with-ofsteds-definition-of-learning/ (Accessed: 13/09/2024)

Report of the ALCAB Panel on Geography (July 2014). https://alevelcontent.wordpress.com/wp-content/uploads/2014/07/alcab-report-of-panel-on-geography-july-2014.pdf (Accessed: 13/09/2024)

Caviglioli, O. & Goodwin, D. (2021) *Organise Ideas: Thinking by Hand, Extending the Mind.* London: John Catt Educational Ltd.

Chiles, M. (2021) *The Sweet Spot: Explaining and Modelling With Precision.* London: John Catt Educational Ltd.

Tharby, A. (2018) *How to Explain Absolutely Anything to Absolutely Anyone: The Art and Science of Teacher Explanation.* Carmarthen: Crown House Publishing.

Tayler, L. (2024) *Visualising Physical Geography: The How and Why of Using Diagrams to Teach Geography 11–16.* London: Routledge.

Willingham, D.T. (2021) *Why Don't Students Like School?: A Cognitive Scientist Answers Questions About How the Mind Works and What It Means for the Classroom*, 2nd Edition, Hoboken, NJ: Jossey-Bass.

29. EXPLORING GENDER IN THE SECONDARY GEOGRAPHY CLASSROOM
BRILEY HABIB
@MAP_ADDICT

'Everywhere, women are worse off than men, simply because they are women. Migrant and refugee women, those with disabilities, and women members of minorities of all kinds face even greater barriers. This discrimination harms us all. Just as slavery and colonialism were a stain on previous centuries, women's inequality should shame us all in the 21st.' (United Nations, António Guterres, speech 27 February 2020)

WHY DO WE NEED TO TEACH ABOUT GENDER IN THE GEOGRAPHY CLASSROOM?

According to the latest figures published by the Geographical Association (2023a), geography remains a relatively gender balanced subject at A-level, where 54.1% of entries are male and 45.9% are female. When analysing A-level results, it was found that girls outperform boys by an average of 3 points (Brace, 2020) which is equal to a third of a grade. GCSE figures also show a similar trend in England (Geographical Association, 2023b). As an international teacher, I also recognise a similar pattern in my classroom at both IGCSE and the International Baccalaureate qualifications. Entry patterns have remained consistent throughout my teaching career and, as a reflective practitioner, I began to question why there is still a higher number of boys opting to choose geography as a subject, post 14.

Feminist geographies are not a new concept. Authors such as Massey (1994) and McDowell and Sharp (1997) have all produced earlier work which centred around feminist geography and yet we see very little feminist geography being identified or taught in English curriculums, most likely due to its absence from the geography national curriculum of England. When designing schemes of work, teachers may investigate briefly how gender equality can increase development (Gardner et al, 2018). Gender equality has a dedicated United Nations Sustainable Development Goal (UNSDG) as well as being integrated throughout the remaining UNSDGs due to intersectionality. According to the latest snapshot of Goal 5 (Gender Equality), progress shows that the world is not on track to achieve gender equality by 2030 (United Nations, 2022).

A polling company in the United Kingdom conducted a survey for BBC radio and BBC Bitesize in September 2023. One thousand girls were surveyed, with 27% of respondents saying that they had experienced sexual harassment and 44% that they felt unsafe when walking on the street (BBC, 2023). This 'geography of fear' (Valentine, 1989 and Khurana, 2020) can be used as a starting point for encompassing gender equality into the geographical ideas of space and place.

This chapter will explore how I have integrated various issues that girls and women may face into the geography curriculum, as well as how I have used the idea of girls feeling unsafe in public spaces to create fieldwork opportunities. (See also chapter 8.)

DATA GAP

Caroline Criado Perez (2020, pp.11–13) has argued that a gender gap exists in news, science, city planning and economics which creates an absent presence known as 'gender data gap'. She suggests that the data gap is not malicious but is a product of 'not thinking' and that when we fail to include the perspective of women this can be an unintentional driver of male bias.

If we consider Perez's view on unintended male bias, where can we overcome this in our curriculums? One area of the geography national curriculum in England (DfE, 2013) that can include more equitable knowledge using a female perspective is the topic of urbanisation. A single story has evolved about favelas in Brazil as being substandard slums, crime-ridden areas of poverty and lawlessness, where cowed residents live in fear of prowling gangs (Perez, 2020, p.41). After I visited Rocinha Favela in 2022, I was keen to show a different side to the single story that is presented in textbooks at key stage 3. The informal settlement is one that is a thriving and bustling area of commerce in Rio de Janeiro where women work in shops, have art galleries and their children attend local schools. The pacification of favelas is also taught by some teachers as part of some GCSE geography courses in England in order for students to answer questions about urban planning. One example of a question that has appeared in a GCSE examination is '*Evaluate the effectiveness of an urban planning strategy in helping to improve the quality of life for the urban poor. Use an example of a city in a LIC or NEE.*' (AQA, 2017) With this question in mind, it is important to refer back to Perez where she discusses the project of Minha Casa, Minha Vida (My house, My Life), a project which was setup by the Brazilian government to move those from inadequate housing (Perez, 2020, p.41). Perez uses the voices of the women who have been affected by the housing project. Using an extract of this book, a 'guided reading' activity was created to enable students to understand that being resettled from a bustling favela might not always be better for women and children. (See also chapter 14.)

USING POPULATION STATISTICS ALONGSIDE A WOMAN'S PERSPECTIVE

Teaching population pyramids and choropleth maps at key stage 3 is a popular activity for teachers who follow the national curriculum of England because of the requirements to teach models and theories alongside population. Using population pyramids or maps can only tell us one side of the story: the data of ageing or youthful populations. It cannot tell the stories behind the data. According to the United States 2020 census, one sixth of people are aged over 65. This is a ticking time bomb for greying women.

A 2019 study by the University of Pennsylvania did not differentiate between females and males homelessness; supporting Perez's idea of unconscious bias in the data gap (University of Pennsylvania, 2019). Bearing this in mind, I felt it was important for students to 'hear' from a woman who has experienced homelessness as an ageing person.

Using an article from Vox (2016), which was an autobiographical piece of writing, students read and discussed the reasons why women over the age of 65 are growing in the homeless population. To support the high number of students who are English-as-a-second-language speakers (over 90% in my school), the article is translated into different languages. I share the article with students and they use the translate button on their devices. The reason that I enable students to read the article in their own language is because I want the students to be able to use English in the classroom in geographical discussions. By allowing translation of longer pieces of prose, students are able to work at the same speed as their peers. Having undertaken training from Bard College (Institute for Writing and Thinking) in the USA, I now use a technique involving students using the following steps after reading the article:

1. Choose a paragraph that resonates (about women's homelessness).
2. Choose a sentence that would make people more interested in this issue (women's homelessness).
3. Choose a word that you do not understand about the topic.

By using techniques such as these in class discussions, it gives all students (irrespective of English language ability) the opportunity to have a starting point from where they can share their opinion.

WOMEN AND DISASTERS

Disasters is another area where women's voices can go unheard in the geography classroom. Too often we can brush over the effects of natural disasters by categorising the effects of the disaster into short or long term with little regard being given to how women are specifically affected.

The first gendered data of natural disasters did not come about until 2007. Neumeyer (2007) highlighted that women are more likely to die in natural disasters than men. Professor Maureen Fordham, Director of the IRDR Centre for Gender and Disaster at University College London, has carried out extensive research into gender disparities caused by disasters. Fordham found that in Bangladesh, cyclone shelters were built by men for men (Perez, 2020, p.301) and that women were often afraid to enter this space because of the fear of sexual violence and harassment.

One such example that can be explored in the classroom is how maternal mortality rates increase when a disaster strikes. Examples from the Philippines typhoon Yolanda (Haiyan) in 2013 and the Ebola pandemic (2014) were investigated to answer the enquiry question: *Why is it not the disaster that kills women?* As part of this enquiry, students are given a worksheet of evidence (Figure 29.1) and asked to write their own factsheet (in the style of a geo factsheet) which answers the enquiry question.

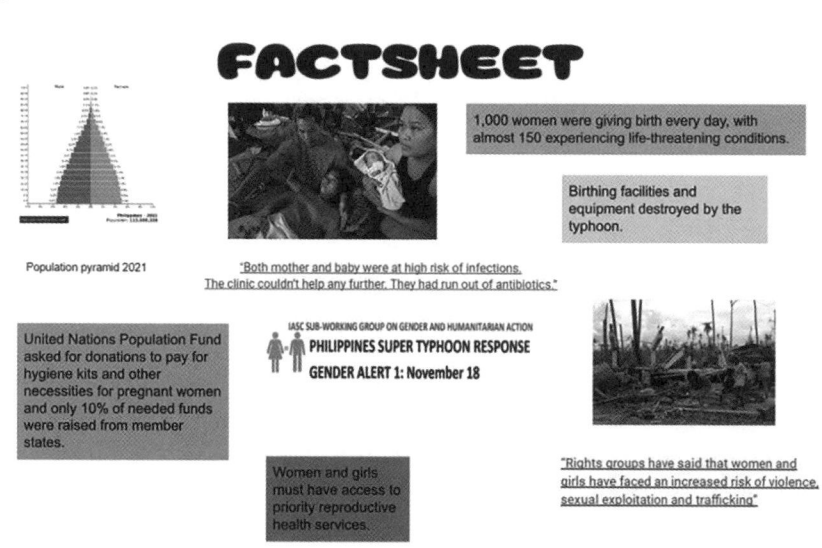

Figure 29.1 Geo factsheet which contains hyperlinks for articles

HIDDEN GEOGRAPHERS

We have all most likely taught about Esther Boserup's theory of population policies, agriculture and technology at some point. Rachel Carson, the author of the famous *Silent Spring*, is another female geographer who may make it into the geography curriculum when examining the origins of climate change in the media. Marie Tharp probably not so much. A pioneering geologist, Tharp was the first person to map the Atlantic Ocean floor, which enabled the confirmation of the Mid-Atlantic Ridge. This supported Wegner's theory of continental drift and led to Wegner's theory being included in textbooks for the last 40 years. Tharp, a hidden figure of the geography classroom, can be easily added to a lesson of the theory of plate tectonics by using videos on YouTube from The Royal Institution, such as: 'Marie Tharp: Uncovering the Secrets of the Ocean Floor – with Helen Czerski' (YouTube, 2016). This video would also align with the national curriculum requirements of teaching students about maps. Greta Thunberg, like Boserup, is not a hidden figure in geography, but for children of colour there is a lack of representation of climate activists of colour in our textbooks. The stories of many women activists of colour can be found using the book *Questions for Rebel Girls* (Sher et al, 2021). My students research women activists of colour such as Isatou Ceesaymm, Hu Weiwei and Autumn Peltier, and produce either a presentation or other form of media about the activists' work to educate the school community about different voices in environmental activism.

PERCEPTION OF SAFETY IN FIELDWORK FOR WOMEN

After reading Kit Marie Rackley's 'Geogramblings' post in conjunction with Diverse Educators about their experience as a trans child carrying out fieldwork, I began to think about the children in my school who identify as trans. Overnight fieldwork or trips can be an absolute joy for students to undertake. Students can make lasting memories of not only the geographical landscape but also the social connections that young people will make with their peers. For trans students like Kit, or students questioning their identity, their fieldwork experience, where after the lights go out they cry themselves to sleep because they don't feel safe, is not one that any child should experience.

It has been suggested that perceived safety relates to people's perception of their safety because of their fear and anxiety, whereas actual safety relates to being a victim of crime (Allessie, 2022, p.7). The University of Sheffield hosted an online workshop, 'Dealing with violence: women doing fieldwork' (University of Sheffield, 2022). Through this conference it was ascertained that female geographers experienced gender-based violence in the field which encompassed physical and verbal street harassment, threats and discrimination and mansplaining in male-dominated research teams. With this in mind, it is imperative that any female student feels safe when collecting data in the field, and not just for perceived levels of safety. Hamilton (2018, p.40) outlines the following suggestions as to why it is important for participants to feel safe in the field:

- Subconscious or conscious actions by females to mitigate safety risks may result in bias in the fieldwork. This can include selecting and interviewing participants which results in possibly biased approaches to data collection.
- Avoiding controversial topics to avoid confrontation.

Before students approach fieldwork, they should be briefed on safety in the field. While this should be carried out in a classroom discussion, it is also an important part of risk assessment.

USING LOCAL AREAS TO ASSESS PERCEPTION OF SAFETY

Dr Anna Barker (2023) has led research for the University of Leeds and has found out that women in Britain are three times more likely to feel unsafe in a park during the day, with this figure rising to four out of five women feeling unsafe after dark. This means that women and girls are less likely to use parks than boys and men. Using this research as a starting point, I decided to create a fieldwork enquiry for Year 9 students which asks them to investigate: *To what extent is Emirgan Park safe for women and girls?*

Emirgan Park is one of Istanbul's most used parks. It is famous for its tulip display which takes place in April each year and people can travel internationally to see the tulip festival. The park is owned by the Istanbul Municipality.

This fieldwork followed a traditional format:
- Aims and introductions
- Methods and risk assessment
- Data presentation
- Analysis
- Conclusions and evaluations

As part of this enquiry it was important to give students a booklet with clear instructions on what is required of them. The fieldwork also included a mark scheme which was adapted from the International Baccalaureate Internal Assessment requirements and the Cambridge IGCSE coursework. As my students undertake fieldwork post 14 and are externally assessed I feel that it is necessary to give them as many opportunities of completing fieldwork as possible at key stage 3 but that will also not be repeated post 14-years old. This fieldwork was chosen as it was a cumulative enquiry at the end of the topic on gender in geography.

Students are given three mini-hypotheses to include in their introduction:

1. Emirgan Park is a safe space because the toilets are in an open area and are well lit.
2. Emirgan Park is a safe space because of the presence of security.
3. Emirgan Park has a 'safe perception' by women and girls due to how many women and girls visit the area.

It is expected that students will explain why they think these hypotheses will be true, as well as including appropriate maps of the park area that they will be investigating.

There are various methods that students will use in the park to be able to answer their hypotheses.

An example of student work is shown in Table 29.1:

Table 29.1 Student response

Method type	Description of how to carry out method	Justification as to why this is the most suitable method and how it relates to a hypothesis
Pedestrian count	Stand in position at the investigation site. Start the timer for 5 minutes. Count and tally how many pedestrians walk past your right.	It shows how popular Emirgan Park is without talking with others or moving around too much. You can see how many women walk past as well. It also requires a set time so slower groups will have time to catch up. It is also useful for understanding the park's popularity among women and girls. This is useful for the third hypothesis.
Toilet observation and environmental quality survey	On the map, identify if there is a toilet at your investigation site. Observe the environmental factors outside of the toilet, e.g. if there are no lights outside of the building you would mark this as −1 on the survey sheet.	You can observe the quality of the toilets at Emirgan Park. It also doesn't require a set time so groups will have the opportunity to work at their own pace. This is useful for the first hypothesis.
Security count	Stand in position at the investigation site. Start the timer for 5 minutes. Count and tally how many security people move past your right.	It is useful for figuring out the security level at Emirgan Park and how the security is distributed and without external communication. It also has a set time so slower groups will be able to catch up. This is useful for the second hypothesis.
Perception survey	Interview at least one person. The survey is in both Turkish and English. You only need to fill out one sheet. Ask all questions and record the answers on the sheet. Alternatively, give the interviewee the sheet to fill out themselves.	It helps improve communication by letting students interact with their environment. It is also useful for understanding the opinions of the people, in particular the safety-related opinions of women and girls and their perceptions of the park, which are major factors of the park's safety. This is useful for the third hypothesis.
Photographs	Take photographs at each site. These should show the environment and pedestrians.	It is useful for capturing a still image of how people, in particular, women and girls, interact with the park environment. This is useful for the third hypothesis.

As part of the method section, students also included a risk assessment. As I work in an international school, I also made sure that each group had a Turkish speaker and the surveys were translated into Turkish so that students felt safe and that everyone could participate. As part of the data presentation method, students also took photographs, making sure that they did not include people's faces.

FINAL REFLECTION

At the beginning of this chapter, I used the introductory quote from Secretary-General António Guterres which discussed that women are worse off because they are simply women. If we highlight the intersectionality of women and how they are treated differently because of their gender, then hopefully more students would be agents of change. There are many areas of the curriculum which can be adapted to include a woman's, and indeed a young girl's, perspective. We cannot talk about equality in the curriculum if we do not highlight that there is an issue for girls and women across all aspects of life.

Five reflection questions

1. Audit your curriculum. Are there areas where you have unintentionally created a scheme of work where women's voices and experiences are not used but could be?
2. Create a survey in school asking where girls feel unsafe in school or on the way to school. Students could create community presentations about the results.
3. Carry out your own fieldwork in your own local areas about the reasons why women and girls may feel unsafe.
4. Research how other girls in other schools in other countries have felt about their safety. Carry out a comparison to see if it is the same reasons why.
5. Contact your local MP to ask what is being done to ensure the safety of women and girls in your local area.

DIGGING DEEPER – THREE RESOURCES TO DELVE FURTHER

- Seager, Joni (2018) *The Women's Atlas*. Myriad Editions.
- Caroline Criado Perez (2019) *Invisible Women: Data Bias in a World Designed for Men*. New York, NY: Abrams Press.
- Greenflag award: https://www.greenflagaward.org/news/new-guidance-launched-to-create-safer-parks-for-women-and-girls/

REFERENCES

Allessie, B. (2022) Individual Safety Perceptions of Women in Public Space Towards Socio-Spatial Indicators for Improving the Environment. Unpublished MA thesis.

AQA. (2017) Specimen Question for AQA Geography Exam. https://filestore.aqa.org.uk/resources/geography/AQA-80352-EX.PDF (Accessed: 21/05/2024)

Barker, A. (2023) Making parks safe for women and girls. https://www.leeds.ac.uk/news-society-politics/news/article/5295/making-parks-safe-for-women-and-girls (Accessed: 28/05/2024)

BBC. (2023) Many Teens Feel Unsafe and Anxious'. https://www.bbc.co.uk/news/uk-66855386 (Accessed: 21/05/2025)

Brace, S. (2022) *Geography of Geography Report*. Royal Geographical Society. https://www.rgs.org/geography/key-information-about-geography/geographyofgeography/report/geography-of-geography-report-web.pdf/ (Accessed: 21/05/2024)

Caplan, Z. (2023) 2020 Census: United States Older Population Grew From 2010 to 2020 at Fastest Rate Since 1880 to 1890. United States Census Bureau, 25 May 2023. https://www.census.gov/library/stories/2023/05/2020-census-united-states-older-population-grew.html (Accessed: 24/10/2023)

Department for Education (DfE). (2013) Secondary National Curriculum - Geography. https://assets.publishing.service.gov.uk/media/5a7b8699ed915d13 1105fd16/SECONDARY_national_curriculum_-_Geography.pdf (Accessed: 21/05/2024)

Geographical Association. (2023a) A-level Geography Results Analysis. https://geography.org.uk/wp-content/uploads/2023/08/A-level-Geography-Results-Analysis-2023-final.pdf (Accessed: 21/05/2025)

Geographical Association. (2023b) Geography GCSE Results 2023 Report. https://geography.org.uk/wp-content/uploads/2023/08/Geography-GCSE-results-2023-report.pdf (Accessed: 21/05/2024)

Hamilton, J. & Fielding, R. (2018) 'Safety First: The Biases of Gender and Precaution in Fieldwork' in *Femininities in the Field: Tourism and Transdisciplinary Research* (B.A. Porter and H.A Schänzel eds).

Khurana, Nalini V. (2020) Geographies of Fear: Sexual Harassment and Women's Navigation of Space on the Delhi Metro. *South Asian Journal of*

Law, Policy, and Social Research. https://ssrn.com/abstract=3671579 (Accessed: 28/05/2025)

Perez, Criado C. (2020) *Invisible Women: Exposing Data Bias In a World Designed For Men*. London: Vintage Books.

Porter, B.A. & Schänzel, H.A. (2018) *Femininities in the Field: Tourism and Transdisciplinary Research*. Bristol: Channel View Publications.

United Nations. (2020) 'Secretary-General António Guterres' Remarks at The New School: 'Women and Power'. https://www.un.org/sg/en/content/sg/speeches/2020-02-27/remarks-new-school-women-and-power. (Accessed: 15/10/2023)

United Nations. (2022) Progress on the Sustainable Development Goals: The Gender Snapshot 2022. https://www.unwomen.org/sites/default/files/2022-09/Progress-on-the-sustainable-development-goals-the-gender-snapshot-2022-en_0.pdf (Accessed: 15/10/2024)

University of Pennsylvania. (2019) The Emerging Crisis of Aged Homelessness. https://aisp.upenn.edu/wp-content/uploads/2019/01/Emerging-Crisis-of-Aged-Homelessness.pdf (Accessed: 21/05/2025)

University of Sheffield. (2022) Dealing with violence: women doing fieldwork. https://www.sheffield.ac.uk/geography/news/dealing-violence-women-doing-fieldwork (Accessed: 28/05/2025)

Valentine, G. (1989) The geography of women's fear. *Area*, 21(4), pp. 385–390.

Vox. (undated) Homeless Over 50: The Statistics. https://www.vox.com/first-person/2016/9/29/12941348/homeless-over-50-statistic.

YouTube. (2016) Marie Tharp: uncovering the secrets of the ocean floor. https://youtu.be/TgfYjS0OTWw?si=8lLVfBdvYuw0X4KM (Accessed: 21/05/2025)

30. GEOGRAPHY ... IS IT QUEER?
JACOB JAMES PROFITT (THEY/THEM)
@JACOBJAMESGEOG

'Openness may not completely disarm prejudice, but it's a good place to start.' (Jason Collins, see Huffington Post, 2013)

Teaching queer identities within the geography curriculum is vitally important. However, it can be a complex conversation to have, especially for teachers who may not be as confident in their knowledge of the queer community. This chapter will explore queer identities across the globe and some potential topics that can be explored within the geography curriculum. It will also look at ways to ensure that your geography classroom is an inclusive environment for queer students, especially when it comes to exploring geography outside the classroom.

QUEER RIGHTS ACROSS THE GLOBE
Since the start of the twenty-first century, queer rights have undergone a transformation, with 33 countries recognising marriage equality as of 2023 (Council on Foreign Relations, 2022) and more countries announcing plans to legalise marriage equality each year. As the century moves forward, many countries are also abolishing laws that targeted queer people, with countries such as Singapore, St Kitts and Nevis, Barbados and the Cook Islands becoming some of the most recent nations to decriminalise same-sex activity (Human Dignity Trust, undated).

Figure 30.1 A map showing where LGBTQ+ rights are progressing
Source: ILGA World Database (database.ilga.org)

However, while successes in queer rights need to be celebrated, there is also a darker side that needs to be recognised equally and that is the areas where queer people face discrimination for their identities. Even in 'developed' countries where queer rights are seen are being more equal, there has been a growing cloud of anti-queer hatred. In the UK alone there has been an increase of 41% in hate crimes against people based on sexual orientation and 56% based on gender identity (Home Office, 2022).

Across the globe, 66 jurisdictions criminalise private, consensual, same-sex activity between two men via 'sodomy', 'buggery' and 'unnatural offences' laws. Nearly half of these countries are in Commonwealth jurisdictions. Of those 66 jurisdictions, 41 criminalise same-sex relations between two women; 12 countries across the globe have laws that could result in the death penalty for same-sex activities – Iran, Northern Nigeria, Saudi Arabia, Somalia, Yemen, Afghanistan, Brunei,

Mauritania, Pakistan, Qatar, UAE and Uganda; 14 countries criminalise gender identity and the expression of transgender people; and, in many other countries, transgender people are targeted under the laws that criminalise same-sex activities (Human Dignity Trust, 2025).

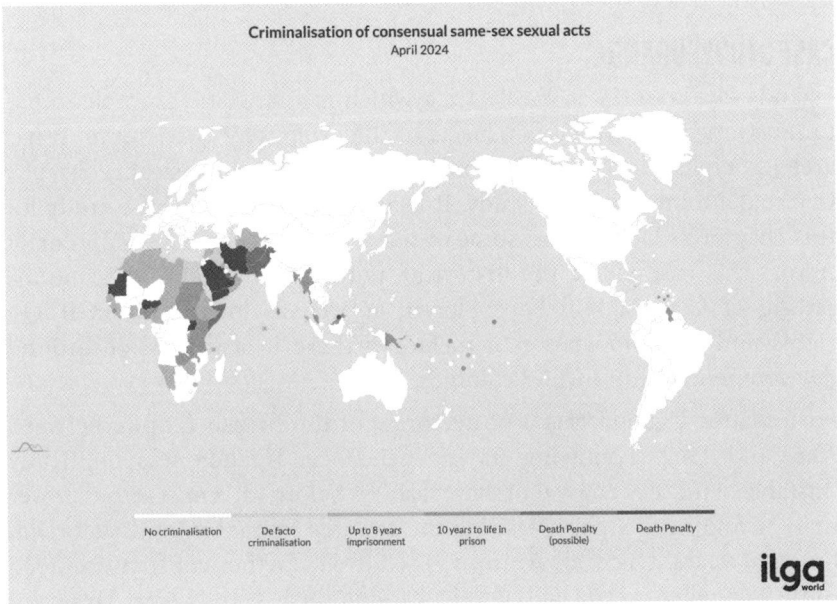

Figure 30.2 A map showing where LGBTQ+ rights are regressing
Source: ILGA World Database (database.ilga.org)

There is one clear link to anti-LGBTQ+ laws across the globe; it comes down to colonialism, more specifically British colonialism. Following the spread of the British Empire in the sixteenth and seventeenth centuries, many places that were colonised were forced to adopt laws that had been imported from Great Britain. One of the most prominent examples is India, where in 1860 the first prohibition of sodomy was written into a penal code. Eventually, this law was exported to other British colonies over Asia, the Pacific and Africa. Despite independence of these nations, often these laws remain. Many of the 66 countries where criminalisation of relationships between two men remain have links to the legacy of colonialism and the British Empire.

Homophobia on the African continent is sad to see, as history provides us with many examples of societies which accepted homosexual members. Many Egyptian deities are portrayed androgynously, and tombs have been found containing men's bodies in an embrace (Buckle, 2020). How did this change happen and what impact is it having today?

CASE STUDY: UGANDA

Uganda is a country in East Africa which is a fantastic example to use to illustrate many different aspects of the geography curriculum. It has areas to explore in its social, economic, political and physical geography that make it an ideal case study. It is also a very relevant case study for this chapter as Uganda has some of the worst LGBTQ+ rights on Earth. Across this case study, the historical, political, physical and economic setting of Uganda will be explored, as well as looking at LGBTQ+ rights and the importance of including these in a discussion around development within former colonies.

Historically, Uganda was a protectorate of the British Empire between 1894 and 1962. Following its independence, Uganda was politically unstable, with the removal of many leaders before Idi Amin seized power in 1971 following a coup, leading the country into a brutal dictatorship for eight years. Once Idi Amin was removed, further conflict occurred until 1986 when, at the end of the Ugandan Bush War, Yoweri Museveni rose to power and became the president of Uganda. Museveni remains the autocratic leader of Uganda at the time of writing, with recent elections in the country condemned for a lack of transparency and abuses against opposition candidates and supporters (House of Commons Library, 2021). Economically, the combination of historical, political and physical influences in the country mean Uganda has faced many challenges in developing. With a human development index (HDI) of 0.525 in 2021 (166th out of 191) it is one of the least developed countries in the world. Uganda not only has poor economic development with a gross domestic product (GDP) per capita of $1,073 but is also poor in its social development, especially with LGBTQ+ rights which are currently regressing at a significant rate.

Anti-LGBTQ+ laws were first introduced in Uganda during British colonial rule and the change in the culture from accepting to homophobic was exacerbated by the spread of Christianity by white missionaries. Stories say that Kabaka Mwanga II of Buganda, who reigned from 1884 to 1897, was queer, but with many records from this time coming from white Christians it is hard to ascertain the truth. Mwanga II is famous for killing the 22 Ugandan Martyrs, but was this because they had converted to Christianity or because they refused his advances? He was against British colonial rule and was exiled to the Seychelles; was his alleged homosexuality used to demonise him and help remove him? Laws from colonial times have been maintained and strengthened over the years to limit the rights of queer people in Uganda further. As well as the influence of agents of formal social control such as laws, Ugandans are also subject to informal social control through socialisation. As successive generations have been brought up to be homophobic, this has become a dominant aspect of the culture.

Queer people in Uganda therefore have a difficult journey; same-sex activity is illegal for both males and females in Uganda and many political leaders have used openly anti-queer rhetoric. This has permeated the general population in Uganda, with polling over the years showing a majority of the country's population opposing the advancement of queer rights in Uganda. This has led to high-profile attacks on queer people in Uganda, with LGBTQ+ rights activists facing severe harassment and, in some cases, being killed. An example is David Kato, referred to as 'Uganda's first openly gay man' in *The Economist*, who was a founding member of Sexual Minorities Uganda (SMUG). His address was published in a Ugandan paper under the title 'Hang Them!' and after months of harassment he was found dead at his home in 2011 (University of York, undated).

Uganda made world news in 2023 for the passing of its 'anti-homosexuality bill'. The law was met with condemnation from world leaders as it increased the punishment for homosexual acts to life imprisonment, and in some cases of 'aggravated homosexuality' for repeat offenders the death sentence could be handed out. The law has also been condemned for including homosexuality, rape and paedophilia in the same legislation, which could lead to people thinking that they are associated. This legislation has led to further anti-LGBTQ+ sentiment across the country.

Following the passing and implementation of the 'anti-homosexuality bill', Uganda has faced a backlash from many countries and organisations including the World Bank – which announced Uganda would no longer be able to borrow money – and the USA, which expelled Uganda from tax breaks for sub-Saharan countries. There has also been condemnation from political associations including the European Union. Weakening links with such countries and organisations could well affect Uganda's development in the future. In August 2023, a 20-year-old man became the first person to be prosecuted for 'aggravated homosexuality', for which he faces the death penalty (*The Guardian*, 2023).

However, is it right for western countries and organisations to put pressure on Uganda to change laws made by Ugandans? Is this yet another form of neo-colonialism? Buckle (2020) argues that:

> 'Rejecting pro-LGBT legislation is rejecting neo-colonialism and is in favour of African nationalism, self-determination and self-worth. Unfortunately, African homophobia is a tricky mix of anti-neo-colonialism, politics, and religion, made worse by the HIV/AIDS crisis... So what is the future for LGBT rights in Africa? In many countries, despite the legacy of colonisation, citizens are taking a more autonomous stance on LGBTQ+ legislature, with the queer communities taking the lead, instead of external pressures from the West. Across the world, countries that have improved their LGBT rights records have done so because of the hard work, organising and leadership of local LGBT groups and communities, and the case of Africa is no different.'

Uganda is not unique in its position on anti-LGBTQ+ rights. Countries you may study as part of your geography curriculum may have similar approaches. When we explore the physical and human geography of such countries, we also have the opportunity to consider their queer geography.

QUEER GEOGRAPHY – FROM ACADEMIA TO THE CLASSROOM

Koh (2022) explores queer geography, stating that 'The value of thinking geographically about queer communities can be considered through the lens of academia, where it can add to progressive discourse within the discipline of Geography and provide a more comprehensive understanding of spatial ontologies' (Knopp 2007). He suggests that studying the way queer people interact with space adds a new layer of meaning to geographical understanding, giving the example of the home often being seen as a refuge, but not necessarily for queer individuals. In fact, Koh presents research which suggests that public places may become more homelike for queer people. Koh states that 'thinking geographically about the queer community adds value to the discipline of geography by introducing alternative voices into academic discourse and deepening our knowledge of various geographical phenomena.'

As classroom teachers, we can learn from this work. Do we consider the queer experience of places when studying topics such as 'Changing Places', a popular A-level geography topic. With sexual orientation featuring in the 2021 Census, we can explore qualitative data such as that shown in the Office of National Statistics census map (https://www.ons.gov.uk/census/maps/choropleth/identity/sexual-orientation/sexual-orientation-4a/lesbian-gay-bisexual-or-other-lgb), see QR code for link. Which areas have highest number of LGB+ respondents? What could explain these patterns?

We could also explore quantitative data, using resources such as Queering the Map (https://www.queeringthemap.com/). We could explore diverse experiences of being queer in different places with our sixth form students. Could our students work with other students to gather anonymous experiences of being queer in our local area and map them in a similar way? If our schools have PRIDE groups or similar, could geographers help them map the school site to show which parts feel welcoming and which not, then present this to the leadership team? Could this help increase the safe spaces for queer students in our schools?

CREATING A SAFE SPACE FOR QUEERNESS

There are many ways to create an inclusive geography classroom, including through the incorporation of queer geography into schemes of learning (as shown with the case study above); this is better than just random lessons related to a key date within the queer calendar as it shows it is part of the learning journey for all students and related to the topics they are learning within their geography classroom.

The key thing around creating an inclusive classroom is to not be scared about the topic you are teaching, or the questions you may get asked. It is ok to not know everything – students within the classroom would rather you were honest and learn the answer to a question together rather than hear you attempt to answer, but not feel that confidence in your answer. Your classroom should be a safe space for all questions to be asked (as long as they are done respectfully). Aim to create a culture of respect, caring and being open without fear of ridicule.

Taking students out of school for geography fieldwork (see chapter 8) can make it more difficult to ensure a safe space. It is important to have conversations as teachers and with students and their parents/carers to make sure all fieldwork is inclusive. Of course, it is also crucial to ensure that queer staff are also safe. There are some key questions to think about when planning a geography trip with queer students:

- Are student going to an area with accessible/inclusive facilities (gender neutral toilets/changing spaces)?
- Are students going to an area where they may feel uncomfortable expressing their identity?
- Will the dress code be inclusive for all students, however they identify?
- Will the place of visit be welcoming for all students?
- If an overnight trip, how will gender non-conforming students be roomed? Would individual rooms be a possibility?
- If the trip is international, are there any anti LGBTQ+ laws in place that could cause stress and anxiety for the student?

- How will you have sensitive conversations with students, parents/carers around any areas of difficulty identified from the questions above?

Some of these questions may seem basic when it comes to planning a trip, but these are very important to think about with queer students, and especially non-gender conforming or trans identifying students. Owen (2022) considers some of these questions and also issues such as the dilemmas people face about whether to reveal their sexuality/identity in the field.

When I was at university, we did a trip to Guatemala, which doesn't have anti-LGBTQ+ laws, but can have hostile attitudes towards LGBTQ. I wasn't anxious due to the conversations that the university had with me ahead of the trip which helped reassure me that I would be well looked after, regardless of my gender identity and sexual orientation. This helped make the trip feel more inclusive to queer students; although I could have faced hostility, I knew would be made to feel safe for the duration of the trip (and I didn't face a single issue).

A crucial element of making a safe space available is communication – we need to listen to each other and learn from each other. Together, geographers can enrich their curriculum and their queer students' experiences of geography.

Five reflection questions

1. Are there any areas in your current curriculum where queer geography could be explored?
2. How can you show a deeper understanding of queer identities across the globe and link variations to other aspects of the geography curriculum?
3. Why is it important to ensure that when discussing these topics a safe space is created within your classroom?
4. What issues could you foresee when trying to ensure that fieldtrips are inclusive to all queer identities and what are some potential solutions to these?
5. How can you celebrate important queer dates within your department?

> **DIGGING DEEPER - THREE RESOURCES TO DELVE FURTHER**
> - Envisioning Global LGBT Human Rights: (Neo)colonialism, Neoliberalism, Resistance and Hope on JSTOR.
> - LGBTQ+ Inclusive Fieldwork - Geography Directions.
> - LGBTQ+ Inclusive Fieldwork – https://queermemorials.leeds.ac.uk/2021/03/18/lgbtq-inclusive-fieldwork/ (Accessed: 28/05/2025).

REFERENCES

Buckle, L. (1 October 2020) African sexuality and the legacy of imported homophobia. https://www.stonewall.org.uk/news/african-sexuality-and-legacy-imported-homophobia

Council on Foreign Relations. (2022) Marriage Equality: Global Comparisons. cfr.org (Accessed: 28/05/2025)

Home Office. (2022) Hate crime, England and Wales, 2021 to 2022. www.gov.uk (Accessed: 21/05/2025)

House of Commons Library. (2021) Uganda: reactions to the 2021 election. https://commonslibrary.parliament.uk/research-briefings/cbp-9206/ (Accessed: 21/05/2025)

Huffington Post. (2013) Jason Collins' Courage in Coming Out. https://www.huffpost.com/entry/jason-collins-courage-in_b_3193476 (Accessed: 21/05/2025)

Human Dignity Trust. (2025) A History of LGBT Criminalisation. https://www.humandignitytrust.org/lgbt-the-law/a-history-of-criminalisation/ (Accessed: 21/05/2025)

Human Dignity Trust. Map of Countries that Criminalise LGBT People. https://www.humandignitytrust.org/lgbt-the-law/map-of-criminalisation/ (Accessed: 21/05/2025)

https://theafricanroyalfamilies.com/2023/06/03/kabaka-king-mwanga-ii-of-buganda-gay-bisexual-or-queer/

Koh, B. (9 January 2022) Is Geography coming out of the closet? Why queer geography matters. https://bloomsburygeographer.com/2022/01/09/is-geography-coming-out-of-the-closet-why-queer-geography-matters/ (Accessed: 21/05/2025)

Owen. (2022) How can we make geography more inclusive? https://geogmum.wordpress.com/2022/01/03/how-can-we-make-geography-fieldwork-more-inclusive-part-1/ (Accessed: 21/05/2025)

The Guardian (via Agencee France Presse). (2023) Ugandan man charged with "aggravated homosexuality" under new law. https://www.theguardian.com/world/2023/aug/29/uganda-prosecutors-charged-man-aggravated-homosexuality-anti-lgbtq-law (Accessed: 21/05/2025)

University of York. The Life and Legacy of David Kato. https://www.york.ac.uk/colleges/david-kato/life-and-legacy/ (Accessed: 21/05/2025)

31. WHY VOICES MATTER NOW
CHANTAL MAYO-HOLLAWAY, EDITED BY AKHERA WILLIAMS (VOICESPROJECT CO-CREATORS)

'Do your little bit of good where you are; it's those little bits of good put together that overwhelm the world.' (Desmond Tutu)

We are living through a time of great change; in a shattered nation (Dorling, 2023) with an increasingly divided society (Bhattacharyya et al, 2021). We not only face internal change, but multiple global 'crises - economic, social, environmental' which impact 'vulnerable and racialised groups' most (Mondon and Winter, 2020, p.3). In addition, across Europe we are seeing the rise of extremes in 'Alt Right' and 'Alt Left' approaches to politics (Winter and Mondon, 2018), creating further division at national scales. These combine to create increasingly unstable futures for our students. (See chapter 38.)

In times of instability 'individuals ... segregate themselves socially' (Duffy et al, 2019), creating neo-tribalism (Lammy, 2020). This segregation involves splitting populations into 'in groups', assigned 'positive attributes' and 'out groups', assigned 'negative attributes' (Johnston (Eds), 2021, p.15). Such division is accentuated by our virtual environments, where students are increasingly receiving their information from algorithm-based applications and find themselves in echo chambers not only of their geographic communities, but of their online subcultures (Dorling, 2023). In these spaces people often speak their own voice, rather than listening to that of others, making it a space where people see opinions that may be harmful to others as 'free speech' rather than potential hate speech (Surviving Society Podcast, 2019). Combine this with the current 'post truth, anti-intellectualism and ... mistrust of expertise' (Nichols, 2017) and a rise in disinformation is enabled. As a consequence, there has never been a more important time to show students alternative, authentic voices and how to respect the voices of others.

The purpose of this chapter is to share ways marginalised voices can be ethically incorporated into our lessons. Here we are taking marginalised voices to mean 'unheard voices' (Stoecker and Tyron, 2009), ones which are not included, historically excluded, voices who struggle to be heard, and those who are unable to access the tools to be heard (VoiceProject, 2020). These voices may be similar internationally and nationally, but will be very different at a local scale. We suggest ways of making safer spaces for students to have their own voices heard and learn to listen the voices of others, as well as sharing ways which we can support students in reflecting on the teaching and learning of their own communities, cultures and identities. Blood (2007) commented that action to do this is necessary as such division is a 'problem [which] must be tackled at a very early stage before students grow up ... with no real understanding' of other communities. Therefore, we, as educators, have a responsibility to the communities we serve, to help their future denizens and leaders to see that 'we have far more in common than that which divides us' (Jo Cox, MP).

PRACTICAL STEP 1: INCLUDING AUTHENTIC VOICES

'Everyone deserves to have their lives elevated through the beauty of truthful representation.' (Rohit Bhargava, writer, in *Beyond Diversity: 12 Non-obvious Ways to Build a More Inclusive World*, 2022)

WHY SHOULD WE DO THIS?

The inclusion of a variety of voices from the precise location or an exact issue being taught is crucial to avoiding the 'single story' narrative (Adichie, 2009). Even our textbooks can create extreme binaries (Taylor, 2014) and are often 'problematic and unrepresentative' (Rackley, 2020). Showing students that more than just powerful and privileged voices matter, and the inclusion of indigenous knowledge, is an essential part of the decolonisation process (Anderson, 2021), which can often lead to the platforming of voices whose truths have been historically sidelined.

Here, it is argued that the use of marginalised voices should be uncut and contextual. Although the power relationships leading to such marginalisation make it difficult for these voices to be accessed: For example, who chooses the voices being heard? Within your classroom, your school and your community you will have some voices that are more marginalised than others and it is argued here that these are the voices to platform. Hammond and McKendrick (2020) also emphasise the need for teachers to make 'connections between the geography they are teaching and children's everyday lives'. If students do not feel that their learning represents them, speaks to them, then how relevant is it? And how well will they connect with and remember it?

HOW CAN WE DO THIS?

Below are four strategies used to facilitate the inclusion of marginalised voices.

School-to-school projects: Making connections with schools in other parts of the world enables students to explore their own and other young people's geographies. 'Our World is a Class' is a Brazilian project aiming to 'explore new cultures and broaden our horizons'. Students communicate directly through lesson-based video calls where there is an opportunity to exchange learning, as well as 'ask questions … about … culture and the area we live'. Most importantly, this communication empowers students to be facilitators by shaping the interaction and asking questions themselves in the design stage of the process (Murilo Czaika, project founder).

Video interviews: Hope and Healy's 'Geographies of Sustainable Development: What does Bolivia teach us?' is based on academic geography research and the co-production of knowledge on sustainability and development (Geographies of Sustainable Development, 2020; Healy et al, 2023). The resource includes interactive ArcGIS StoryMaps, text extracts and full video interviews with Indigenous leaders and development workers (Figure 31.1). These resources can be used to encourage students to critically engage 'with related local, national and global debates and provides them with the opportunity to make connections to their everyday lives. [The resource draws] on Indigenous territorial politics, conservation and resource conflicts to ask: "What does Bolivia teach us about sustainable development?"' (Geographies of Sustainable Development, 2020). Students are able to engage with the perspectives and expertise of indigenous leaders and development workers through 'their own words and modes of talking about sustainability and development within and for Bolivia' (Healy et al, 2023, p.68). There are, however, limits of this sort of resource. For example, Healy et al (2023, p.68) 'acknowledge there could be greater potential to explore how students could enter into a two-way dialogue with Bolivian Indigenous leaders and development workers', such that 'the films could be developed to span over time to highlight changes in the significance of ideas tied to specific ongoing local, national and global debates of importance in Bolivia'. The impact of this is that teachers have access to often marginalised voices from Indigenous leaders. This enables teachers to support students in engaging with people with expertise about sustainability and sustainable development that runs counter to other dominant narratives about sustainable development (including SDGs).

Figure 31.1a and b Screenshots from Dr Jess Hope's recordings featured on https://sites.google.com/view/geographiesofsustainabledev/what-does-bolivia-teach-us

Using speeches: Instead of selectively quoting marginalised voices, or worse creating fictional voices, it is possible to include videos of young people directly linked to the topics you are teaching. These 'real' examples are a crucial approach to anti-racist pedagogy (Puttick and Murrey, 2020; Puttick, 2017). Two examples of contributions from Indigenous Peoples are Brazilian environmental activist Txai Surui's speech to COP 26 in 2021 and Maxida Marak's video for Amnesty Internation in 2017 (see Surui, 2021 and Märak, 2017). Not only are their words in print and unabridged for students to interpret, but videos of them telling their own story, uncut, enables students to feel the emotion and intent of the words being used. Honesty and representation matters. These resources enabled students to understand how Indigenous People often have deeper, spiritual and more interconnected ties to their land compared to a more exploitative vision of national governments and businesses. The impact this had on students was one of surprise and astonishment. Teachers commented that students were shocked about what they heard and many felt a sense of injustice having listened to the emotion behind the voices.

Audio recordings: The VoicesProject is a free, anti-racist, critical and inclusive learning resource which was created to: promote learning which responds to structurally marginalised voices in our classrooms, actively challenging colonial legacies in schools – from pedagogy to curriculum – and amplify the voice of marginalised communities in the UK.

The resources include audio recordings which discuss issues related to geographical themes of space/place, identity, migration and empire. The listening activities linked to audio recordings of intergenerational or friendship conversations encourages active listening.

Voices can be used with guidance from the VoicesProject website, or integrated into existing schemes of learning. At Ormiston Shelfield Community Academy, Malikah and Sharmin's audio was used with key stage 4 to consider the impact of past migrations from the Indian sub-continent to Britain, post partition. This enabled students to understand the reality of the complex factors creating this movement and the implications of it. As part of the task, students were asked to focus on five key moments (of their choice) from the audio, as well as considering how this reflected a moment in their own life. One student responded that 'this reflects to my life as I had to move from my home country to here'. This powerful moment led to follow up conversations with staff, and a greater sharing of the student's own personal experiences, helping them feel more 'in place' in their classrooms. In a Birmingham school with a large and diverse catchment, the resource also generated 'important conversations ...[and] some of the students were surprised by the experiences of their friends and classmates and started to understand their own privilege' (Ruth Till, geography subject lead and PGR Universities of Birmingham and Warwick).

PRACTICAL STEP 2: MAKING SPACE FOR STUDENT VOICES

'Nothing about them, without them.' (Anonymous)

WHY SHOULD WE DO THIS?

Making space for student voices reflects the changing nature of geography as an academic discipline, with a move towards greater participatory involvement. It is also Article 12 of the United Nations Convention on the Rights of the Child, 'The right to be heard' (UNICEF, 2016). The curriculum making model, initially proposed by Lambert and Morgan (2010) and further developed through the Geographical Association is one way school geography can do this. This is something which has been seen historically in projects like the Young People's Geography Project (Biddulph and Frith, 2009) and in recent projects such as Young People at a Crossroads (Walker et al, 2022). When 'children [are seen] as active agents in the creation of their own lives and geographies' it leads to a more realistic representation of the world the students experience (Yarwood, 2012). Pike (2020) also argues that 'geography pupils are also citizens in local, national and global communities'. Not only is this important in classrooms, it is also important in our case studies. According to Cresswell, place meanings are 'both individual and shared' (2019, p.134), therefore, in order to truly understand and represent a place in the classroom, we need to know how others see, feel and experience it, as well as considering our own positionality (Burnett et al, 2021). Working together, with students and our communities, we can make something greater than the sum of our parts; with each voice contributing towards the rich tapestry of our own classrooms.

HOW CAN WE DO THIS?

Inclusive classrooms: Offering opportunities for students to add their voice in safe classrooms and the empowerment to question resources, is something Sinclair has written about (2022 and 2023). In school, a climate is set up to guide students to 'respect the **diversity** of the world; represent people and places **equitably**; champion the **inclusion** of different thoughts and experiences and to provide **justice** for communities harmed by misrepresentation and minimisation.' Students are encouraged to ask questions such as: *Have we included the **voices**,* *actions, words* and **knowledge** *of the communities we are learning about?* and *Have we considered if the data sets are* **accurate**, *who collected them and why?* when conducting project-based learning (Sinclair, 2023). When this is applied as a consistent classroom habit, students become more synoptic and reflective (key geographical skills) and more engaged and empowered. One student opted 'to present the "real" Haiti, not Haiti from the textbook'. Sinclair tells us that this type of learning 'values the student as an engaged participant in the creation of knowledge subsequently empowering the student in a way that destabilises the norm' while also giving students the tools they will need for their future involvement in society (Sinclair, 2022).

Listening projects: The VoicesProject has a listening programme for schools. This is aimed at encouraging schools to listen to the marginalised voices in their classrooms, schools and communities. Such a project was trialled by students at Ormiston Shelfield Community Academy (Figure 31.3). Here students found it important to share the issues they faced and put forward ideas to reduce misconceptions through opportunities to share and celebrate their culture, for example through the food in the school canteen. This has been taken further at Preston Manor School, Wembley, where sixth formers are leading their own VoicesProject. 'This is focussing on anti-racism and diversity in the curriculum. Until this project, much of the work was teacher led and we are now in a position to empower students to capture their own voices and experiences of school, and our curriculum, so that we can incorporate this back into our own work and make our school as inclusive as possible through this listening exercise' (Hamda Sheikh, Preston Manor School).

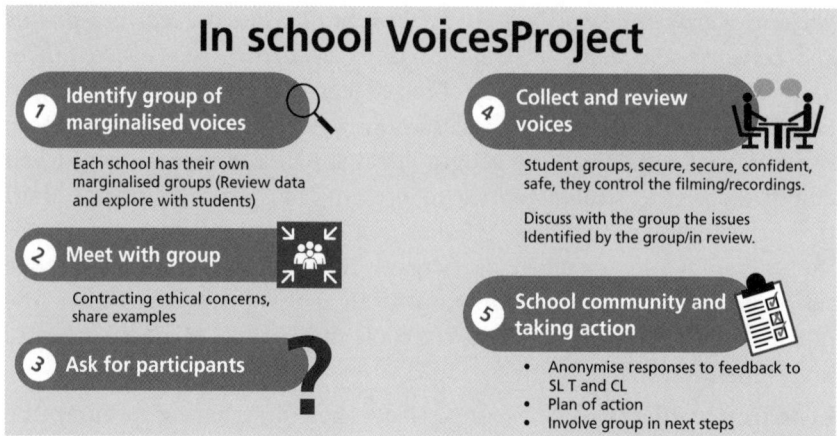

Figure 31.2 In school VoicesProject

Source: www.theVoicesProject.co.uk

Valuing the voice of others: By showing students how to ethically collect, interpret and protect voices through fieldwork, we can help them value and understand the views of others. In addition, students can acquire a key qualitative fieldwork skill. One example is the VoicesProject NEA resource (Figure 31.4). A second example is a homework activity at St Ambrose College, Manchester, where students held an intergenerational listening exercise about family migration and heritage. With permission, this was then brought into the classroom to enrich discussions and links between students' geographies and theoretical learning. Finally, at Kingsthorpe College, Northampton, students took inspiration from the VoicesProject website to create their own recorded interviews. The project had a massive impact on the confidence of students in approaching the local community, knowing that people would be willing to share their stories. What more powerful way to involve your local community and make positive links between our schools and the communities they serve?

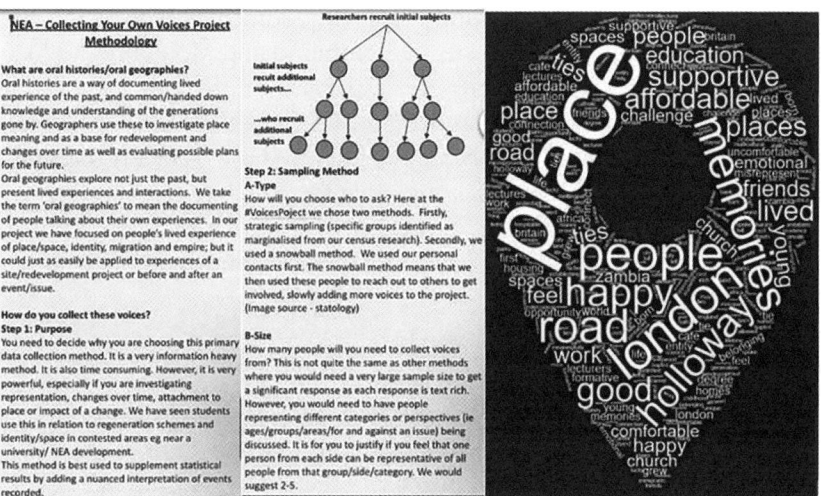

Figure 31.3 NEA Collecting your own voices

Source: www.theVoicesProject.co.uk

Super case studies and curriculum review: Lambert and Biddulph (2020) acknowledge 'the power, and potential, of drawing on children's geographies to enhance geography education' (Lambert and Biddulph, 2014, p.215 in Hammond and McKendrick 2020). Their focus is on a 'knowledge exchange' in the curriculum-making process, not a one-way flow of information.

One area where this is particularly relevant is in the representation of place. The *Geography Subject Review* (2023) acknowledged that place was 'often poorly planned … typif[ied] just one issue or phenomenon' was 'at times … outdated and inaccurate … [leading students to] … learning "single stories".' The co-creation and use of super case studies, on which to hinge lesson content, is another way to do this. Adding student voice into super case studies helps to interweave narratives, images, artefacts and lived experience of places (Cresswell 2019).

At Ulidia Integrated College, Carrickfergus, staff and students worked together to create a super case study for Belfast linked to urban issues (key stage 4) and 'ethnically diverse city' (Sixth Form). 'Working together to create the super case studies has been incredible powerful for students as we explored the history together to develop a wider understanding of different perspectives on the city.'

At Ormiston Shelfield Community Academy, students involved in super case study creation for India echoed Hammond and McKendrick (2020) reflecting that 'students' voices are important as they are part of the education system'. The final student commented that 'if we had more lessons on this, I would have been more likely to have picked [geography] as an option', proving that making space for student contributions has a direct impact on perception, especially by marginalised communities who do not always see themselves represented in what and how we teach. (See also chapter 16.)

CONCLUSION

> *'Geography should not ... simply be about witnessing the world but, rather, taking action to change it.'* (Yarwood, 2021, p.3)

Every classroom, every school, every community has marginalised voices. Those who are marginalised in my classroom may not be the same people who are marginalised in yours. Everyone is somewhere along the journey of including voices in their work, but hopefully this chapter has enabled you to identify ways in which you can include globally marginalised voices through school-to-school links, interviews, videos and audio; as well as how to include your own students' voices in their lessons and in the creation, review and editing of materials as participatory agents.

Five reflection questions

1. How do you reflect on the worldview and assumptions embedded in the curriculum, resources and narratives you teach?
2. What steps could you take in curriculum review to ensure it is more pluralistic and reflective of the world your students inhabit?
3. Who are the marginalised peoples and voices within your classroom, school and community?
4. How do you seek to overcome marginalisation through the resources you use and your classroom practice?
5. If you could do one thing to make a change for the students in your classroom, what would it be?

> **DIGGING DEEPER – THREE RESOURCES TO DELVE DEEPER**
> - Adichie, C. N. (2009) The Danger of a Single Story TEDGlobal 2009. https://www.ted.com/talks/chimamanda_ngozi_adichie_the_danger_of_a_single_story?language=en
> - The VoicesProject (2020) Making Voices Matter: Including Marginalised Voices in your Classroom Geography. TeachMeet 2020 presentation. https://x.com/voicesproject_/status/1912584260650774904?s=46
> - Burnett, L., Brack, S. and Anderson, S. (2021) Teaching about a place stop and think first. decolonisegeography.com, 9 April 2021. https://decolonisegeography.com/blog/2021/04/teaching-about-a-place-stop-and-think-first/

REFERENCES

Adichie, C. N. (2009) The danger of a single story. TEDGlobal 2009. https://www.ted.com/talks/chimamanda_ngozi_adichie_the_danger_of_a_single_story?language=en (Accessed: 22/10/2023)

Anderson, N. (2021) Why do we need to decolonise geography? decolonisegeography.com, 10 February 2021. https://decolonisegeography.com/blog/2021/02/why-do-we-need-to-decolonise-geography/ (Accessed: 22/10/2023)

Blood, M. (2007) *Watch My Lips, I'm Speaking!* Dublin: Gill and Macmillan Ltd.

Bhargava, C. & Brown, J. (2021) *Beyond Diversity: 12 Non-obvious ways to build a more inclusive world.* London: Ideapress Publishing.

Bhattacharyya, G., Elliott-Cooper, A., Balani, S., Nişancıoğlu, K., Koram, K., Gerbrial, D., El-Enany, N., de Noronha, L. (2021) *Empire's Endgame: Racism and the British state.* London: Pluto Press.

Biddulph, M. & Firth R. (2009) Young Peoples Geographies. *Teaching Geography,* 34(1).

Burnett, L., Brack, S. & Anderson, S. (2021) Teaching about a place stop and think first. decolonisegeography.com, 9 April 2021. https://decolonisegeography.com/blog/2021/04/teaching-about-a-place-stop-and-think-first/ (Accessed: 22/10/2023)

Cresswell, T. (2019) *Maxwell Street: Writing and thinking place.* Chicago: University of Chicago Press.

Dorling, D. (2023) *Shattered Nation: Inequality and the geography of a failing state*. London: Verso.

Duffy, B., Hewlett, K., McCrae, J. & Hall, J. (2019) *Divided Britain? Polarisation and fragmentation trends in the UK*. London: The Policy Institute Kings College London.

Geographical Association. (2019) Creating an inclusive geography classroom. https://geography.org.uk/ite/initial-teacher-education/geography-support-for-trainees-and-ects/learning-to-teach-secondary-geography/inclusion-and-adaptation/creating-an-inclusive-geography-classroom/ (Accessed: 23/10/2023)

Geographical Association. Curriculum Making. https://geography.org.uk/curriculum-support/support-guidance-curriculum-planning/curriculum-making/ (Accessed: 28/05/2025)

Hammond, L. & McKendrick, J. H. (2020) Geography teacher educators' perspectives on the place of children's geographies in the classroom. *Geography*, 105(2), 86–93.

Healy, G., Laurie, N. & Hope, J. (2023) Creating stories of educational change in and for geography: what can we learn from Bolivia and Peru? *Geography*, 108(2). Summer 2023.

Geographies of Sustainable Development. (2020) Geographies of sustainable development: what does Boliva teach us? https://sites.google.com/view/geographiesofsustainabledev/what-does-bolivia-teach-us (Accessed: 17/07/2025)

Hopkins, P., Botterill, K., & Sanghera, G. (2018) Towards inclusive geographies? Young people, religion, race and migration. *Geography*, 103(2), pp.86–92.

Johnston, L. (ed) (2021) *Bias Busting for Beginners: An introduction to anti-bias in education*. Belfast: The Northern Ireland Council for integrated Education: Belfast.

Lambert, D. & Morgan, J. (2010) *Teaching Geography 11-18: A conceptual approach*. Maidenhead: Open University Press.

Lammy, D. (2020) *Tribes: A search for belonging in a divided society*. London: Constable.

Märak, M. (2017) Amnesty International. https://youtu.be/0KRZz8dc-4s?feature=shared (Accessed: 25/10/2025)

Märak, M (2023) featured in *Author Ann-Helén Laestadius on reindeer herding*, BBC, The Arts Hour, 4 February 2023. https://www.bbc.co.uk/programmes/w3ct391f (Accessed: 25 October 2023)

Marshall, T. (2018) *Divided: Why we're living in an age of walls*. Elliott and Thompson.

Mondon, A. & Winter, A. (2020) *Reactionary Democracy: How racism and the populist far right became mainstream*. Verso.

Ofsted. (2019) Education inspection framework: Equality, diversity and inclusion statement. May 2019. https://assets.publishing.service.gov.uk/media/5d39d3e1e5274a400af8142a/Education_inspection_framework_-_equality__diversity_and_inclusion_statement.pdf (Accessed: 22/10/2025)

Ofsted. (2023) Getting our bearings: geography subject report. 19 September 2023. https://www.gov.uk/government/publications/subject-report-series-geography/getting-our-bearings-geography-subject-report (Accessed: 22/10/2025)

Nichols, T. (2017) *The Death of Expertise: The campaign against established knowledge and why it matters*. Oxford: Oxford University Press.

Pike, S. (2020) 'GIS for Young People's Participatory Geography' in Walshe, N. & Healy, G. eds (2020) *Geography Education in the Digital World: Linking theory into practice*. London: Routledge.

Puttick, S. (2017) Should we only teach about real people and real places? *Geography*, 102(1), pp. 26–32. https://doi.org/10.1080/00167487.2017.12094006 (Accessed: 28/05/2025)

Puttick, S. & Murrey, A. (2020) Confronting the deafening silence on race in geography education in England: learning from anti-racist, decolonial and Black geographies. *Geography*, 105(3), pp.126–134.

Rackley, K. (2020) Decolonising geography. geogramblings.com, 1 August 2020. https://geogramblings.com/2020/08/01/decolonising-geography/ (Accessed: 22/10/2023)

Royal Geographical Society. Dr Jess Hope and Grace Healy on sustainable development: what does Bolivia teach us? https://www.rgs.org/schools/resources-for-schools/dr-jess-hope-and-grace-healy-on-sustainable-development-what-does-bolivia-teach-us (Accessed: 15/10/2025)

Royal Geographical Society (n.d.) The case for qualitative fieldwork. https://www.rgs.org/schools/resources-for-schools/the-case-for-qualitative-fieldwork (Accessed: 25/10/2023)

Sinclair, D. (2022) How getting your students to teach can increase their success and support decolonising your classroom. https://dsinclairwriting.com/2022/02/27/how-getting-your-students-to-teach-can-increase-their-success-and-support-decolonising-your-classroom/ (Accessed: 22/10/2023)

Sinclair, D. (2023) DEIJ resources questions to ask about our learning today. https://dsinclairwriting.com/2023/04/23/deij-resources-questions-to-ask-about-our-learning-today/ (Accessed: 22 October 2023)

Stoecker, R. & Tyron, E. (eds) (2009) *The Unheard Voices: Community organizations and service learning*. Philadelphia, PA: Temple University Press.

Suruí, T. (2021) Indigenous activist Txai Suruí's full speech at COP26. https://www.youtube.com/watch?v=TP5Nbc5P0GM (Accessed: 25/10/203)

Surviving Society Podcast (2019), EO69 Aurelien Mondon and Aaron Winter, Racism and the populist far right. https://on.soundcloud.com/EVvyL. (Accessed 25/10/2023)

Taylor, L. (2015) Research on young people's understanding of distant places. *Geography*, 100(2), pp.110–113.

The VoicesProject. www.thevoicesproject.co.uk (Accessed: 25/10/2023)

The VoicesProject. (2020) Making voices matter: including marginalised voices in your classroom geography. TeachMeet 2020 Presentation for Geography Teachmeet Share Learn Teach. https://www.youtube.com/watch?v=8wxCrtpDBbo (Accessed: 25/10/2023)

UNICEF. (2016) *United Nations Convention on the Rights of the Child*. UNICEF UK. https://www.unicef.org.uk/wp-content/uploads/2016/08/unicef-convention-rights-child-uncrc.pdf (Accessed: 28/05/2025)

Walker, C. with Rackley, K. M., Summer, M., Thompson, N. & Young Researchers. (2022) Young people at a Crossroads: Stories of climate education, action and adaptation from around the world. www.sci.manchester.ac.uk/research/projects/young-people-at-a-crossroads (Accessed: 25/10/2023)

Winter, A. & Mondon, A. (2018) Understanding the mainstreaming of the far right. 26 August 2018. https://www.opendemocracy.net/en/can-europe-make-it/understanding-mainstreaming-of-far-right/ (Accessed: 22/10/2023)

Yarwood, R. (2021) Geography citizens: from witness to action. *Geography*, 106(1), 2–3.

32. THE POWER OF STORYTELLING IN GEOGRAPHY: A DECOLONIAL PERSPECTIVE
IRAM SAMMAR
@IRAMSAMMAR

'...when a story survives in folklore, it expresses in some way a region of the "local soul".' (Fanon, 1952, p.45)

We all tell stories in the geography classroom, some about our holidays and others about our everyday geographies 'near and far'. Telling a good story is to enter an enchanting realm of learning through imagination, where we discover the extraordinary capability of narratives to transcend the familiar and captivate new learning. I begin with Fanon's quote above, as it has revolutionary implications for much of the discourse on antiracism and decolonial thinking in geography today. His work on colonialism, racism and the psychology of oppression is reflected in his perspective on the significance of folklore in expressing the essence of a particular cultural or regional identity through oral traditions of storytelling. Geography teachers can use storytelling as a key with which we unlock the door to a world of understanding and discovery. A well-told story possesses the mystical power to ignite curiosity, establish connection, and breathe life into the geography we explore with our students.

As a geography teacher, I included stories as an integral part of teaching, as I aspired to engage my students with decolonial and antiracist concepts rarely found in the geography curriculum. British geography, as a subject, has long been entwined with narratives that reflect historical biases and perpetuate colonial perspectives which require an understanding and implementation of antiracism (Esson and Last, 2020). Puttick and Murrey (2020) also highlight that the key issue here is the 'remarkably persistent whiteness' and a 'deafening silence' of 'race' within geography education in the UK, which is coupled with the lack of diversity in student bodies and faculty members in higher education geography departments. As a result, the profound silence on 'race' in English school geography stands as an important strand of thought in the discipline, especially as there is a potential for important lessons from antiracist and decolonial conversations unfolding.

In this chapter, a decolonial exploration of three interwoven concepts that hold profound implications for our understanding of education, pedagogy and storytelling will be offered. First, I will demonstrate how I use stories to engage with students in the geography classroom through personal geographies. Secondly, I will explore storytelling as a decolonising tool for geography, which involves dismantling these ingrained narratives and paving the way for a more inclusive, diverse and antiracist approach to understanding this complex and dynamic world we share and learn about. The final section will explore how diversifying pedagogy through storytelling is central to decolonising and antiracism and can foster transformative learning, offering an empowering medium through which educators can retell the story of geography and how it came to be understood.

BRINGING IN STORIES THROUGH PERSONAL GEOGRAPHIES

On a bus ride home one day, I had an encounter that would leave a lasting impact on my perspective on how I approach teaching about race and racism in the geography classroom. Through the power of storytelling, I engage my students with this particular event, to think beyond the curriculum presented to them in the form of existing textbooks and extensive online software tools. When I introduce ideas and concepts around migration and development studies, I retell a particular story which sparks a unique interest in my audiences, however young or mature:

> *"Don't let her on. Get off! Go on, go back to where you came from,"* said an elderly man looking at me with a walking stick and wool-blended cap. *"You're all over the place, I tell ya ... GET 'ER OFF!"* This elderly man really wanted my attention.
> (Sammar, 2022)

As I boarded this London bus, an elderly gentleman made racist remarks to me and the driver, asking him not to allow me on the bus simply because I am Muslim and South Asian. I felt deeply upset and hurt by this display of discrimination. Instead of ignoring the situation, I decided to confront him about his racist comments. Initially, our interaction was tense, but as I engaged him in conversation, something extraordinary happened. Our conversation moved from conflict to reason. As I explained I am a geography teacher, we began a new conversation where we began discussing geography from what we eat, wear and use in our daily lives, to connections with people and places in the wider world.

I first told this story to a key stage 3 class, for there were many comments being made about migration. Many comments were being made about how 'immigrants' come over to this country, the United Kingdom, and take 'our' jobs and create tensions between themselves and the host country. It is this kind of dialogue that encouraged me to share stories from my own experience to shed light on misconceptions of immigration of South Asians, Africans and those from the Caribbean who have settled in the UK. Discussions on race and racism are absent in the geography curriculum and sharing lived experiences invites students to engage in critical thinking and analysis, essential skills for understanding complex geographical issues. This particular story helps my students understand the importance of 'place' for me as a teacher with a heritage rooted in Pakistan, where most of my extended family lives. An opportunity is created to talk about how my own parents migrated to the UK, and how Pakistan was created in 1947. Unfortunately, in textbooks and much of the resources available on Pakistan, there is an over emphasis on negative imagery and representation. Little is taught about how British colonisers, such as Lord Mountbatten, were deeply involved in the tragic events leading up to the partition between Pakistan and India, which also left important territories under dispute, such as Kashmir, Bangladesh and many other communities (Sammar, 2022). By examining multiple perspectives through stories, students learn to question dominant narratives and recognise the biases inherent in traditional geographical education. This critical lens empowers students to challenge stereotypes, question historical injustices and advocate for a more inclusive and equitable representation of geographical knowledge.

STORYTELLING AS A DECOLONISING TOOL

As a geography teacher, I use storytelling as a decolonial tool to encourage students to develop a passion for geography and to enjoy the whole experience of learning through personal geographies. Puttick and Murrey (2020) have addressed the deafening silence of race in the geography curriculum in England, and recognised the need to address issues of race and racism within the context of geography education. They suggest that the absence of such discussions presents a challenging and 'racist' impression of a post-racial Britain, which thus reinforces negative connotations of race in societies that foster meritocracy and privilege. It is thus important to take heed of decolonial and antiracist learning. The silence of race can only be broken if we embrace a more engaging and inclusive learning environment, where incorporating storytelling can serve as a powerful tool to introduce decolonial conversations about these sensitive topics.

To decolonise is to remove the continual 'colonial domination', which calls for the dismantling of racist social constructs guised through 'Eurocentric world power ... to Indigenous-led demands for radical restructuring of land, resources and wealth globally' (Esson et al, 2018, p.385). Decolonising and antiracist storytelling can therefore address important contemporary global concerns such as environmental (in)justice and racial capitalism, illustrating how historical patterns of colonisation and racism contribute to unequal access to resources (see chapter 39) and environmental degradation. By connecting these narratives to current events, educators can inspire students to become agents of change in addressing global challenges related to climate change, resource distribution and environmental sustainability.

Post-colonial and decolonial geographers have used stories to help personalise historical narratives, allowing readers to connect emotionally with the experiences of individuals, similar to the one shared earlier. Jazeel (2019, pp. xi-xii) narrates his parents' stories by unravelling the realities of empire through 'its colonial geography and post-colonial remains; the colonial migrations and diasporas that cannot be uncoupled from imperial histories' which his mother embodies in a photograph he describes. The photograph is of his mother and aunt from the 1950s, both wearing saris ready to vote for the first time as British citizens with a powerful representation of South Asian women no longer subjugated to colonised rule. This depicts the struggles former colonised people endured before finally being accepted in British society as citizens of the land that once colonised them. Jazeel acknowledges the complex nature of identity, including factors like race, class, gender and geography, offering a more comprehensive understanding of how these aspects intersect and influence one another. By weaving together postcolonial and decolonial stories, geography teachers who may identify with a heritage in a former British colony, can embrace the intersectionality inherent in such experiences. This personalisation of narratives makes abstract concepts more tangible and relatable. Storytelling thus allows educators to connect personal narratives to broader global issues, fostering a sense of interconnectedness. By exploring stories that transcend national borders and highlight shared human experiences, students can develop a more empathetic and globally aware perspective. This interconnected view of the world challenges ethnocentrism and encourages students to consider the implications of geopolitical decisions on a global scale.

DIVERSIFYING PEDAGOGY THROUGH STORYTELLING

Each lesson taught in a geography classroom is the beginning of a story, whether it is a tale from a holiday or something that happened in the past. My own pedagogical approach to teaching geography is largely dependent on storytelling, especially as I enjoy engaging students in conversations about personal geographies linked to my own postcolonial realities. Incorporating storytelling in geography is an innovative way to encourage decoloniality and antiracism within the teaching and learning, as it is crucial for fostering a more inclusive and equitable geography classroom. Storytelling requires a deep engagement with diverse pedagogical approaches to teaching and learning geography, such as culturally responsive pedagogy (CRP).

Some may argue that stories may seem abstract in theory, however they can be very meaningful if told through CRP. Diversifying your educational approach through CRP helps teachers recognise and value the diverse cultural or spiritual backgrounds of students, as it aims to make education more inclusive and relevant (Pirbhai Illich et al, 2017). In the context of geography classrooms, incorporating CRP can enhance storytelling by considering antiracism, decolonisation and indigenous knowledge. Traditional geography education has often been complicit in perpetuating colonial narratives that marginalise and erase the voices and perspectives of indigenous communities and people of colour (Esson, 2018; Craggs, 2019). Maps, textbooks and curricula have historically centred on the experiences of colonial powers, reinforcing a Eurocentric worldview. Decolonising geography seeks to address this imbalance by acknowledging the rich diversity of human experiences and recognising the importance of local and indigenous knowledge. Beyond our personal and geographical histories, Esson and Last (2020) argue that antiracism in geography education involves actively dismantling discriminatory practices, challenging stereotypes, and fostering an environment that promotes recognition of an individual's humanity and 'saying the unsayable'. By incorporating antiracist principles into storytelling, educators can address systemic racism embedded in geographical education and contribute to a more just and equitable society.

Storytelling becomes a powerful tool in promoting antiracism by highlighting the stories of resilience, resistance and empowerment within marginalised communities. Through narratives that focus on the impact of racism on geographical spaces and the lived experiences of Black and Global Majority people, students can better grasp the intersectionality of race and geography. This approach encourages students to critically examine how historical and contemporary factors contribute to the uneven distribution of resources, power and opportunities across different regions. Figure 32.1 is an example of how I used the narrative of my own experience of racism to get my students thinking about ideas around race, (anti)racism, empire, colonialism and decolonising, all through CRP.

Blog example: **Bus Ride Home: 'Go back to where you came from!' by Iram Sammar**

Read an extract:

So there I was, alone, me and just this elderly man sitting directly in front of me. He leaned forward a little, so people couldn't catch his comment: 'You know what I mean, go back to where you are from – where your people are from. The same colour as you, the same smell!'

'Where do you think I'm from?' I said, after regaining courage after onlookers rolled their eyes at his arrogant behaviour. Somehow that gave me courage, I took it as a hint of solidarity. I leaned into his territory to replicate the intimidation. Slowly I spoke, 'Take a guess.'

The intense pressure of the confrontation switched his mood from conviction to passiveness. He was getting uncomfortable, so he continued, 'Oh, I don't know Tuvalu … I don't care. Just get out of here'. His shouting subsided as he was confused as to where I was actually from, which I found extremely amusing.

In my teacher soft tone I said, 'Sir, do you know where Tuvalu is?'

Read the full blog here:

https://salaamgeographia.com/2022/04/15/bus-ride-home-go-back-to-where-you-came-from/

Key discussion points:

1. What are your first feelings upon reading this extract?
2. Why do you think the elderly man is upset with the person getting on the bus?
3. Imagine you are the person getting on the bus. How would have you reacted?
4. How was geography used to change the elderly man's perspective on migration, race and racism?

Write a response to the author:

Figure 32.1 Engaging with a blog post and making sense of stories

CONCLUSION

This chapter has reflected on how I infused stories with the principles of decolonisation and antiracism together, to engage students in a journey that not only disseminates geographical knowledge but also challenges and dismantles stereotypes and dominant narratives about the wider world. Storytelling stands as a potent vehicle for decolonising geography and promoting antiracism education. By weaving diverse narratives into the curriculum, educators can dismantle entrenched colonial perspectives, challenge stereotypes and foster a more inclusive understanding of the world. Through our stories, students can connect their own personal experiences to global issues, develop critical thinking skills, and become advocates for justice and equity. As we embrace the power of storytelling in geography education, we take a significant step toward creating a more just, equitable and antiracist society.

Five reflection questions

1. Do you have a story of your own you share with your students, or could share in the future?
2. How do you currently address issues of decoloniality and antiracism in your geography lessons, and do you see storytelling as a valuable tool for this purpose?
3. How can storytelling help you include diverse perspectives and knowledges within the geography curriculum, particularly those that have been marginalised or omitted in traditional materials?
4. Can you think of how storytelling might help students connect with their own personal geographies to foster a deeper understanding of the impact of colonialism and racism in different contexts?
5. What barriers stop you from using storytelling in lessons and how could these be overcome?

> **DIGGING DEEPER – THREE RESOURCES TO DELVE FURTHER**
> - Eddo-Lodge, R. (2020) *Why I'm no longer talking to white people about race.* London: Bloomsbury Publishing.
> - Sammar, I.R. (2022) Unorthodox Geography Teacher. Available at: https://salaamgeographia.com/blog/ (Accessed: 28/05/2025)
> - Yancy, G. and Sharpe, C. (2018) *The fire now: Anti-racist scholarship in times of explicit racial violence.* London: Bloomsbury Publishing.

REFERENCES

Craggs, R. (2019) Decolonising the geographical tradition. *Transactions of the Institute of British Geographers*, 44, 444–446. https://doi.org/10.1111/tran.12295 (Accessed: 28/05/2025)

Esson, J. (2018) The Why and the White: racism and curriculum reform. *British Geography*, 1–8. Area Special Section.

Esson, J. & Last, A. (2020) Anti-racist learning and teaching in British geography. *Area*, 52(4), pp.668–77.

Fanon, F. (1952; 2008) *Black Skin, White Masks.* London: Penguin.

Jazeel, T. (2019) *Postcolonialism.* London: Routledge.

Pirbhai-Illich, F., Pete, S. & Martin, F. (2017) *Culturally Responsive Pedagogy.* Springer.

Puttick, S. & Murrey, A. (2020) Confronting the deafening silence on race in geography education in England: Learning from anti-racist, decolonial and Black geographies. *Geography*, 105(3), pp.126–34.

Sammar, I.R. (2022) Bus Ride Home: "Go back to where you came from!". https://salaamgeographia.com/2022/04/15/bus-ride-home-go-back-to-where-you-came-from/ (Accessed: 28/05/2025)

33. WHAT IS GEOGRAPHY TEACHING TODAY AS A BLACK BRITISH GEOGRAPHY TEACHER?

SHANIQUE HARRIS
@GEOGSHANIQUE / SHANIQUE.GEOGRAPHY@GMAIL.COM

'Ubuntu: I am because we are, and since we are, therefore, I am.'
(African proverb, southern Africa)

Today, there is a known shortage of geography teachers; the UK Government target for recruitment was not reached in 2021 or 2022 despite the reintroduction of teacher training bursaries (School Teachers' Review Body report, 2023). This situation is worse for the recruitment of teachers from ethnic minority groups and is a complex issue that has the potential to have significant implications for both geography education and the students who are being taught. Of course, the issues will be more specific and more profound in areas where there are larger groups of students from minority ethnic backgrounds and in this chapter, I would like to share my own perspectives and a classroom practice that I have used, as well as share some of the positive initiatives that have been taken by organisations and individuals.

It is important to address potential critics who may dismiss the specific area of teaching discussed here as a minor issue that only affects a very small proportion of the students and teachers in the UK. However, I want to emphasise that my intent is to be a voice of and for the under-represented teachers and students from ethnic minority backgrounds. Drawing on my unique experience as being one of those and having had the privilege of teaching and working with many, I seek to shed light on some of their perspectives and challenges. I also feel privileged to contribute to the discourse on geography education, providing insights that can benefit other educators and academics in the field.

REPRESENTATION

The significance of representation cannot be overstated. Having diverse geography teachers in schools not only serves as a beacon of positive role models for students from diverse backgrounds but also contributes to breaking down barriers that may impede their academic and professional aspirations. Statistics show that there is a notable under-representation of Black individuals both teaching and studying geography in UK universities. Furthermore, Desai (2017) makes reference to an observable 'awarding' gap that is present, indicating that Black students studying geography tend to graduate with grades that are proportionally lower compared to their white counterparts.

Personally, reflecting on my own experience as a Black student passionate about geography, I never had the privilege of being taught by a geography teacher who shared my cultural background. This lack of representation led to frequent questioning from peers and parents about my enthusiasm for a subject not commonly associated with individuals from ethnic minorities. Aisha Thomas (2022), in her book, underscores that positive representation transcends mere diversity in the room which can often be tokenistic. It necessitates transforming attitudes, behaviours, language and cultural norms while dismantling prejudiced narratives. As Thomas rightly emphasises, when students encounter educators who mirror their identities and experiences, it fosters a deep sense of belonging and serves as a powerful inspiration for them to pursue both academic and professional goals with unwavering confidence.

CULTURALLY RESPONSIVE TEACHING

Geneva Gay (2000) defines culturally responsive teaching as 'using the cultural characteristics, experiences and perspectives of ethnically diverse students as conduits for teaching them more effectively'.

I now teach in an international school in Turkiye with students from various backgrounds and was able to leverage the broad diversity of the students in a Year 7 class to enhance the learning experience for all in their initial geography lesson. In my Year 7 class of 11 students, I had students from eight different countries. In one of their first geography lessons, we used Google Earth to *fly* to each of their home cities and have a look at what the place was like. This activity was in part unplanned and was initiated by a student requesting to see what their home city looked like as they hadn't lived there for very long. The cities we explored included Addis Ababa, Baku, Bern, Baghdad, Cardiff, Denver, Dushanbee and Hanover. Of all the cities, the students were notably surprised by the presence of modern skylines in places like Addis Ababa and Baku, and the lack thereof in Hanover, a post-war industrial city located in north-central Germany. This opened up discussions about what influences places to look the way they do. I was able to introduce geographical terminology such as rural and urban, distinguish between human and physical features, and students discussed how the stereotypical view we may have of a place may hold true in certain parts of a country but is not always universal. During this 45-minute lesson, students shared their experiences of their cities, many of which they had only lived in for a short period of time, and the classroom atmosphere allowed for students (and myself) to hear and learn from each other and merge personal experiences with what we could see on Google Earth.

Through this activity we were able to explore 'place' as a concept and use the technology to increase students' geographic awareness while also incorporating critical thinking and enquiry through the discussion that came from it (Todd, 2007). This was all before they had any theoretical knowledge of geography at KS3. It meant that students began to hold an understanding that there can be multiple perspectives in understanding the world around us which I believe is a very important foundation for young geographers at this phase of their learning journey as it exposes them to an idea of multidimensional ideas of places and encourages curiosity and open-mindedness.

INFLUENCES OUTSIDE OF THE CLASSROOM

In 2022, I had the privilege of leading a project entitled 'Geography for All' which was funded by the Royal Geographical Society (RGS). The primary objective of the project was to increase the representation of students from diverse backgrounds choosing to pursue geography beyond GCSE, A-level and into university. The genesis of this project was rooted in the stark recognition from previous research undertaken by the RGS of the existing lack of diversity within the discipline at university level. The project provided opportunities for first-generation students from under-represented backgrounds in KS3 to visit universities and partake in workshops with students and professors (University of Northampton, 2022). As a student whose parents did not attend university, I can personally attest to the value of such experiences. They provide a crucial space for asking questions and gaining insight into what university life entails. These initiatives not only offer practical support but also serve as powerful motivators, inspiring students to consider higher education despite potential barriers.

We must remember that in the realm of British education, universities for years have played a pivotal role as a source of learning and research. The Secretary of State for Education has the authority to decide what is taught and studied in UK schools through the national curriculum, but for this to be done well they will draw on expertise and knowledge of the subject communities. These will likely be recognised subject bodies such as the Geographical Association (GA) and the RGS, but also academic researchers working on geographical study at higher levels. An absence of diverse representation in these influential spheres inevitably casts a shadow on the national curriculum, potentially overlooking perspectives, ideas and knowledge from non-white communities. This oversight may inadvertently hinder a comprehensive understanding of diverse viewpoints within the geography education.

QUESTIONS OF 'RACE' AND 'ETHNICITY'

Questions around identity include the notions of 'race' and 'ethnicity' and are integral to getting to know the students we teach as individuals (Sammar, 2024). Since the tragic death of George Floyd there has been heightened discourse on Black lives and Black experiences in different spaces. In education, discussions of decolonising the curriculum have gained prominence (Sammar, 2024). Despite encountering some criticism, the fact remains that this discourse exists and is growing. Decolonisation provides an alternative theoretical framework and methodology for critique through students' and teachers' lived experiences (Winter, 2022). There are teachers, academics and even students who are taking the time to articulate their perspectives and, as professionals in geography education, I believe it is imperative that we try to engage with this dialogue and not dismiss it.

Ultimately, our shared goal as educators is to prioritise the academic success and well-being of our students. While terms like 'decolonising' and 'anti-racist teaching' may be intimidating to some, they are not meant to instil fear; rather, they serve as invitations to embrace and learn from those with valuable experiences and insights. It is crucial to recognise the importance of listening to individuals with diverse perspectives and informed recommendations, to create teaching and learning environments where both students and teachers can excel and achieve their best potential.

FUTURE OF BLACK STUDENTS AND TEACHERS

Esson and Last (2020) highlight the importance of actively embedding Black staff and students in decision-making processes that influence the development of pedagogical approaches and curricula. They make it clear that it is vital to acknowledge and appreciate the subjectivities, epistemologies and ontologies specific to Black communities within such educational frameworks. This is essentially saying that recognising and valuing the diverse perspectives, knowledge systems and fundamental beliefs held by individuals within Black communities is crucial for designing an education system that is truly inclusive and equitable. By actively involving Black individuals in shaping educational policies and practices, you not only honour our unique experiences but also enrich the educational experience for everyone involved. This approach ensures that the subjectivities, epistemologies and ontologies inherent to Black communities are not only acknowledged but integrated into the very fabric of the educational system, creating an environment where all voices are truly heard and respected.

What is most profound is the advocacy for proactive anti-racist measures to guarantee equitable opportunities for Black staff and students, particularly when it comes to accessing, participating in and succeeding in geography. Such initiatives should be pitched at fostering change, ensuring equity for Black students and staff within British geography at a structural level.

For students, there have been a number of initiatives that have been taken by universities to expose under-represented students to geography at university level and also to actively support those who are already studying at higher levels. These include but are not limited to the Fi Wi Road project (Black Geographers and RGS), Encompass Project (York St John University) and the Equator Research school, showcasing institutional commitment to widening participation in geography education.

Within the landscape of schoolteachers, it is clear that supporting teachers from minority ethnic backgrounds, especially within the UK context, should involve the recognition and understanding of the challenges they may face within the school system. By implementing strategies to enhance the professional experience of these teachers, whether it is through improved methods to progress with equitable opportunities, tailored professional development programs or mentoring programmes. Specific strategies like these can contribute to the broader goal of fostering a sense of belonging, professional growth and empowerment among teachers. In turn, this comprehensive approach not only uplifts individual educators but also enriches the educational landscape as a whole, promoting true diversity, equity and a more vibrant learning experience for students across the UK.

CONCLUSION

In conclusion, as educators and particularly geography teachers, we are entrusted not only with the dissemination of knowledge but also with the responsibility of shaping the future. The shortage of geography teachers, particularly from ethnic minority backgrounds, underscores a critical gap in our educational system. Embracing the spirit of the 'Ubuntu' quote — recognising that 'I am because we are' — compels us to proactively address this imbalance. With regards to decolonial and anti-racist practices, the Ubuntu philosophy implies that the collective wellbeing and success of teachers and students regardless of their background are intrinsically linked, hence interconnectedness within the educational community is imperative.

While commendable efforts are being made by some in classrooms to elevate the significance of geography education for students from diverse backgrounds, to see a shift, teachers, schools and institutions must continue to go further in order to truly embody the diversity inherent in geography's essence. Deliberate and collective efforts are required that not only acknowledge but actively embrace non-white perspectives. Only with such proactive steps can we ensure that the subject is taught in a manner reflective of its inherent diversity and claim that we are authentically creating an inclusive space where every student feels seen, heard and valued.

Five reflection questions

1. Who in your class, especially students from diverse backgrounds, might benefit from better representation in geography lessons, and how can you address this?
2. What specific strategies can make geography lessons more interesting and relatable for students, taking into account diverse backgrounds?
3. Where in your class can you create a more inclusive environment, fostering a sense of belonging and inspiration for students from diverse backgrounds?
4. Why is it important to engage with discussions on decolonising the curriculum in geography, and how might it impact your students' learning experiences?
5. How can you actively connect and collaborate with teachers who bring different cultural perspectives to ensure a more diverse and inclusive approach to geography education in your school community?

> **DIGGING DEEPER - THREE RESOURCES TO DELVE FURTHER**
> - Book: *Representation Matters* by Aisha Thomas - explores the significance of representation in education and its impact on students, providing practical insights for teachers.
> - Activity: Who Am I? - exploring students' personal geographies through a classroom activity (Sammar, 2024).
> - Podcast: 'Have You Heard George's Podcast?' by George the Poet - thought-provoking discussions on culture, identity and education, offering valuable insights for teachers and students, particularly from multicultural cities like London.

REFERENCES

Desai, V. (2017) Black and minority ethnic (BME) student and staff in contemporary British geography. *Area*, 49 (3): 320-323. https://rgs-ibg.onlinelibrary.wiley.com/doi/10.1111/area.12372 (Accessed: 28/05/2025)

Equator Research School. https://equatorresearchgroup.wordpress.com/ (Accessed: 28/05/2025)

Gay, G. (2000) *Culturally Responsive Teaching: Theory, research and practice.* New York: Teachers College Press.

Patterson, T. C. (2007) Google Earth as a (not just) geography education tool. *Journal of Geography*, 106(4), 145-152.

Royal Geographical Society (n.d.) Geography for all. https://www.rgs.org/schools/geography-for-all (Accessed: 28/05/2025)

Royal Geographical Society (n.d.) Geography of geography: the evidence base. https://www.rgs.org/about-us/what-is-geography/geography-in-schools/geography-of-geography-the-evidence-base (Accessed: 28/05/2025)

Sammar, I. (2024) Decolonial and anti-racist pedagogy through personal geographies. *Teaching Geography*, 49(1), 22-25.

School Teachers' Review Body 33rd report, available at: https://assets.publishing.service.gov.uk/government/uploads/system/uploads/attachment_data/file/1170121/STRB_33rd_Report_2023_Web_Accessible_v02__1_.pdf

Thomas, A. (2022) *Representation Matters*. London: Bloomsbury Publishing.

University of Northampton. (2022) Young people chart careers in geography at UON. University of Northampton. https://www.northampton.ac.uk/news/young-people-chart-careers-in-geography-at-uon/ (Accessed: 7 February 2024)

Winter, C. (2023) The geography GCSE curriculum in England: a white curriculum of deceit. *Whiteness and Education*, 8(2), 313–331.

York St John University, Encompass. https://www.yorksj.ac.uk/working-with-the-community/schools-and-colleges/outreach-projects/encompass/ (Accessed: 17 September 2024)

34. HOW CAN WE MAKE OUR UK GEOGRAPHY CONTENT LESS ENGLAND-CENTRIC?

ALICE MCCAUGHERN
@TEACHWITHALICE

'If geography itself has any significance it is that we are made to lift our eyes from our small provincial selves to the whole complex and magnificent world.' (Richard Burton, 1821-90, to the Royal Geographical Society)

In recent years, tremendous efforts have been made to make our geography curriculums more representative and inclusive, all with the aim of increasing depth of knowledge. It is important to portray the complexity and nuance of different places, so to not reduce them down to a single story such as poverty of Africa, but instead a place of stark contrasts of physical landscapes and rapidly growing economies (Faloyin, 2022). This work can be incredibly challenging to do well and within the restrictions of contact time with students and available time for teachers to plan properly.

CREATING TOPICS WHICH TEACH WHAT THEY SAY ON THE TIN

WHAT STORY ARE WE CURRENTLY TELLING ABOUT THE UK?

As a Northern Irish geographer, who was trained and only ever worked in English schools, the UK case studies taught all seemed to stem from one place – England. This was despite there not needing to be an English example, simply a UK example. And if we did venture beyond the English border, it was a focus of rural decline in the Outer Hebrides, with the response from one student being, 'Who on earth would want to live there?' (see Figure 34.1). Students also express genuine intrigue about where I am from and what it is like as a place. Their basic knowledge of wider UK geography was fairly limited! With our eyes (and curriculum) fixed on England, we are missing out on the spectacularly beautiful and equally interesting physical and human geography of not just Northern Ireland but Wales and Scotland too.

GCSE topic	UK-based case study
Rivers/Coasts	River Tees Banbury river management Dorset Walton on the Naze
Urban	London – urban growth, Stratford/Olympic Park, Docklands, traffic management
Resources	England – water redistribution
Ecosystems	A British woodland – Epping Forest
Hazards	UK weather – Beast from the East – impacts on England
Changing Economic World	UK economy focus North–South divide (England) Rural decline – **Outer Hebrides** Suburban growth – Cambridge Improving transport (of England) Science park – Cambridge Sustainable industrial development – Sunderland car plant

Figure 34.1 Summary of UK-based case studies for GCSE specification. (Representation of one English school)

Look at Figure 34.1. This England-centric choice of case studies isn't isolated to one scheme of work, school or multi-academy trust, but is common across different topics, in different schools.

As there is so little room to teach more, we have to critically examine the topics we already teach (see chapter 16) and make sure they have valuable geography within them, and that the content is actually what is said on the tin. For example, if we want to teach about valuable UK geographical landscapes, then we need to make sure that is what we are teaching. If we need to teach about extreme UK weather events, we need to include impacts from all over the UK, rather than just the facts and figures about what happened in England – then it's just an English weather event. If we are to examine economic inequality within the UK, we should look beyond the North-South divide (of England) and examine the economic divide between England and the rest of the UK. This helps students, particularly those growing up in England, to have a less England-centric view of the UK. (See also chapter 16.)

WHAT MIGHT BE STOPPING US, REALLY?

So, what is stopping us for making our case studies for UK-focused topics not focused on England? • We already have resources planned and available. • It's familiar to students already. • It's in the textbook.	But, • Northern Irish, Scottish and Welsh teachers have resources around their countries too. Time to collaborate? • With an increasingly international, non-local intake of students, are England-centric case studies still always familiar to your class?

In a time-pressured, heavy-workload job, it is much easier to keep doing what we are doing, with resources ready to go and familiar. But there are geography teachers in and from Northern Ireland, Scotland and Wales who have resources about these UK countries too. The Geographical Association Annual Conference 2023 was all about collaborating and this is an area where collaboration could help teachers find materials and knowledge to make our UK-focused topics as rich and authentic as possible.

Another reason why we might keep to the status quo is that students supposedly know their regional geography already and a head start is always an advantage. It is so important for students to have an understanding of their local geography. In designing schemes of work, there is a real challenge in getting the balance right between local, national and international case studies which add sufficient variety and nuance to ensure we avoid telling the single story. However, as class populations have become more diverse, multicultural and multinational, there may be a misconception of pre-existing awareness and knowledge.

It's in the textbook. Textbooks, if used appropriately can be an excellent resource to point the overstretched new early career teacher (ECT) or seasoned pro in the right direction of where to start and so it's tempting to pick the example from the textbook or exam board scheme of work (SoW). Therefore, exam specifications need to lead the way in exemplifying a more diverse range of case studies and content to allow teachers to teach a much more inclusive curriculum, even until Year 13.

HOW CAN WE STRIKE THE BALANCE OF SPECIFICATION REQUIREMENTS AND MAKE THE BEST CHOICES OF CASE STUDIES?

Changing schemes of work, even just a case study, can take considerable time in decision making and remaking resources and assessments to fit your new choice. One also doesn't want to change a case study to tick boxes or for more tokenistic reasons. Therefore, pick areas where you can genuinely get an equally good or even better geographical example or case study that will, without doubt, enrich your curriculum and aid students' understanding of the theory and of the place. Gain your colleagues' perspectives as they may have examples that you can tap into.

You might not need to change your case study at all. For example, a case study for an extreme weather event for GCSE geography which is commonly used is Beast from the East (February 2018). Despite this being a weather event that not only impacted all four countries of the United Kingdom and also Ireland and parts of mainland Europe, many resources just focus on what happened in England (BBC News, 2018). Give students a few more facts about what happened in Wales, Scotland, Northern Ireland as well as England and they can discuss impacts at a local, national and even international scale (if your specification requires), not only aiding their understanding of the event itself but supporting development of higher tier-skills to write like a geographer by considering events in regard to scale.

Local case studies can offer opportunities for fieldwork. However, using case studies from further afield can offer opportunities for GIS and virtual fieldwork (see chapter 11), with students using Google Earth to conduct land use and environmental quality surveys. The online collection of digitised maps from the National Library of Scotland can support development of map skills. Within the topic of Changing Places (AQA A-level) this can be used for applying theory or investigating a new case study.

It is impossible to fit everyone and everything into the curriculum even if we start in Early Years Foundation Stage (EYFS), however, if we make small tweaks and changes here and there, it helps us strive in the right direction to help 'to lift our eyes ... to the whole complex and magnificent world'.

Five reflection questions

1. Do your topic titles cover what you aim to teach (e.g. UK landscapes which focus on more than just England)?
2. What regions does your school have a focus on?
3. What case studies have potential to highlight another area of the UK?
4. What case studies does your school need to keep?
5. Where within the curriculum could you as a teacher gain greater knowledge of other examples?

DIGGING DEEPER - THREE RESOURCES TO DELVE FURTHER

- Google Earth case studies - Andy Funnel - https://earthcasestudies.com/ (Accessed: 28/05/2025)
- National Library of Scotland maps - https://maps.nls.uk/ (Accessed: 28/05/2025)
- GA Welsh Special Interest Group - https://geography.org.uk/welsh-special-interest-group/ (Accessed: 28/05/2025)

REFERENCES

Faloyin, D. (2022) *Africa is not a country: breaking stereotypes of modern Africa*. London: Harvill Secker.

BBC News. (2018) Europe freezes as "Beast from the East" arrives. BBC News. [online] 28 February 2018. https://www.bbc.co.uk/news/world-europe-43218229 (Accessed: 30/12/2023)

35. LATINX VISIBILITY
IZZY WOOD
@GEOGRAPHYJOG

'A lot of stereotypes that I've faced is that I sell drugs or that I am somehow related to a drug cartel and like it's a joke that gets very boring very quickly because people say it so much. I guess it's something people think because of the movies that they watch and the way that the media portrays us ... Latinos have the right to be seen.' (Alvarado, 2018)

Did you know that most UK residents originating from the entire continent of South America, Central America and some Caribbean islands don't have a box to tick for the ethnic group section on the UK census? There have been various campaigns for the term Latin American/Latinx/Hispanic to be included and formally recognised before the 2021 census, but they were not approved. In comparison, the options for Asian people include Asian Bangladeshi, Asian Indian, Asian Chinese, Asian Pakistani, Other Asian. There are also limited options for other ethnicities, such as the binary of Black African or Black Caribbean. What if you are African but not Black, or are Latinx and Black? Do we explore these limitations of quantitative data with our students?

This chapter discusses Latinx history and diaspora, with specific focus on Colombian people in London. I suggest three curriculum artefacts that can be used in the classroom and school as an institution. My intention is not to suggest finite resolutions to the issues, but rather acknowledge them and raise awareness.

LATINX TERMINOLOGY

In Spanish, Latino refers to men, Latina refers to women and Latinos refers to both men and women. Latinx is an inclusive gender-neutral term that rejects the notion of gender as a binary. I use Latinx 'with the appreciation of Latine as an alternative gender neutral term' (Andrade et al, 2023). I have opted to use Latinx as the least offensive, rather than the correct term (Sinclair, 2023). I appreciate people born prior to the 1990s, on the whole, use the terms Latino/a, whereas younger generations, including those in our classroom, use Latinx.

I am White British and recognise the associated privileged positionality and mistakes I could make by not being Latinx. My writing is informed through research and lived experience. I want to make space for Latinx voices, with the Latinx voice, not as an authority on it (Sinclair, 2023).

Please note I am not using the terms 'South America' and 'Latin America' interchangeably.

LATIN AMERICA COLONIAL PAST AND PRESENT

Latin America is a place, an identity, an ethnicity, a culture, but none of these things are homogenous. Even within one country, Colombia, there is a vast difference between Rolos, Paisas, Costeños and Caleños, people from Bogotá, Medellín, Northern Coast and Cali, respectively. In the limited space here I deconstruct a small amount about the term 'Latin American' and encourage you to dig deeper.

South America is often considered a monolingual continent. The assumption is that everyone speaks Spanish, with Portuguese-speaking Brazil often an afterthought. The reality is more complex. The Latinx identity can also include Central American countries that are politically classified as part of North America, as well as some Caribbean islands like Cuba, the Dominican Republic and Puerto Rico (a territory of the USA). French Guiana is still governed by France and the official language is French. Suriname was a Dutch colony; Dutch is the official language, but English is widely spoken (Suriname, cia.gov, 2023). Guyana was initially a Dutch colony, later a British colony. It is the only South American country in which English is the official language and has shared 'cultural and historical bonds with the Anglophone Caribbean' (Guyana, cia.gov, 2023). The official languages of Paraguay, Peru and Bolivia include both Spanish and Indigenous languages – Guarani in Paraguay, Quechua in Peru and Bolivia. Colombia's official language is Spanish but there are 65 Amerindian languages, which 'can be grouped into 12 language families, including Arawakan, Cariban, Tupian and Quechuan' (translatorswithoutborders.org, 2023). Due to the wording on census questions there is no room to verify whether sections of the population are monolingual in languages other than Spanish (translatorswithoutborders.org, 2023).

Latinx identities are complex in other ways, beyond multilingual experiences; one other aspect is due to the transportation and enslavement of African people. Cartagena, Colombia was by far the largest single port of enslaved debarkation in the Spanish Americas from as early as 1537 (Landers et al, 2015). Records between 1573 and 1640 document 463 ships with 73,000 enslaved Africans, but many more were smuggled or escaped. Estimates go up to 250,000 enslaved people during Spanish colonisation which ended around 1819 (Landers et al, 2015; theworld.org). The Spanish, seeking many things including power, land, empire and gold, killed hundreds of Indigenous people, burned their villages and forced enslaved people to extract gold from tombs of Sinú Indians (Landers et al, 2015). Today, 30 per cent of Afro-Colombians live in 'multidimensional poverty', meaning they lack access to basic needs such as adequate housing, secure employment and schooling (theworld.org). This is partly due to institutional and structural racism in Colombia. Between 1540 and the 1860s, over 5.5 million African people were forcibly taken to Brazil, making it the country where the highest number of enslaved people arrived (brazillab.princeton.edu, 2023).

Brazil and Colombia have recognised the historical and civic neglect of their Afro-Colombian and Afro-Brazilian communities and now offer legal and cultural recognition. Irrespective of these attempts, racism is still evident in both nations (Landers et al, 2015). Francia Márquez, the first black female Vice President in Colombia, has been subject to discrimination and racism, putting this down to the legacy of colonialism and slavery (nbcnews.com, 2023). Akala speaks of his experience of racism in Brazil in his book *Natives*. He explains: Brazil is 'where racial slavery lasted the longest, and where by far the largest number of Africans were taken' (Daley, 2018, p.54). He says, 'like all of the other former slave colonies of the Americas, [Brazil] worked to extend and maintain white supremacy long after slavery had ended, despite claims to being a racial democracy' (Daley, 2018, p.54).

Regardless of this evidence, it is often felt in Latinx communities that their experience of the colonial past and the racialised colonial present has been overlooked even when seeking to decolonise. Latinx people suffered as a result of colonialism. We must understand this in order to deconstruct history.

The above demonstrates two things. First, the complex history of the Latin American colonial past and the impact it had on Indigenous people, their lives and their territories. Secondly, it outlines the diversity of the Latinx community. A Latinx individual could have African, European and multiple Indigenous heritages and identities. As such, adding one box on the UK census for Latin Americans would be reductive. One box, however, would provide some visibility to the community, and this is what campaigners have asked for.

LATINX IN LONDON

While the UK census does not collect the data, statistics from countries of origin can give some indications regarding numbers of Latin Americans abroad. There have been many waves of Latin Americans arriving in the UK and London is now home to one of the largest Latin American populations in Europe. From the 1970s onwards, relatively large numbers of Latin Americans arrived either as political refugees and/or with work permits. In the late 1990s and early 2000s, the Latin American population in London grew significantly as complex political and economic upheavals coincided with US border controls tightening, and in Colombia specifically, the armed conflict intensified (McIlwaine, et al, 2011, p.13). Trust for London suggests a quarter of a million Latin Americans live in the UK, with over half living in London, around 145,000 people (TrustforLondon.org.uk). Colombians make up around a fifth, with large concentrations in Lambeth and Southwark

(TrustforLondon.org.uk). In 2009, the Ministry of External Relations in Colombia stated that some '4 million Colombians lived abroad', with 19.4 per cent in the UK (McIlwaine et al, 2011, p.14). Many Latin Americans arrived from Spain after the global financial crisis but before Brexit, using the EU settler status route. Therefore, a significant proportion of Latin Americans will be un- or under-recorded as they are instead registered as Spanish (McIlwaine et al, 2011).

While the UK census does not record Latinx people specifically, the 2021 census can tell us something. Compared to 2011 data, in the 2021 census those who chose to specify their ethnic group through the 'Other ethnic group' almost tripled. The largest of these ethnic groups includes Hispanic or Latin American which amounts to 76,000 people and 0.1% of the 'Other ethnic group' (ons.gov.uk). Despite this data, the Latinx community and their lived experience, is not visible or officially accounted for through the census, NHS, Universal Credit, schools, universities or workplace data collection. We should insist our schools add a Latinx box on ethnicity surveys because they are heavily data driven; while there are many things wrong with this, if there is no data, Latinx students get forgotten or deprioritised.

We know that a quarter of Latin Americans in the UK work in low-paid elementary jobs such as cleaners, kitchen assistants, waiting staff and security guards (trustforlondon.org). Many individuals work two or three jobs with short sleeping windows to make ends meet. In addition, Latinx access to health services remains low. It is estimated that 1 in 6 are not registered with a GP and nearly 7 in 10 have not accessed dental services in the UK (trustforlondon.org). This became a particularly serious issue during the pandemic when people were concerned about missing out on vaccinations (Johnston, 2021).

Additionally, when an ethnic group or cultural community is not recorded as residing in a specific location, they are at risk of being rendered invisible. Governments and local councils rely on statistics and quantitative data. With no census data on the number of people existing in a place, and then being displaced via gentrification, it is easy to remove them as if they were never there in the first place.

LATIN ELEPHANT

Gentrification has been and is a threat to Latin Americans living in both South and North London (Seven Sisters) for over 10 years. The project Latin Elephant has been gathering research and support for the community facing displacement in Elephant and Castle, Southwark, London; interviews here are taken from their research. Regeneration is occurring in deprived boroughs where diverse ethnic populations have settled and now fear displacement. At the start of the 1990s, Elephant and Castle was relatively derelict. Latin American businesses were attracted to the cheap rent and breathed life into the area. The Latin American community in Elephant and Castle built an economic hub, home to more than 80 businesses (London's Latin Quarter, 2015).

The growth of the Latin American population in Elephant and Castle snowballed over the last 30 years and people have expressed an emotional attachment to the place. One interviewee says, 'I find everything that brings me closer to my homeland' (London's Latin Quarter, 2015). Another interviewee explained that there was a 'Colombian cafeteria where we could find *our* products: natural juices, buñuelos, arepa, chicharron ... and it became a meeting point for us Colombians' where they could 'help each other find work' (London's Latin Quarter, 2015). A shop owner explains that people don't just go to his shop to buy things but also 'because they don't speak English and they have just arrived. They're looking for information about National Insurance, the GP, work, accommodation, translation of work contracts, to fill out forms, to make a phone call' (London's Latin Quarter, 2015). This is evidence that genuine Latinx social capital has been built in Elephant and Castle. Each Latin American country has their own style of haircuts, music, clothing, jokes, slang and food and to find the style you like from your homeland is an important part of maintaining, producing and reproducing personal identity.

These collective cultural community centres are vital and should be protected by the city. The regeneration, however, has made things more expensive, and the fear is that 'within 10 years we might not exist commercially in this area ... I'd call it commercial displacement where the ethnic minorities are displaced by economic development'. Latin Elephant and the community are asking that they 'continue being part of Elephant and Castle's history because we helped to write that history and we oppose any attempt to move us from this area that we helped to build.' (London's Latin Quarter, 2015).

For one fieldtrip I take my students on a walking tour of Elephant and Castle and Brixton to see the fortress developments, the post-modern architecture and the gentrification, but mostly to eat Colombian snacks, see Colombian fashion, food and music, and listen to so many people speaking Spanish in London.

CURRICULUM ARTEFACT 1: *MY UNCLE IS NOT PABLO ESCOBAR*

Bringing art and culture into the classroom can shed light on invisible and displaced communities. A trilingual play called *My Uncle Is Not Pablo Escobar* was supported to be produced by LatinXcluded, The Advocacy Academy, Arts Council England, Kings College London, among others. The play blurb reads: 'Latinx Women from South London take centre stage and dare you to call them invisible ... From not having a box to tick to challenging toxic stereotypes ... they confront the audience with what it means to be both Londoner and Latinx' (Andrade et al, 2023).

The opening quote of this chapter is by Elizabeth, the co-creator of this play, about how she feels her community is misrepresented. This play has also been a campaign success: 'Elizabeth managed to secure Latin American as an ethnicity on all Arts Council forms' which is the 'first example of a major organisation diverting from the census' (Andrade et al, 2023). The play gives voice to anyone facing visa complications, stereotypes and having dual identity. The aim is for girls to be able to say 'that's who I am. I've gone through these problems. I've gone through this, she's very much like me, she's very much like my mum, like my sister, we have experienced the same hardships and the same complications' (Andrade et al, 2023).

This is what we should aim for in our classrooms too – to give voice and space to marginalised groups (see chapter 31). The immigration interview in the Appendix of the play can be coded or used as a curriculum artefact. One part in particular illuminates the stress the visa applications can bring on individuals: 'I'm always worried that for some reason one of us will get our visa renewed and the other will be denied' (Andrade et al, 2023, p. 72).

This may be how some of our students, or indeed colleagues, feel on a day-to-day basis. Awareness of this can build empathy and understanding between your own school community and not simply be a tool to study migration. This is particularly important when considering bringing the Home Office into a school or college for a careers fair, which could make some students or colleagues feel unsafe if their family are dealing with visas and status to reside in the UK. This is an aspect of inclusivity and safeguarding some institutions or individuals do not consider. In their 'On Co-Creation' section, they discuss the system of calling in and calling out, as well as Johari's window model (see Figure 35.1), (Andrade et al, 2023, Tommy on co-creation). These methodologies are central to creating a democratic or decolonised space, classroom or institution. Such methods can be used between colleagues and students. For example, if I teach about Pakistan with a Pakistani student, I am in the box blind spot where there will be some knowledge and experiences *not known to self, known to others* and can acknowledge this student as an expert or leader. In addition, it could be used as a way to discuss curriculum or school policies with other teachers, managers or senior leaders, particularly when advocating for decolonisation. As a teacher or leader, decolonising curriculum may be *known to self, not known to others*, so in the hidden area box. The use of this model levels discussions when trying to navigate institutional hierarchy. This play has many valuable lessons, and I hope you can implement some in your own classrooms.

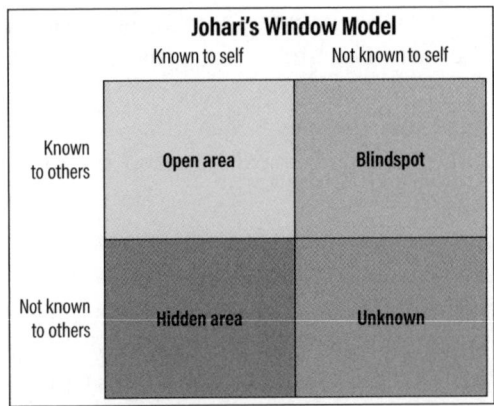

Figure 35.1 Johari's Window Model

CURRICULUM ARTIFACT 2: MEDALLO CITY

Maluma is a Colombian, international singer, model, actor and the voice of Mariano in Disney's *Encanto*. *Medallo City* is his song about Medellín. The lyrics and music video can be used as curriculum artefacts to teach about the city of Medellín. I use Medellín as an example of a sustainable and innovative city as well as an example of an informal representation of place. I teach about Medellín at A-level when studying how past and present connections within and beyond a locality can shape places and the lives of people who live there. Lastly, I use it to show how artistic sources such as art, poetry, photography, song lyrics and other qualitative sources can effectively represent the character of a place over time.

Mariano's aim in this song is to tell the world about the city he is from; each verse suggests a place to visit. He sings in Spanish but translates as: 'I'm telling you about my city, I'll take you from the north to the south and on the graffiti tour, expressing myself with the local slang, I represent Medallo [Medellín] on a global level' (Lodoño Arias, 2020). Mariano is proud and inviting the world into Medellín to see it for themselves. The music video showcases different shots from the city, which has been historically stigmatised as a city of drug lords and crime. It also features important monuments; the Metro, street football, art and many exotic flowers (rmadesign.tv, 2020). There are many street food sellers featured at different times of day and for my students, many of whom have never left the UK or Europe, it is a cultural feature of Colombia demonstrated far better here than in any picture or sentence in a textbook. The music video brings the day to day to life in a way which challenges preconceived notions of the 'Global South'. Stereotypes about the 'Global South' land heavy in my classroom as a college teacher. Students pick these up through non-specialist (and probably specialist) teachers in primary and secondary school as well as parents and the media. It is a challenge to have students for only two years, get them to A-level standard and deconstruct and decolonise their minds before they go off into the world with these notions more or less fixed. For this reason, this is not the only music video I use in the classroom.

The last verse, however, is a clear signal to the world to stop talking about the Colombian 'single story' (Adichie, 2009). Mariano sings: 'Medellín isn't Pablo Escobar, there's a better story to tell'. In his music, Mariano demonstrates that Medellín, the city of his birth, is not synonymous with violence and drug trafficking (salsaeslacura.com, 2020). Maluma says: 'I feel that in this moment I have a big responsibility to continue to change the face [perception] of the country [Colombia]' (salsaeslacura.com, 2020). Despite Maluma's best efforts, Pablo Escobar has been 'reincarnated as a jest or as a brand' globally since his death in 1993 (Hernández, 2020). To give this some context, Maluma was born in 1994, three days before Harry Styles. Could you imagine a Harry Styles song asking the world to stop talking about a UK historical figure active around the same time as Pablo Escobar?

CURRICULUM ARTEFACT 3: JBALVIN, SCHOOLS BBC 1XTRA

When teaching about cultural diversity, social segregation and economic inequality in London, I use an interview from BBC 1Xtra between the British presenter Ace and JBalvin to demonstrate the misrepresentation of Colombians. JBalvin is a Colombian, international singer who has collaborated with Beyonce and Cardi B and has clothing lines with Guess and Nike.

This interview demonstrates and challenges a major stereotype about Colombia and Colombian people. JBalvin spends 7 of the 11-minute interview schooling Ace on why Pablo Escobar should not be glamorised. Ace tells JBalvin that Medellín, JBalvin's home city, is famous because of the Netflix show *Narcos*. Ace asks if it's safe to visit and JBalvin replies: 'I am sorry, but *Narcos* is telling a story that happened 30 years ago and I wasn't even born you know, I was a little baby at the time so the bad thing about that is like, people don't get that that was 30 years ago.' JBalvin explains that there are many good Colombians and questions why people constantly refer back to Pablo Escobar and glamorise his life. JBalvin says: there's a lot of people like James Rodriguez, Falacao and Shakira who 'are representing real well but then they pull out with this Netflix thing and its taking us back to 30 years ago'. He continues: 'the way you see it is as if you are watching a movie eating popcorn but what if you were right there, you know, with the guns all over and bombs and your mum crying because your brother die. You get me? It's not fun' (JBalvin, BBC1Xtra, 2018).

Colombians like JBalvin work hard to break out into the global music scene. They deserve dignity and freedom from negative representations of their community. This is just one high-profile example of how Colombians are routinely associated with Pablo Escobar, his violence and drug use. It causes their validity and legality to be doubted and questioned. Such a stereotype contributes to individual, institutional and structural racism. Visibility of this issue is vital in order to tackle it: glamorising and glorifying murderers is not OK.

OTHER IDEAS FOR CURRICULUM ARTEFACTS

- 'Ya Rayah' by Rachid Taha – 'Hey, immigrant', an Arabic song about the hardships and regret of migration.
- 'Latinoamérica' by Calle 13 – a song in Spanish which discusses the essence and struggles of being from the continent.
- 'Sonny's Lettah' by Linton Kwesi Johnson – an anti-sus poem about Brixton.

CONCLUSION

This chapter briefly discusses Latinx history and status in the UK. It looks specifically at London and one of the many stereotypes faced by the Colombian community. It presents three curriculum artefacts that can be used to demonstrate and challenge this. I acknowledge what I have written is only a part of and not the full story, and that the examples used could reinforce a negative stereotype rather than challenge them.

The key takeaways are, first, encourage students and teachers alike to investigate the complex Latinx history in the Americas and the UK. Secondly, advocate for the use of qualitative sources and to be critical of quantitative ones. Lastly, and most importantly, emphasise the importance of including the Latinx voice in the classroom, on administrative paperwork and in decolonial narratives.

You can download my lesson from TES (https://www.tes.com/teaching-resource/-12930408).

> **Five reflection questions**
> 1. Whose voices are invisible in the stories you are telling?
> 2. What resources or curriculum artefacts are most interesting and accessible for students and you?
> 3. Are you counterbalancing the limitations of quantitative data with qualitative sources?
> 4. Do you or your colleagues speak another language which can shed light on an invisible community through curriculum artefacts?
> 5. How can you deconstruct and decolonise throwaway comments about marginalised communities in textbooks?

> **DIGGING DEEPER – THREE RESOURCES TO DELVE FURTHER**
> - The Motorcycle Diaries – Walter Salles' 2004 film
> - Colombia Calling – podcast
> - Oblivion – memoir by Hector Abad (2006)

REFERENCES

Adichie, C. N. (2009) The danger of a single story. TEDGlobal 2009. https://www.ted.com/talks/chimamanda_ngozi_adichie_the_danger_of_a_single_story?language=en (Accessed: 22/10/2023)

Alvarado, E. (2018) The stimulus for *My Uncle is Not Pablo Escobar* filmed in 2018 by Toby Lloyd.

My Uncle Is Not Pablo Escobar. https://brixtonhouse.co.uk/shows/my-uncle-is-not-pablo-escobar/ (Accessed: 26/10/2023)

Andrade, V., Alvarado, E., Wray, L. & Ross-Williams, T. (2023) *My Uncle Is Not Pablo Escobar*, London: Bloomsbury Publishing.

JBalvin interview for BBC 1Xtra (31 January 2018). https://www.bbc.co.uk/programmes/p05wpb6j (Accessed: 28/05/2025)

Brazillab.princeton.edu. (2023) *Racialized* Frontiers: Slaves and Settlers in Modernizing Brazil. https://brazillab.princeton.edu/research/racialized_frontiers#:~:text=Courtesy%20of%20Firestone%20Library.,1540%20and%20until%20the%201860s (Accessed: 29/10/2023)

Daley, K. (2018) *Natives: Race and Class in the Ruins of Empire.* London: Two Roads.

Hernández, D. The Second life of Pablo Escobar: Through the hippopotamus walk. *Casqui*, November 2020, 49(2) pp.125–43.

McIlwaine, C. J., Juan Camilo, C. & Linneker, B. (2011) No Longer Invisible: The Latin American Community in London. Trust for London.

London's Latin Quarter. (2015) https://www.youtube.com/watch?v=DRC2cyhpzAM (Accessed: 28/05/2025)

Landers, J., Gómez, P., Polo Acuña, J. & Campbell, C. J. (2015) 'Researching the history of slavery in Colombia and Brazil through ecclesiastical and notarial archives' in Kominko, M. (ed) *From Dust to Digital: Ten Years of Endangered Archives Programme.* Cambridge: Open Book Publishers.

Lodoño Arias, J. L. (Maluma) (2020) Medallo City

Salsa es la cura (28/12/2020) Medallo no es Pablo Escobar, mensaje de Maluma al mundo. https://salsaeslacura.com/medallo-no-es-pablo-escobar-mensaje-de-maluma-al-mundo/ (Accessed: 25/10/2023)

Rueda, M. (31 August 2021) https://theworld.org/stories/2021-08-31/reclaiming-colombia-s-black-history-one-tour-time

Guyana .(11 October 2023) www.cia.gov (Accessed: 25/10/2023)

Suriname. (11 October 2023) www.cia.gov (Accessed: 25/10/2023)

Navarro, L. (November 12 2013) Photos reveal harsh details of Brazil's history with slavery. https://www.npr.org/sections/parallels/2013/11/12/244563532/photos-reveal-harsh-detail-of-brazils-history-with-slavery (Accessed: 26/10/2023)

NBCNews.com. (11 April 2023) Woman is convicted after racist rant against Colombia's first Black vice president. https://www.nbcnews.com/news/latino/woman-convicted-racist-rant-colombias-first-black-president-rcna79097 (Accessed: 29/10/2024)

Office for National Statistics (ONS). (released 29 November 2022) ONS website statistical bulletin, *Ethnic group, England and Wales: Census 2021.*

TrustforLondon.org.uk (Accessed: 26/10/2023)

Johnston, I. (15 Feb 2021) Low GP access rates among Latin Americans in UK raise vaccine concerns. *The Guardian*.

Sinclair, D. (2023) How Getting Your Students To Teach Can Increase Their Success And Support Decolonising Your Classroom. https://decolonisegeography.com/blog/2022/01/how-getting-your-students-to-teach-can-increase-their-success-and-support-decolonising-your/ (Accessed: 26/10/2023)

36. ENVIRONMENTAL EDUCATION: AN INTEGRATED APPROACH
SANDRA ZOE PATTERSON
@SANDYZPATTERSON

'Alone we can do so little; together we can do so much.' (Helen Keller, n.d.)

Environmentalism has always been important to me; it is one of the reasons that I love geography so much. As a newly qualified teacher (NQT) working in Wales, I found an outlet for my passion when a Sixth Form student came to me to ask me for career advice and to help them with an environmental project in the community; my Eco Team was born. I was delighted to realise that I could make a difference in young people's lives simply by being myself. The student in question went on to further study and is now an environmental campaigner.

Later I had the opportunity to lead my own department when I moved to Ulidia Integrated College in Carrickfergus, Northern Ireland. Here I developed and led a whole-school environmental education programme. When I returned home, I chose to work in the integrated sector as inclusion is central to my personal educational philosophy.

Within this chapter, I will explain the context of my school setting, contextualise my educational philosophy and outline my approach to environmental education within my school setting.

INTEGRATED EDUCATION IN NORTHERN IRELAND

'The system of education in NI is fundamentally divided' (Taggart and Roulston, 2022, p.8). In Northern Ireland there are four categories of schools for families to choose from: denominational, non-denominational, integrated and Irish medium. There are currently over 70 integrated schools, in a context of around 1080 schools in total. In an integrated school there is an intentional balanced mix of students, staff and school governors from the Catholic, Protestant and other or no faith backgrounds. This is intentional and purposeful. Every integrated school was campaigned for by parents; I am currently a parent campaigner for a new integrated school, but that is a story for a different day.

Integrated education is intentional in striving for a mix in schools and creates an environment that, not only acknowledges the diversity in its community, but recognises that intentional and balanced mixing is an opportunity to learn with, from and about each other (www.nicie.org). This ethos flavours the school and is a reminder that integration infiltrates everything from what is taught, how occasions are marked and how we are with one another.

Within an integrated school students learn together every day; this becomes our normal. Students and staff come together to intentionally learn about each other's identity together, while also appreciating that through our daily lives, we have so much more in common than we have differences.

There are four pillars to integration. These are equality, parental involvement, faith and values and, finally, social responsibility. The four pillars are central and actively evident in every integrated school.

The Four Principles of Integrated Education

Equality

The integrated school promotes equality in sharing between and within the diverse groups that compose the school community.

Faith and Values

The integrated school provides a Christian based rather than a secular approach. It aspires to create an environment where those of all faiths and none are respected, acknowledged and accepted as valued members of the school community.

Parental involvement

The support and commitment of parents is a fundamental element of Integrated Education and historically, parents have been central to the development of integrated schools.

Social Responsibility

The integrated school delivers the curriculum on all ability and inclusive basis to all of its pupils. It respects the uniqueness of every pupil and acknowledges his/her entitlement to personal, social, intellectual and spiritual development in the attainment of individual potential.

Figure 36.1 The four principles of integrated education Source: NICIE, 2022, with permission

Integrated schools are different from other educational establishments in Northern Ireland as integration is a people-led movement committed to peace and reconciliation. Within integrated schools, educators are committed to intentionally engaging with issues such as human rights, faith, culture, politics and equality by embracing and celebrating differences such as ethnicity and world religions 'where those of all faiths and none are respected, acknowledged and accepted as valued members of the school community' (Taggart and Roulston, 2022, p.7). 'In societies emerging from conflict, education plays an important part in instituting peace and reconciliation' (Roulston et al, 2023); the anti-bias lens which is utilised within an integrated school permeates into everyday life to engage with and challenge the contested socio-political identities which are at the root of the conflict within the region. This redress reframes difference to shift responses to these socio-political identities to build a peaceful, more conciliatory future.

SPACES OF INCLUSION

While studying as an undergraduate I was very fortunate to have Tim Cresswell as my dissertation supervisor; his ideas and guidance on the meaning of space and place have been integral to my work within the integrated sector and to my practice as a teacher.

Cresswell (1992) argues that 'value and meaning are not inherent in any space or place' (p.9), but that practice is simultaneously a form of consumption and that meaning of place can only be developed using space in an intentional and meaningful manner. 'Expectations about behaviour in place are important components in the construction, maintenance, and evolution of ideological values' (p.4). This consumption of place meaning therefore results in the ability to create the attribution of meaning by meaningful intentional action. Space and place are used to structure a 'normative landscape' (p.60) in which ideas about what is right, just and appropriate are transmitted through space and place. These ideas are central to my educational philosophy and are central to my approach to integration.

WHOLE-SCHOOL ENVIRONMENTAL EDUCATION WITH AN INTEGRATED LENS

My team have identified our vision as being: *'Highest standards of global education for all to create tomorrow's sustainable society'*. We review our vision regularly and have it displayed prominently in our teaching and learning spaces. Our vision guides and focuses our actions.

To achieve our vision, I have devised a series of key actions. I will highlight three of these to illustrate my whole-school approach to environmental education through the integrated lens.

ECO CODE

It is important to meet the vision that all members of the school community are part of the direction setting process. To this end, an environmental vision in the form of our Eco Code is devised by the school community. The code is conceived bi-annually by our whole school community. In keeping with our integrated ethos, our code is created together. When choosing each line, it is important that it reflects the school community, so contributions from a range of year groups are chosen. The final code is then translated into Spanish and Irish to reflect languages spoken in my school community (see Figure 36.2).

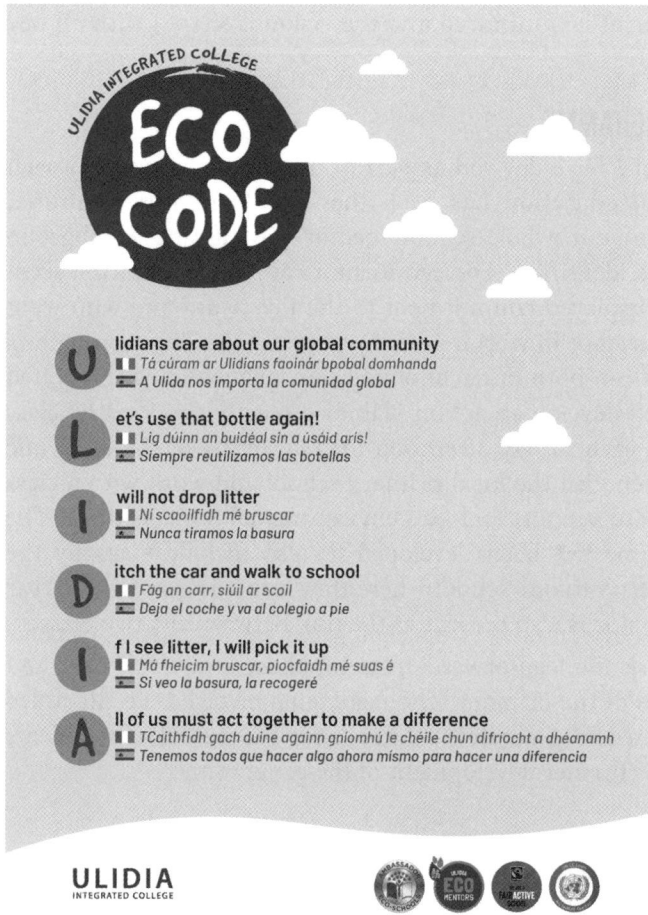

Figure 36.2 Whole-school eco code. There is space at base of the poster for both students and staff to 'sign up' to the code. It is presented in three languages, reflecting our school community

Reflecting the principles of integrated education, the statement is shared with parents, governors, students, all staff and our wider community. Students 'sign up' to the code each year and a copy is displayed in each form room. All staff (referring to teaching, auxiliary and learning support staff) in the school also sign a paper copy which is displayed in the staff room. Parents, governors and the wider school community also sign up to the code using digital signatures to ensure that both the carbon footprint is minimised and the vision is shared with all our school community.

ECO MENTOR PROGRAMME

A key action that I have devised as part of our whole-school approach to environmental education has been the 'Eco Mentor Programme'. Embedded within our whole-school mentor programme a subgroup of eco mentors is identified. The eco mentors are students in the sixth form who have displayed commitment to the Eco Team and who want to gain vital experience in working with young people. The eco mentors work with staff from both our school and our neighbouring integrated primary school to develop an action plan of eco activities for a P4 class to participate in each Friday afternoon for two terms of the academic year. Students then visit the local primary school and work with a class and their teacher to support and lead environmental projects there. The mentor programme has been developed to also include visits by the primary school class to our school where they join a lesson with a Year 8 class. Their mentor is also present as the link between the two classes.

Each year, we ask for feedforward from all those who took part and from the families of the primary school students involved. Feedforward information is used to generate a reference for the sixth form students involved and aids further development of the programme.

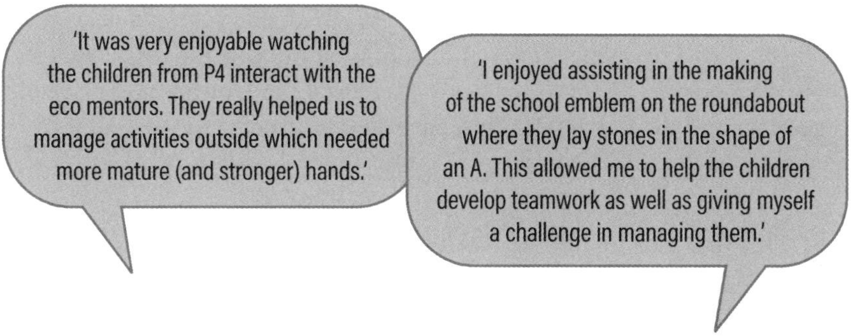

> 'It was very enjoyable watching the children from P4 interact with the eco mentors. They really helped us to manage activities outside which needed more mature (and stronger) hands.'

> 'I enjoyed assisting in the making of the school emblem on the roundabout where they lay stones in the shape of an A. This allowed me to help the children develop teamwork as well as giving myself a challenge in managing them.'

Figure 36.3 Feedforward from students and staff involved in the eco mentor programme

The impact of the eco mentor programme has been substantial. Primary students gain a role model sixth form student who they can learn from; the primary school receives help with eco projects; sixth form students gain valuable work experience with young people; mentors gain references. This project also builds links between two integrated schools in our wider community to increase collaboration.

CURRICULUM MAPPING

Social responsibility is a core value of the integrated ethos. As such, every department in our school contributes towards our whole-school environmental education programme. This is mapped bi-annually using a whole-school curriculum audit. The 'Social Responsibility' pillar of integrated education is emphasised through this vein of whole-school curriculum. The whole-school bi-annual audit provides clear evidence of coverage for evaluation alongside opportunity to identify areas to be built upon and developed further.

Environmental education is also mapped within the geography spiral curriculum. A spiral curriculum has been developed to build on prior learning to amplify learning of core concepts prior to external assessments. Lupton and Whaites (2023) state that for a curriculum to be sustainable it must include biodiversity, climate change and living sustainably as both indiscreet and discreet strands. They also argue that these areas must be taught in sequence to ensure that students develop curiosity and care for the environment, build on that knowledge to learn about why it is under threat, then sequentially build to develop both sense of responsibility and activism that they can protect the biodiversity. I achieved this by using a curriculum map to build these three stages of the sequence into each year of study. The three stages were mapped both discretely and indiscreetly within each year of study.

The four core goals of anti-bias in education are also mapped within the curriculum. Opportunities to embed identity, diversity, justice and activism within the curriculum are intentionally sought out and identified explicitly.

EVERY DAY IS A SCHOOL DAY

Along the journey I have learned some key lessons about my teaching and leadership.

- **Vision is important; it is central to planning that a vision is both clear and shared with stakeholders.** The process of defining my vision was central to me to identify what I wanted to achieve and why. The vision also helped my team to identify key actions and not go off on a tangent!
- **Build your community.** The team that I have built have been sources of inspiration, opportunity, sounding boards and my most important critical friends. I have built my own networks through attending events, training and simply keeping my ear to the ground. My networks have opened doors and opportunities in many ways. My advice would be to take opportunities to talk to people and always exchange contact details. Building a community network has been one of the most important things that I have done and has helped me, in turn, to support others.
- **Lead with passion and confidence.** I have found that it is important to lead while being unashamedly passionate about my ethos. As Marian Wright Edelman stated, 'you can't be what you can't see'. Don't be afraid to be passionate about the area of leadership. I am a keen environmentalist and I shout this from the rooftops as I believe in the importance of our environment. In turn, my students have been confident to also follow their passion and stand up for what they believe.

- **Visible leadership is important.** As Donald McGannon stated, 'Leadership is an action, not a position'; be active and be approachable by your team.
- **Expect the unexpected.** It is important to be prepared for anything that can happen; I have built a bank of 'go to' activities so that if a speaker cancels at the last moment or if there is a weather emergency during a planned activity, we have a backup plan.
- **Be open to opportunities as they arise; opportunities to gain experience, opportunities to work with others and opportunities to develop the programme.** And if you cannot find the opportunity, make one!
- **Finally, have fun!** We spend so much of our lives working, we should be enjoying what we do!

Five reflection questions

1. What are you unashamedly passionate about?
2. How do you bring the wider community into your role?
3. How can you build connections with feeder schools?
4. What ethos is embedded within your curriculum map?
5. How can you inspire young people to follow their passions?

DIGGING DEEPER - RESOURCES TO DELVE FURTHER

- Council for Integrated Education article 'Anti-bias in Education', available at: https://nicie.org/what-we-do/supporting-integrated-schools/ethos-development/anti-bias/
- Ulster University's Transforming Education Project, available at: https://www.ulster.ac.uk/research/topic/education/our-research/current-research-projects/transforming-education (Accessed: 28/05/2025)

REFERENCES

Cresswell, T. (1992) *In Place/Out of Place: Geography, ideology, and transgression.* NED-New edition, University of Minnesota Press.

Milliken, M., Roulston, S. & Cook, S. (2021) *Transforming Education in Northern Ireland: Stimulating debate among teachers, educationalists, decision-makers and the wider public. Developing the education system in Northern Ireland to meet the needs of young people in an inclusive society.* Transforming Education, Integrated Education Fund.

Milliken, M., Roulston, S. & Cook, S. (2021) *Transforming Education in Northern Ireland.* Ulster University, Coleraine.

Northern Ireland Council for Integrated Education. (2016) Big Small Stories. https://nicie.org/what-is-integrated-education/history-of-integrated-education-2/big-small-stories/ (Accessed: August 2023)

Northern Ireland Council for Integrated Education. (2021) Bias Busting for Beginners. https://nicie.org/who-we-are/publications/other-publications/ (Accessed: August 2023)

Northern Ireland Council for Integrated Education. (2022) Statement of Principles. https://nicie.org/what-is-integrated-education/integrated-ethos/integration-in-practice/ (Accessed: August 2023)

Lupton, E. & Whaites, H. (2023) A time for change: Developing a coherent sustainability curriculum. *IMPACT*, Issue 18, Chartered College for Teaching.

Roulston S., McGuinness S., Bates J. & O'Connor-Bones U. (2023) School partnerships in a post-conflict society: addressing challenges of collaboration and competition. *Irish Educational Studies*, 42(2), pp.257–274.

Taggart, S. & Roulston, S. (2022) *School Ethos in Northern Ireland.* University of Ulster Transforming Education, Integrated Education Fund.

37. CLIMATE CHANGE, SUPER WICKED PROBLEMS AND ANTIFRAGILITY
CATHERINE OWEN AND SEBASTIAN WITTS
@GEOGMUM AND @4WARD2MARS

'Perhaps the key to sustainability lies not just in resilience but in embracing the nature of ecosystems, allowing them to not only endure but flourish amid the dynamic forces of change.' (Taleb, 2013)

It is hard to overstate how important climate change education is in schools today. With the World Meteorological Organisation (WMO) warning us to expect record temperatures and the breaching of the 1.5°C target for warming above pre-industrial levels in the next five years, there is an urgency to tackle not only the climate crisis, but also to support students in coping with the uncertain future (see chapter 38).

TEACHING ABOUT CLIMATE CHANGE AS A 'SUPER WICKED' PROBLEM
Levin et al (2012) characterise climate change as a super wicked problem because:
- time is running out
- those who cause the problem also seek to solve it

- the central authority needed to tackle it is weak or non-existent
- policy responses discount the future irrationally.

They use an 'applied forward reasoning' approach to 'trigger *sticky* interventions that, through progressive incremental trajectories, *entrench* support over time while *expanding* the populations they cover' (Levin et al, 2012).

As teachers, we can't shy away from the complexity of climate change and risk limiting our students' understanding. We need to seek ways to support students in learning about causes, consequences and responses to climate change fully, but without being overwhelmed. Owen and Taylor (2023) suggest practical strategies for teaching different elements of climate change, as shown in Table 37.1.

Table 37.1 Practical strategies for teaching different elements of climate change
Source: Adapted from Owen and Taylor (2024)

Issue	Explanation	Potential approaches
Teaching tricky concepts	Climate change (CC) is complex, but we can use the systems approach to make it easier to understand.	Use clear and accurate diagrams. Model answers to tasks such as explaining causes of CC. Give feedback that move learners forward.
Critical thinking about climate change	We can empower students to … … become better at thinking. … make better use of information. … learn to think more openly about CC.	Encourage students to ask questions and reflect on their learning. Use different sources to explore fact v. opinion and different views. Challenge assumptions during discussions and consider ethical issues.
Seeing climate change	Students may not make links between visual indicators and climate change; we can make these explicit.	Use images or outdoor learning to help students identify indicators of carbon, its impacts and responses.
Taking a social justice lens	Climate justice means that everyone is able to prepare for, respond to and recover from the consequence of CC.	Explore work on this area from organisations such as the Joseph Rowntree Foundation. Read speeches from young climate activists and make links between their words and their contexts.

Issue	Explanation	Potential approaches
Using analytical models	Models such as the cognitive congruence framework or double loop learning can help students make connections between different aspects of CC.	Challenge students to look at how their actions connect with their understanding of CC. What assumptions underly this? How could changing actions affect CC?
Taking action	Students may feel empowered if they take action at home or in school to tackle CC, but we must make sure they don't feel personally responsible for CC. We also need to keep the focus on the geography and comply with government guidance.	Consider actions students could take, but also consider barriers, such as living in a rural area, flying to see family overseas or living in rented accommodation. Involve students creation and implementation of a whole school climate plan. Look at advertisements to explore examples of 'greenwashing'.
Work with the community	Parents and carers may be more willing to make changes if students are informed and interested. Local organisations may be keen to link.	Keep parents and carers informed about what students are learning about CC. Look for potential links with organisations in the community such as youth groups, religious organisations, etc.

Climate change runs through every topic we teach in geography. Roberts (2023) stresses the importance of teaching climate change through the geography curriculum and advocates for using an enquiry approach to do this. She recognises the need for students to learn about climate change and possible mitigation and adaptation strategies, but adds four more areas to consider:

- Are students aware of different ways in which climate change is framed? This could include catastrophic, economic, environmental, social and social justice framing.
- Are students made aware that views about responsibilities for reducing emissions and issues that divide opinion vary across geographical scales? This ranges from individual, through local and national, to international.
- Are students aware of how issues related to climate change have been influenced by the actions of fossil fuel companies through campaigns of disinformation?

- Are teachers in England aware of the current legal requirements when teaching about climate change?

Source: Adapted from Roberts (2023), pp.116–119

As teachers, we need to keep up to date with developments in understanding of climate change and ideas for climate change education. We also need to keep abreast of different ways to perceive climate change consequences and responses, including concepts such as resilience and antifragility.

RESILIENCE AND ANTIFRAGILITY

Students are likely to hear a lot about resilience – the ability to 'bounce back' – in school in relation to their learning, but this concept is also often used in relation to preparation for and recovery from the consequences of climate change. For example, A-level students may learn that the people of the Sundarbans are resilient, maintaining their way of life despite the threat of storms, sea-level rise and the resulting coastal erosion.

There is a growing body of criticism of the use of the concept of resilience in geography. Cretney (2014, p.1) notes that:

> Despite its widespread use, there remains confusion over what resilience is and the purpose it serves. Resilience can, in some cases, speak to a desire to successfully respond and adapt to disruptions outside of the status quo. However, this conceptualisation of resilience is far from uncontested. Emerging research has shown a lack of consideration for power, agency and inequality in popular and academic use of these frameworks.

We can encourage our students to engage with these ideas when we are considering the resilience of people and places in relation to climate change, asking them to consider who holds the power in the situation, if inequalities are disadvantaging any groups and if a community is being expected to look after itself so that the state doesn't have to.

While resilience is about 'bouncing back' to a previous state, antifragility is about a stressful situation causing an improvement in the situation. Taleb (2013) uses the following definition:

Some things benefit from shocks; they thrive and grow when exposed to volatility, randomness, disorder, and stressors and love adventure, risk and uncertainty. Yet, in spite of the ubiquity of the phenomenon, there is no word for the exact opposite of fragile. Let us call it antifragile.

Taleb (2013) defines the concept of antifragility as systems or entities that not only withstand chaos but thrive in it. Taleb contrasts fragility and antifragility, arguing that some things break under stress, while others grow stronger. Taleb discusses the unpredictability of 'black swan events', emphasising the need for robustness in the face of unforeseen, occurrences that could have a significant impact. He encourages embracing uncertainty as an opportunity for growth and improvement, stressing that antifragile systems learn and adapt through exposure to disorder. He advocates for decentralised systems and local decision-making, highlighting their adaptability and resilience.

In our teaching of geography there is no greater example of stress and disorder than the impact of climate change and its impact on global and local systems. Encouraging our students to explore the possible responses to this from the perspective of antifragility can lead to a unique level of insight. There are broadly seven key themes that can be drawn from the concept of antifragility and applied to the teaching of climate change.

1. **Antifragile systems in ecological resilience:** Ecosystems, often antifragile in nature, exhibit characteristics that can inform climate change management. Research by Folke et al (2004) highlights the importance of biodiversity in enhancing the resilience of ecosystems. The diversity of species within an ecosystem allows for adaptive responses to changing environmental conditions, thereby making the system more antifragile.

2. **Barbell strategy for climate adaptation:** This is a dual approach to risk that combines extreme safety with high-risk. The strategy is characterised by a cautious allocation of resources, with one side dedicated to safe and secure approaches, and the other side to speculative opportunities. Through the barbell approach Taleb

(2013) suggests we can better protect against unforeseen negative events while also capitalising on potential high-payoff outcomes Taleb's barbell strategy, which combines robust and experimental approaches, can be applied to climate change adaptation. The Intergovernmental Panel on Climate Change (IPCC) emphasises the importance of both traditional, well-established measures (such as infrastructure improvements) and innovative, experimental strategies such as ideas advocated in Prince William's earth shot prize in order to address the complex and uncertain nature of climate change impacts (IPCC, 2014).

3. **Decentralised approaches for climate resilience:** Decentralised and community-based initiatives for climate adaptation can enhance antifragility. Ostrom's work on the commons (Ostrom, 1990) suggests that locally managed resources often lead to more sustainable outcomes. This decentralised approach aligns with the antifragility principle of promoting resilience through diverse and context-specific strategies.

4. **Learning from climate events:** The adaptive management framework, advocated by Holling (1978), emphasises the importance of learning from environmental events. Each climate-related occurrence provides valuable insights that can inform future strategies. By adopting an adaptive management approach, decision-makers can turn climate-related challenges into opportunities for learning and improvement.

5. **Convex payoffs in sustainable practices:** Convex payoffs are about finding opportunities where potential gains increase more than rise in risk. As uncertainty grows, the positive outcomes from a strategy or investment get much better, making it more resilient and capable of significant gains in unpredictable situations. Investments in sustainable and renewable practices can be viewed through the lens of convex payoffs. Stern's review on the economics of climate change (Stern, 2007) argues that the benefits of early investments in sustainable practices, though initially perceived as costly, can yield substantial long-term gains by avoiding the economic and environmental consequences of unchecked climate change.

6. **Domain dependence in regional approaches:** Recognising the domain dependence of antifragility, it is crucial to tailor climate change management strategies to specific regional challenges. The regional focus is emphasised by Adger et al. (2007), who argue that climate change impacts are context-specific, requiring region-specific adaptation strategies. What works as an antifragile solution in one area may not be applicable elsewhere due to variations in climate impacts and local conditions.
7. **Skin in the game for climate decision-makers:** Taleb's principle of 'skin in the game' suggests that decision-makers should have a personal stake in the outcomes of their decisions. This concept aligns with research by Pielke Jr. (2007), who argues for greater accountability in climate policy. Decision-makers with personal stakes are more likely to make informed and responsible choices that consider long-term consequences.

Could cities face the challenge of climate change in ways that don't just maintain the standard of living of people, but improve it? Could moves to prevent species loss improve the affected ecosystem? Antifragility is implied by the Park Disaster Response Model, which shows the potential of a disaster event to lead to improvement in quality of life, economic activity and social stability, as shown in Figure 37.1. How can we build back better? Both Taleb's antifragility concept and Park's Disaster Response Model promote the idea of systems that go beyond survival, actively benefiting or improving in the face of adversity. This involves not just bouncing back but using challenges as opportunities for growth and improvement.

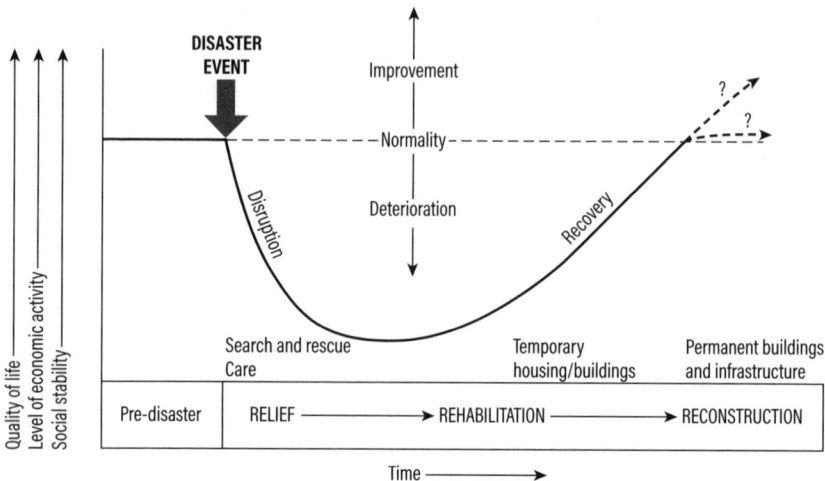

Figure 37.1 Park's Disaster Response Model

Kamranzad (2023) explores the potential of taking an antifragile approach to tackling the challenges posed by climate change and development of ocean renewables. She notes that development of wave farms to generate electricity and protect coastal areas can lead to a change in where sediment is deposited, creating a maladaptation, as this could lead to further changes in the sediment cell. Kamranzad is excited to see projects move from 'maladaptation' to 'antifragility', with systems that 'not only withstand stress and uncertainty, but actually thrive and benefit from them.' She recommends interdisciplinary research aiming to come up with new ways to use existing technology and protection strategies to allow coastal areas to benefit from climate change.

When addressing wicked problems in geography it can be tempting to simplify and fall back on reassuring diagrams. However, we have a duty to meet the problem head on. Teaching the complexity of climate change goes beyond imparting facts; it cultivates a comprehensive understanding, critical thinking skills, and a sense of responsibility that are essential for addressing the urgent environmental issues of our time.

> **Five reflection questions**
>
> 1. How can teachers incorporate diverse framings of climate change, including social justice perspectives, into the curriculum?
> 2. How does the urgency about climate change education align with the current global discourse on climate action?
> 3. In what ways can schools effectively contribute to climate change education and prepare students for an uncertain future?
> 4. How can we balance the complexity of teaching climate change with the need to protect our students from being overwhelmed?
> 5. How is the climate crisis being talked about and tackled in your setting?

> **DIGGING DEEPER – THREE RESOURCES TO DELVE FURTHER**
>
> - Juraszek, D. (2022) Climate Antifragility. Available at: https://www.resilience.org/stories/2022-11-08/climate-antifragility/ (Accessed: 28/05/2025)
> - Owen and Taylor. (2024) *Thinking Critically about Climate Change*. Sheffield: The Geographical Association.
> - Taleb, N. (2013) *Antifragile: things that gain from disorder*. London: Penguin.

REFERENCES

Adger, W. N., Arnell, N. W. & Tompkins, E. L. (2007) Successful adaptation to climate change across scales. *Global Environmental Change*, 15(2), 77–86.

Cretney, R. (2014) Resilience for Whom? Emerging critical geographies of socio-ecological resilience. *Geography Compass*, 8(9). Wiley. https://compass.onlinelibrary.wiley.com/doi/abs/10.1111/gec3.12154 (Accessed: 28/05/2025)

Folke, C., Carpenter, S., Walker, B., Scheffer, M., Elmqvist, T., Gunderson, L., & Holling, C. S. (2004) Regime shifts, resilience, and biodiversity in ecosystem management. *Annual Review of Ecology, Evolution, and Systematics*, 35, pp.557–81.

Holling, C. S. (1978) *Adaptive Environmental Assessment and Management*. Chichester: John Wiley and Sons.

Intergovernmental Panel on Climate Change (IPCC). (2014) *Climate Change 2014: Impacts, Adaptation, and Vulnerability*. Contribution of Working Group II to the Fifth Assessment Report of the IPCC.

Kamranzad, B. (2023) From "Maladaptation" to "Antifragility": Coastal areas under the combined impact of changing climate and development of ocean renewables. *Earth and Environment*, 26. https://earthenvironmentcommunity.nature.com/posts/from-maladaptation-to-antifragility-coastal-areas-under-the-combined-impact-of-changing-climate-and-development-of-ocean-renewables

Levin, K., Cashore, B., Bernstein, S. & Auld, G. (2012) Overcoming the tragedy of super wicked problems: constraining our future selves to ameliorate future climate change. *Policy Sciences*, 45, pp.123. Springer.

Ostrom, E. (1990). *Governing the Commons: The evolution of institutions for collective action*. Cambridge: Cambridge University Press.

Owen, C. & Taylor, A. (2024) *Thinking Critically about Climate Change*. Sheffield: The Geographical Association.

Pielke Jr., R. A. (2007) *The Honest Broker: Making sense of science in policy and politics*. Cambridge: Cambridge University Press.

Roberts, M. (2023) *Geography Through Enquiry*. Sheffield: The Geographical Association.

Stern, N. (2007) *The Economics of Climate Change: The Stern Review*. Cambridge: Cambridge University Press.

Taleb, N. (2013) *Antifragile: things that gain from disorder*. London: Penguin.

38. AT AND BEYOND THE CHALKFACE: THE ROLE OF THE GEOGRAPHY TEACHER IN THE AGE OF CLIMATE CRISIS

KIT MARIE RACKLEY

GEOGRAMBLINGS

'To go back to what we were prior to COVID-19 would only be a testimony to our collective stupidity. We have opportunities now and other pathways ... that hold this world the way it should be, and pathways that sustain and treasure and nurture. Be the generation that said, "No. Enough." We're not lost anymore. We're finding our way home.' (Te Ngaehe Wanikau of the Māori people in New Zealand, 2022)

The climate crisis is effectively an identity crisis. Much of humanity has lost its way by disconnecting itself from nature powered by anthropological processes such as colonisation, industrialisation and capitalism. With the human race essentially a force of nature, this disconnect has disrupted each of the five Earth systems and the interconnections and transfers between them; the warming of our atmosphere being only one symptom. My chapter starts with a quote from a current-day Indigenous leader as it is cultures such as theirs which have demonstrated true sustainable stewardship for thousands of years through connection with the land. Indigenous leaders are storytellers, knowledge keepers and inspiring enablers. These are three traits I believe teachers of *any* subject should recognise in themselves, and along with them the opportunity to empower young people towards a future of *reconnection*.

THE CLIMATE CRISIS AS A MENTAL HEALTH CRISIS

In its Sixth Assessment released in 2022, the Intergovernmental Panel on Climate Change (IPCC) included a summary of studies relating to the impacts of climate change on mental health for the first time (IPCC AR6 WG2, 2022). They conclude, with high or very high scientific confidence, that impacts such as extreme weather, high temperatures and wildfires are having direct detrimental effects on mental health worldwide. While more studies are needed regarding anxiety about the potential risks and awareness of climate change, the IPCC recognise that many national-scale surveys particularly from the USA, Europe and Australia show that the phenomenon of *eco-anxiety* (or *'climate anxiety'*) exists.

> **Eco-anxiety** – 'Feelings of helplessness, anger, insomnia, panic and guilt toward the climate and ecological crisis. Persistent and intrusive worries about the future of the Earth.' (As defined in *The Rise of Eco-Anxiety*, a report released in March 2021 by the charity Force of Nature.)

In 2021, a study by Hickman et al summarised findings from a survey of 10,000 young people aged 16–25 across 10 countries including the UK, Finland, Nigeria and India. 59% were very or extremely worried, with more than 50% saying they felt sad, anxious, angry, powerless, helpless or guilty. Almost half felt that their feelings about climate change negatively impacted their daily lives.

TOWARDS LANGUAGE OF HOPE

Look at the quotes in Figure 38.1. Some of these may be identifiable as headlines from mass media, others from scientific text. Some are phrases that we may all have uttered in the classroom at some point. What has been striking to me is that over the past few years, there has been a convergence in the language used by mass media, scientific articles and education (whether it be verbally uttered in the classroom or written as a curriculum statement). On reflection, I have been guilty of utilising such language in the classroom. My use of language also assumed that because we all live in the UK, we collectively should shoulder most of the blame and that we all should be doing more. Couple that with the fact that climate change is only increasing as a crisis, then this can instil feelings of guilt and inadequacy in our young people.

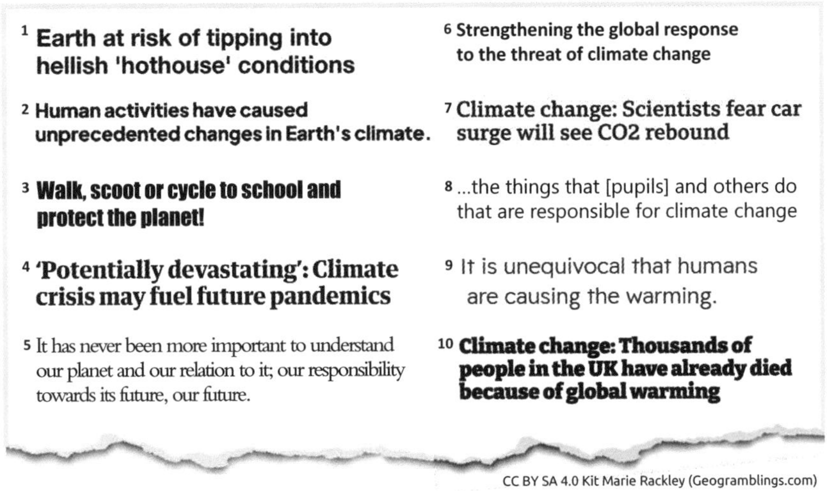

1 **Earth at risk of tipping into hellish 'hothouse' conditions**

2 Human activities have caused unprecedented changes in Earth's climate.

3 **Walk, scoot or cycle to school and protect the planet!**

4 **'Potentially devastating': Climate crisis may fuel future pandemics**

5 It has never been more important to understand our planet and our relation to it; our responsibility towards its future, our future.

6 Strengthening the global response to the threat of climate change

7 **Climate change: Scientists fear car surge will see CO2 rebound**

8 ...the things that [pupils] and others do that are responsible for climate change

9 It is unequivocal that humans are causing the warming.

10 **Climate change: Thousands of people in the UK have already died because of global warming**

CC BY SA 4.0 Kit Marie Rackley (Geogramblings.com)

Figure 38.1 Examples of the type of language used in the topic of climate change

As geography teachers, we therefore must be exceptionally mindful of the language that we use. While we can avoid language used in the popular press (such as statements 1 and 4), we cannot ignore the language that is used in the scientific literature (statements 2 and 9). As teachers, we are in an unenviable position of not shying away from facts, no matter how dire, while not fuelling anxiety. How do we square that circle?

When communicating about what society can do to combat climate change (mitigation), changing just one word can do wonders: *'should'* to *'would'* or *'could'*. Take these sentences for example:

- We *should* be cycling to school rather than getting a lift in a car.

- We are *responsible ['should']* for making sure we don't waste energy and water at home.
- We *should* speak to the local council and challenge them to take *responsibility ['should']* acting on climate change.

How would these be perceived and received differently worded as follows?

- *Would* you cycle to school rather than get a lift in a car?
- How *capable ['could']* are you at making sure energy and water is not wasted at home?
- *Could* we speak to the local council and see if they are *capable* of acting on climate change?

First, this approach doesn't assume that all students are equally responsible or capable. Secondly, in most cases, swapping in words relating to *could* or *would* often changes a statement (or something that sounds like a command) into a question, facilitating powerful discussions.

CLIMATE LITERACY

Reducing anxiety when it comes to irrefutable climate science is trickier. However, Phikala (2020) reasons that eco-anxiety can be a 'component or a dimension' in a person's anxieties, and so addressing factors that drive anxiety in general such as uncertainty, unpredictability and uncontrollability can be helpful. Improving *climate literacy* in ourselves and our students is a synergistic way to do this. On the 'Teaching Climate' section of their education website, the USA's National Oceanic and Atmospheric Administration (NOAA) states in their Essential Principles of Climate Literacy guide:

> 'Reducing human vulnerability to the impacts of climate change depends not only upon our ability to understand climate science, but also upon our ability to integrate that knowledge into human society.'

The 'Essential Principles' are very much worth a read. But what might these principles look like in the classroom? Figure 37.2 has been constructed to demonstrate a systematic approach whereby potential activities in the classroom (bottom) are informed and constructed based on the essential principles for climate literacy (top). The QR code included will allow you to explore the rationale, resources and ideas for application in more detail.

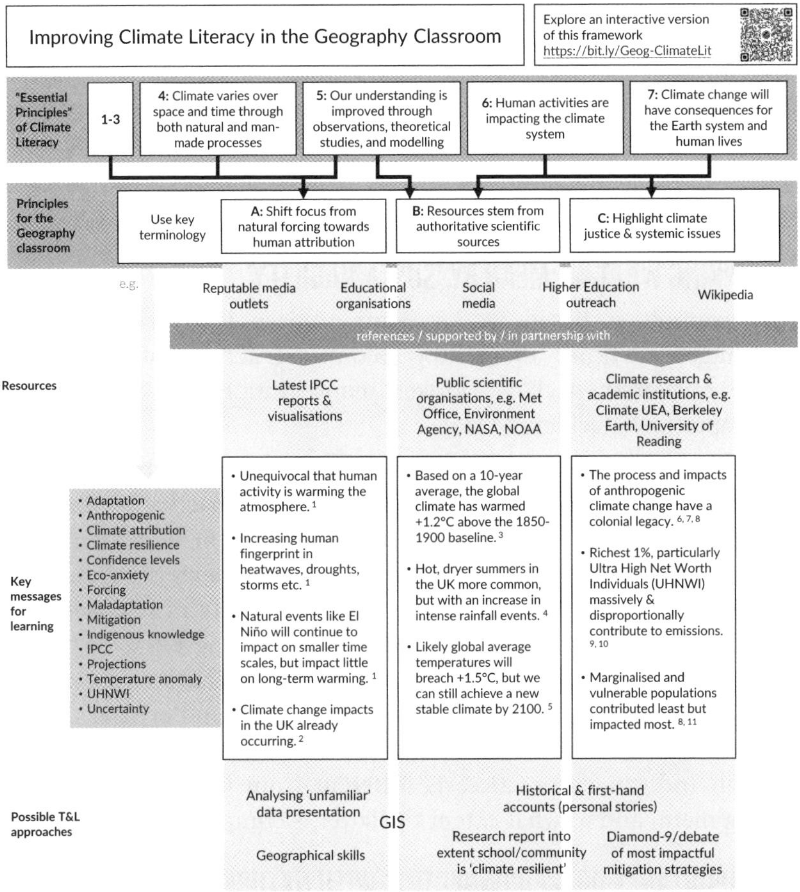

Improving Climate Literacy in the Geography Classroom

Explore an interactive version of this framework
https://bit.ly/Geog-ClimateLit

"Essential Principles" of Climate Literacy	1-3	4: Climate varies over space and time through both natural and man-made processes	5: Our understanding is improved through observations, theoretical studies, and modelling	6: Human activities are impacting the climate system	7: Climate change will have consequences for the Earth system and human lives
Principles for the Geography classroom		Use key terminology	A: Shift focus from natural forcing towards human attribution	B: Resources stem from authoritative scientific sources	C: Highlight climate justice & systemic issues
e.g.		Reputable media outlets	Educational organisations	Social media	Higher Education outreach	Wikipedia

references / supported by / in partnership with

Resources		Latest IPCC reports & visualisations	Public scientific organisations, e.g. Met Office, Environment Agency, NASA, NOAA	Climate research & academic institutions, e.g. Climate UEA, Berkeley Earth, University of Reading
Key messages for learning	• Adaptation • Anthropogenic • Climate attribution • Climate resilience • Confidence levels • Eco-anxiety • Forcing • Maladaptation • Indigenous knowledge • IPCC • Projections • Temperature anomaly • UHNWI • Uncertainty	• Unequivocal that human activity is warming the atmosphere.[1] • Increasing human fingerprint in heatwaves, droughts, storms etc.[1] • Natural events like El Niño will continue to impact on smaller time scales, but impact little on long-term warming.[1] • Climate change impacts in the UK already occurring.[2]	• Based on a 10-year average, the global climate has warmed +1.2°C above the 1850-1900 baseline.[3] • Hot, dryer summers in the UK more common, but with an increase in intense rainfall events.[4] • Likely global average temperatures will breach +1.5°C, but we can still achieve a new stable climate by 2100.[5]	• The process and impacts of anthropogenic climate change have a colonial legacy. [6, 7, 8] • Richest 1%, particularly Ultra High Net Worth Individuals (UHNWI) massively & disproportionately contribute to emissions. [9, 10] • Marginalised and vulnerable populations contributed least but impacted most. [8, 11]
Possible T&L approaches		Analysing 'unfamiliar' data presentation GIS Geographical skills		Historical & first-hand accounts (personal stories) Research report into extent school/community is 'climate resilient'	Diamond-9/debate of most impactful mitigation strategies

CC BY SA 4.0 Kit Marie Rackley (Geogramblings.com)

Figure 38.2 A framework for a systematic approach for the classroom

DECOLONISING CLIMATE CHANGE EDUCATION

A decolonial approach to dealing with climate change and achieving an equitable climate resilient society is not only a moral or historical argument but a scientific one too. The IPCC reports with high confidence that rapid mitigation is only possible where transitions are equitable, inclusive and just, and that 'disruptive changes in economic structure' are needed (IPCC, 2023). Therefore, for educators to be climate literate, a decolonial approach to teaching climate change is necessary. Here are just a few examples of approaches that can be taken.

RETHINKING WHAT WE MEAN BY 'SUSTAINABILITY'

This example is based on an opinion piece I wrote. Students are encouraged to think about how colonial structures of sustainability in the rainforest, such as 'fortress conservation' are an example of maladaptation (Rackley, 2022a).

Facts about deforestation in the Amazon, human habitation and Indigenous use of rainforests, and how fortress conservation has caused conflict can be given out to students in the form of mystery cards. Prompts help them to understand that research shows the Amazon has been inhabited for thousands of years and yet only in relatively recent times the land area covered by tropical rainforests have started to shrink in any meaningful way, and deforestation continues to be an issue despite conservation efforts. Through discussion and critical thinking, students will be able to compare how sustainable use of the forest through Indigenous practices is different from large-scale sustainable development, and to what extent the latter is comparatively ineffective.

THE IMPORTANCE OF (RE)CONNECTING WITH NATURE

Plenty of research shows the natural environment has multiple benefits to mental health (Dillman-Hasso, 2021). In addition, Indigenous peoples strive to remain connected to nature (e.g. Africa Renewal, 2022) and are the world's biggest conservationists (Drissi, 2020). So, climate change education that utilises Indigenous knowledge will be more effective as it is critical, place-based, participatory and holistic (Mbah et al, 2021). Including examples and case studies of Indigenous practices when teaching climate change mitigation and adaptation strategies is an obvious approach. Audio clips from podcasts, online videos and official websites can all be used to provide stimuli.

> 'When you become aware of your space and where you grow and what you're doing, you're going to become more aware of what your community is doing...' (Métis White Raven Woman Candace Lloyd on the Coffee and Geography podcast, 2021)

YOUNG PEOPLE AT A CROSSROADS

The Young People at a Crossroads (YPAX) project involved research into how families composed of first- and second-generation immigrants from the Global South are responding to lived experiences of climate and environmental change. The young people who took part offer unique perspectives which are intercultural and intergenerational, based on the lived experience of their parents and grandparents. A wealth of free-to-access educational resources are available (see 'Digging deeper'), and include:

- a 'creative book' with illustrations, young people's reflections, stories and conversations with family members
- an 8-page educators' guide to the creative book which explores curriculum and whole-school approaches to climate change education
- illustrated videos of some of the young people's stories
- interview training for young people so your students can have their own conversations at home.

TOOLS FOR STORY-TELLING

A child's imagination can be tapped into when it comes to providing narratives and solutions to sustainability-related problems (Caiman and Lundegård, 2017). There are substantial opportunities for cross-curricular approaches when it comes to being imaginative. Here are two examples.

Hot-seating works by having a character who is questioned by a group of pupils about not just the topic of discussion, but also about their personality, motivations, thoughts and feelings. The teacher is usually in the hot-seat (this is called 'teacher in role' (TiR)), or if a teacher is not comfortable, they can use the stimulus and other necessary information to, ahead of time, brief a willing student or colleague to take on the role. Some kind of stimulus is used which involves an identifiable character which could, for example, be a case study of someone involved in climate mitigation or adaptation (Rackley, 2022b).

Scaffolding for pupils to develop literacy through creating stories can come in many forms, whether it be through analysing photographers or focusing on a particular character in an existing story (fictional or not). Introducing an element of randomness can be particularly effective: a random keyword generator; drawing from a pack of flashcards, etc. This example of a story from a primary school student was created using 'Climate Dice' (climatedice.com):

Once upon a time there was a polar bear. She took me on a train all the way to the top of a mountain. The polar bear read me a story, then we had a drink from plastic cups but the polar bear left hers on the ground. A dove flew past and said 'Take your rubbish home with you, don't leave your mess on the mountain!' The polar bear said 'Sorry, I will.'

Figure 38.3 A story constructed using Climate Dice
Source: Climate Dice 'Story Hub', climatedice.com

THE CLIMATE CRISIS *IS* A SCHOOL SAFEGUARDING ISSUE

Based on what has been presented so far, we can now highlight three components which make for a robust argument to treat the climate crisis as a school safeguarding issue:

1. Eco-anxiety exists in our young people and climate change is having a negative impact on their mental health.
2. Increasingly, the impacts of climate change are being felt in the UK, with the most vulnerable feeling the impact first and hardest.
3. Statutory safeguarding duties require teachers to be aware of issues that can put children at harm (Department for Education (DfE), 2023) and be aware of risk and protective factors which can contribute to or mitigate against such harms (such as poor mental health) (DfE, 2018).

Treating the climate crisis as a safeguarding issue can help the school and its community become *climate resilient*, and climate-literate geography teaching staff can lead the charge. One approach is to lead reflective discussions using question prompts explicitly related to safeguarding guidance. For example for 'Early help', questions such as *'Are there students who have directly been impacted by an event attributable to climate change?'* and *'Are there children who are at greater risk to eco-anxiety and what are their risk factors?'* can be explored. Under contextual safeguarding, *'Is the school catchment area situated in an area that is at risk, or increasingly at risk, of events attributable to climate change?'* is a question that can be investigated (Rackley, 2021). There are geographical tools out there that allow for this, such as the flooding and heat maps from Climate Just (climatejust.org.uk). If you are based in London, then there is already a government-led initiative to support schools in becoming climate resilient (London.gov.uk). Other approaches include leading subject areas in mapping their curriculum for both climate change education, to improve subject-specific climate literacy, but also to map safeguarding links (Rackley, 2021). With schools mandated to produce Climate Action Plans by 2025, all the above approaches can support the development of those. A climate-literate educator would be a perfect candidate for the nominated sustainability lead who will oversee their establishment's Climate Action Plan.

Five reflection questions

1. How could you evaluate the 'climate literacy' of yourself, your students and other members of your school community?
2. To what extent does your geography curriculum promote climate literacy? What does the systematic approach of Figure 38.2 tell you that it already does well and what can improve?
3. Does your school have any outdoor classroom facilities or can use outdoor spaces for learning? What opportunities and challenges exist to make learning outside of the classroom a regular occurrence?
4. How can you use your own school's youth voice, lived experiences and cultural backgrounds to decolonise teaching climate change?
5. How could your EHCP/EHC and IEP documents be used to incorporate safeguarding issues relating to climate change?

> **DIGGING DEEPER – THREE RESOURCES TO DELVE FURTHER**
> - The free resources for the aforementioned 'Young People at a Crossroads' (YPAX) project can be accessed at: https://www.sci.manchester.ac.uk/research/projects/young-people-at-a-crossroads/
> - Greer, K. et al. (2023) Teaching climate change and sustainability: A survey of teachers in England. University College London. UK. Accessed at: www.ucl.ac.uk/ioe/departments-and-centres/centres/uclcentre-climate-change-and-sustainability-education - Well-researched insights which contains a reflection section on opportunities for enhancing climate change and sustainability teaching.
> - Carbon Brief (carbonbrief.org) is a fantastic climate journalism website grounded in science, which aims to improve climate literacy for those outside academia. Great for teachers and older students. (Accessed: 28/05/2025)

References for Figure 38.1:

1. https://www.aljazeera.com/news/2018/8/6/earth-at-risk-of-tipping-into-hellish-hothouse-conditions
2. https://www.ipcc.ch/report/ar6/wg1/resources/climate-change-in-data/
3. https://www.youtube.com/watch?v=JTfJmyZJVSA
4. https://www.theguardian.com/environment/2022/apr/28/climate-crisis-future-pandemics-zoonotic-spillover
5. https://virtual-events.stbrn.ac.uk/geography
6. https://www.ipcc.ch/sr15/
7. https://www.bbc.co.uk/news/science-environment-52724821
8. https://www.kingshill.school/geography-curriculum/
9. https://www.ipcc.ch/report/ar6/wg1/resources/climate-change-in-data/
10. https://inews.co.uk/news/climate-change-thousands-uk-deaths-global-warming-1027963

References for Figure 38.2:

1. IPCC (2021). Headline Statements from the Summary for Policymakers. The Current State of the Climate. https://www.ipcc.ch/report/ar6/wg1/downloads/report/IPCC_AR6_WGI_Headline_Statements.pdf. (Accessed: 13 October 2023)

2. UK Office for National Statistics. (2023) Climate change insights, families and households. https://www.ons.gov.uk/economy/environmentalaccounts/articles/climatechangeinsightsuk/august2023 (Accessed: 13 October 2023)
3. UEA Climatic Research Unit. Tim Osborn: HadCRUT4 global temperature graphs. https://crudata.uea.ac.uk/~timo/diag/tempdiag.htm (Accessed: 13 October 2023)
4. Met Office. UKCP headline findings https://www.metoffice.gov.uk/research/approach/collaboration/ukcp/summaries/headline-findings. (Accessed: 13 October 2023)
5. Esri UK Teach with GIS. Modelling Future Climate in the UK. https://teach-with-gis-uk-esriukeducation.hub.arcgis.com/apps/57d73f6d58fb4696a6dea815d2bc8f5d/explore (Accessed: 13 October 2023)
6. Sealey-Huggins, L. (2017) '1.5°C to stay alive': climate change, imperialism and justice for the Caribbean. *Third World Quarterly*, 38(11), pp.2444–2463. doi: https://doi.org/10.1080/01436597.2017.1368013.
7. Douglass, K. & Cooper, J. (2020) Archaeology, environmental justice, and climate change on islands of the Caribbean and southwestern Indian Ocean. *Proceedings of the National Academy of Sciences*, 117(15), pp.8254–8262. doi: https://doi.org/10.1073/pnas.1914211117.
8. Simpson, M.& Alejandra Pizarro Choy. (2023) Building decolonial climate justice movements: Four tensions. *Dialogues in human geography*, p.204382062311746-204382062311746. doi: https://doi.org/10.1177/20438206231174629.
9. Paddison, L. (2021) How the rich are driving climate change. https://www.bbc.com/future/article/20211025-climate-how-to-make-the-rich-pay-for-their-carbon-emissions (Accessed: 13 October 2023)
10. Nielsen, K.S., Nicholas, K.A., Creutzig, F., Dietz, T. & Stern, P.C. (2021) The role of high-socioeconomic-status people in locking in or rapidly reducing energy-driven greenhouse gas emissions. *Nature Energy*, pp.1–6. doi: https://doi.org/10.1038/s41560-021-00900-y.

11. CDP. (2023) *Those on low income and elderly most vulnerable to UK climate change.* (2023). https://www.cdp.net/en/articles/citiesannouncements/those-on-low-income-and-elderly-most-vulnerable-to-uk-climate-change (Accessed: 13 October 2023)

REFERENCES

Africa Renewal (2022) Hindou Ibrahim: Living in harmony with nature. https://www.un.org/africarenewal/magazine/may-2022/hindou-ibrahim-living-harmony-nature (Accessed: 19/10/2023)

Caiman, C. & Lundegård, I. (2017) Young children's imagination in science education and education for sustainability. *Cultural Studies of Science Education*, 13(3), pp.687–705.

Dillman-Hasso, N. (2021) The nature buffer: the missing link in climate change and mental health research. *Journal of Environmental Studies and Sciences*, 11, 696–711.

Drissi, S. (2020) Indigenous peoples and the nature they protect. UNEP. https://www.unep.org/news-and-stories/story/indigenous-peoples-and-nature-they-protect (Accessed: 19/10/2023)

Force of Nature. (2021) The Rise of Eco-Anxiety: A snapshot of how young people in over 50 countries are responding mentally and emotionally to the climate crisis. https://www.forceofnature.xyz/research (Accessed: 5/10/2023)

Hickman, C., et al (2021) Climate anxiety in children and young people and their beliefs about government responses to climate change: A global survey. *The Lancet Planetary Health*, (online) 5(12), pp.863–873. doi: https://doi.org/10.1016/s2542-5196(21)00278-3.

IPCC AR6 WG2. (2022) Chapter 7, Section 7.2.5.1: Observed Impacts on Mental Disorders. https://www.ipcc.ch/report/ar6/wg2/chapter/chapter-7/#figure-7-006#7.2.5 (Accessed: 5/10/2023)

IPCC. Core Writing Team, H. Lee and J. Romero (eds.). (2023) Climate Change 2023: Synthesis Report. Contribution of Working Groups I, II and III to the Sixth Assessment Report of the Intergovernmental Panel on Climate Change. doi: 10.59327/IPCC/AR6-9789291691647.

Lloyd, C. (2021) Coffee and Geography. 31 Aug. https://soundcloud.com/geogramblings/coffee-geography-s01e25-white-raven-woman-candace-lloyd-canada/ (Accessed: 19/10/2023)

London.gov.uk. Climate Resilient Schools. https://www.london.gov.uk/programmes-strategies/environment-and-climate-change/climate-change/climate-adaptation/climate-resilient-schools (Accessed: 20/10/2023)

Mbah, M., Ajaps, S. & Molthan-Hill, P. (2021) A systematic review of the deployment of indigenous knowledge systems towards climate change adaptation in developing world contexts: implications for climate change education. *Sustainability*, 13(9), p.4811.

National Oceanic and Atmospheric Administration. 'The essential principles of climate literacy. https://www.climate.gov/teaching/climate (Accessed: 12/10/2023)

Pihkala, P. (2020) Anxiety and the ecological crisis: An analysis of eco-anxiety and climate anxiety. *Sustainability*, 12(19), p. 7836.

Rackley, K.M. (2021) School safeguarding policy should consider climate change and eco-anxiety https://geogramblings.com/2021/04/25/school-safeguarding-policy-should-consider-climate-change-and-eco-anxiety/ (Accessed: 20/10/2023)

Rackley, K.M. (2022a) Can't See the Wood for The Trees: Rethinking Rainforest Sustainability. https://www.tutor2u.net/geography/blog/cant-see-the-wood-for-the-trees-rethinking-rainforest-sustainability-by-kit-marie-rackley (Accessed: 19/10/2023)

Rackley, K.M. (2022b) Fostering empathy in the teaching of natural hazards. *Teaching Geography*, 47(2), 67–70.

UK Government Department for Education (UK Gov DfE). (2018) Mental health and behaviour in schools. https://www.gov.uk/government/publications/mental-health-and-behaviour-in-schools--2 (Accessed: 20/10/2023)

UK Government Department for Education (UK Gov DfE). (2023) Keeping children safe in education 2023. https://www.gov.uk/government/publications/keeping-children-safe-in-education--2 (Accessed: 20/10/2023)

39. WHO OWNS THE EARTH? PROPERTY RIGHTS AND ENVIRONMENTAL CONFLICT

ANDREW TAYLOR AND CATHERINE OWEN
@GEOGMUM

'Every man has a property in his own person. This nobody has a right to, but himself.' (John Locke)

DISPUTES OVER RESOURCES ARE STRUGGLES OVER PROPERTY RIGHTS

A considerable number of environmental issues and problems are the result of unintended consequences of exploitation of resources. When owners are able to avoid bearing the environmental, economic and social costs of exploiting a resource, this is called an externality. Classic examples might be factory or agricultural pollution going into a river as a by-product of production, which damages fish stocks and drinking water downstream. Externalities occur when the costs of pollution are borne by owners of adjacent resources or society in general rather than the polluters themselves. This happens most often because property ownership regimes do not clearly specify the limits of rights and the necessity of responsibilities. This lack of accountability can lead to increased pollution levels and environmental degradation (Jacobs, 1991; Bromley, 1993). As a result there has been considerable interest in recent times in restructuring property rights and obligations to create more sustainable outcomes.

When teaching about environmental conflicts we can introduce students to the legal regimes and social systems of motivation that surround control of environmental resources, enabling them to develop a deeper understanding of issues. This is particularly relevant in the context of resource management, development and urban change. Our students may accept the way resources are organised, seeing what is done in their spheres as the only option. We can support them in thinking critically about this.

Property rights may be defined legally and/or socially but are almost always central to making sense of environmental issues and developing more sustainable outcomes. We will first explore what we mean by property rights by examining those with which we are most familiar – private property.

WHAT ARE PROPERTY RIGHTS?

Bromley (1993) argues that, contrary to popular belief, to say that you own a property right does not mean that you actually own the thing itself, but rather gain the right to claim a benefit stream protected by the state. Thus 'it is *not* the resource itself which is owned: it is a bundle, or a portion of rights to use a resource that is owned' (Alchain and Demsetz, 1973, p.17). An example is that the rights to exploit land for different activities such as agriculture, hunting, minerals and fishing are often owned by different parties.

The state can also benefit by controlling some of these benefit streams (Taylor, 1998), such as by making laws to restrict an activity, or through taxation to increase its cost. Since the 1970s and rise of neoliberal governments in the UK and USA there has been interest in using free markets to convert negative externalities into potential benefit streams (Gillespie and Thomas, 1989). This process resulted in, for example, emission trading markets with the Kyoto Protocol (Grubb, 1999) and associated European Emission Trading Scheme (Taylor, 2004a). In this scheme companies that emit large amounts of pollution are required to buy permits to pollute. At the end of a defined period they must be able to show sufficient permits to match their levels of pollution or face steep fines. Typically the coordinators of such schemes reduce the number of permits available over time, thereby forcing polluters to either seek their own internal efficiencies or purchase others surplus efficiencies. By attributing a value to reducing pollution it is hoped that the market will seek out the lowest cost pollution reductions. However, all that has really happened is that a liability previously borne by non-owners and future generations has been internalised through the attribution of value to it. The carbon cycle is a core topic in A-level geography specifications in England, including mitigation of climate change through schemes such as emissions trading. A simple roleplay game with groups of students representing different countries and being given varying amounts of cash, permits and pollution to deal with over a series of rounds could stimulate students' thinking about such schemes, followed by a reflection activity challenging them to think about why governments may support this kind of mitigation and its long-term effectiveness.

Drawing from Becker (1977) and Bromley (1993), we can see that property rights are entitlements to enjoy defined benefit streams, which may be broken up. They are not a thing in themselves. When we say that we own something, we don't actually own the thing itself, but rather the right to enjoy the benefits of using the thing in certain ways. With rights come duties and most thinkers, over the centuries, have accepted that the enjoyment of rights is dependent upon fulfilment of duties. These duties can be expressed explicitly through law and regulation, or implicitly through codes of conduct or stewardship.

STEWARDSHIP

From the beginning of time, the benefits of rights have always been balanced by duties and people's relationships to their environment and those around them, based upon some sense of virtue that goes beyond pure economic calculation to place limits upon exploitation (Shaiko, 1987). As Scruton puts it, 'the good things we inherit are not ours to spoil, but ours to use wisely and pass on' (Scruton, 2012, p.216). Cherishing a place and its people as a source of identity, that Scruton calls 'oikophilia', is 'a call to responsibility, and a rebuke to calculation, it tells us to love, and not to use; respect and not to exploit' (Scruton, 2012, p. 253); 'it comes from believing that this problem is our problem, and therefore my problem, as a member of the group' (Scruton, 2012, p. 171). When our students analyse the costs and benefits of extraction of a particular resource, we could explore this concept with them – to what extent are their conclusions based upon economic factors and how much importance are they assigning to cherishing a place? Across the world there exist local stories that communicate stewardship across the generations. For example, in parts of Romania there is a belief that every generation should plant a woodland during their lifetime. Perhaps our students are aware of such stories in their own communities.

Recent thinkers, such as Schumacher (1993) and Scruton (2012, 2017), emphasise the importance of scale and proximity in putting aside economic pressures so that the interests of non-owners become real. We could test this concept with our students, investigating whether their perceptions of issues change according to whether their impact is close to home or far away and/or whether the scale of the issue makes a difference. Energy resources provide a good context for this; would students be concerned about a wind turbine being built near their home? What if it is to be built on the edge of the next town? Is it preferable to build large nuclear power stations such as Hinkley C in remote locations rather than develop smaller sources of power across the country?

Following the industrial revolution, stewardship began to lose its local character. As the scale of human endeavour has grown, personal accountability has often been replaced by some form of collective or corporate interest (Kohr, 1957).

THE PLANNING SYSTEM

In response to both industrial and housing pressure upon the British countryside, the Town and Country Planning System was created to regulate property rights on behalf of society as a whole. Areas that the system has expanded to cover include housing development and access to the countryside, both of which have become contentious issues. For example, rural land is a productive space for farmers, yet for others it is a space to be enjoyed for its natural beauty, so just how much access to rural land is reasonable is highly political (Newby, 1980, 1987; Marsden et al, 1993).

Similarly there are significant tensions between housebuilders and rural residents over construction in rural areas. Here we have a tension between landowners' rights to realise the value of their asset, a desire to protect the character of rural spaces, the rights of local residents to a particular way of life and the needs of the country for more housing. Thus in, often very local disputes, we see what at heart are struggles to redefine property regimes. Planning officers have the often unenviable task of trying to balance private interests with public needs. Given that the English system of government is based upon a contract between individuals and government (dating back to the philosopher Locke's ideas) where the primary responsibilities of government are the protection of life, liberty and property, this is not easy. Every restriction upon development is perceived as a failure to fulfil this contract and a shift in power from private to public authority.

It is likely that housing development will be taking place in areas near to our schools, giving us a great opportunity to unpick these issues with our students. Students could look at local planning documents and decide if they would make the same decisions as the planning officers involved. If there is an example of a development being turned down by the local council, they could consider why and relate this back to needing to balance the rights of local people with the demand for more housing.

COMMON PROPERTY

Elionor Ostrom won a Nobel prize for her work which demonstrated that common property rights were a sustainable economic alternative to private property rights. Ostrom (1990, 2015) suggests that there are eight key criteria for a functioning commons regime:

1. Define clear group boundaries.
2. Match rules governing use of common goods to local needs and conditions.
3. Ensure that those affected by the rules can participate in modifying the rules.

4. Make sure the rule-making rights of community members are respected by outside authorities.
5. Develop a system, carried out by community members, for monitoring members' behaviour.
6. Use graduated sanctions for rule violators.
7. Provide accessible, low-cost means for dispute resolution.
8. Build responsibility for governing the common resource in nested tiers from the lowest level up to the entire interconnected system.

In other words, there must be an agreement, that is enforceable, between a defined group of stakeholders about how the resource is to be managed. Common property regimes are far more widespread than is often realised; from the management of grazing land across large swathes of the world (including the UK), to fishing stocks, to the management of apartment buildings, they have proven to be a means of escaping the prisoner's dilemma (described in the next section).

The challenge with common property regimes is to develop socio-legal mechanisms that balance the interests of rights holders fairly while conserving the resource for the future (Boiler and Ostrom, 1990). However, many such regimes successfully combine a history of managing the shared resource and a proximity of owners that encourages the establishment of norms of behaviour designed to discourage conflict. Community enterprises, from wind farms to community shops, pubs and transport schemes are growing rapidly in the UK, encouraged by an emerging ecosystem of renewable energy, technology and finance firms. You may be aware of local examples of community enterprises in your area – developing students' awareness of these may increase their appreciation of such schemes and may even stimulate them to consider setting up their own.

PUBLIC RESOURCES: STATE-OWNED PROPERTY AND OPEN ACCESS

The problem with State property is that, just like private property, it removes personal responsibility for outcomes. 'Unlike private property where there is a defined owner, or common property where there are a defined group of owners, state property is, in any sense, other than a bureaucratic one, effectively ownerless.' (Taylor and Bronstone 2019, p. 117).

Some, mostly global, resources are, in the absence of a world government or regulatory authority, ownerless entirely. These are usually called open-access regimes. Examples include the oceans and the atmosphere.

With both State-controlled and open-access regimes, the ownership structure means that no one has any personal interest in protecting the sustainability of the resource. As a result, users are incentivised to obtain whatever short-term gains they can for themselves, since the negative consequences for doing so will be shared among all other users (Peterson, 1993; Dobson, 1990; Taylor, 2004b). Effectively, all externalities arising from usage of the asset are socialised from the individual to either the rest of society or to future generations of the whole world.

A-level geography students following English specifications explore the global commons of the oceans, Antarctica and overfishing. The dumping of sewage off the coast of England could be used to develop understanding of private companies exploiting a shared resource for their own benefit, with the negative consequences left to others and ongoing problems being created for the future.

CONCLUSION

Property rights are at the centre of almost all environmental conflicts since they define human interactions with resources and one another. They define the limits of exploitation, duties of care and responsibilities to one another, wider society and the natural world. While there has been considerable interest in using property rights to create markets for environmental goods since the 1970s (Jacobs, 1991) in more recent times, and especially since the awarding of a Nobel prize to Elinor Ostrom in 2009, growing interest has emerged in common property. Not only has this found expression in debates over natural resources, but the emergence of open-source technologies has created renewed interest in community platforms as the basis for sustainable actions (Mason, 2015; Taylor and Bronstone, 2019, 2022; Varoufakis, 2023). Encouraging our students to explore these issues in their own contexts will allow them to develop their critical thinking skills and start to get to grips with ideas which are likely to become more and more important in the future.

> **Five reflection questions**
> 1. Which issues in your geography curriculum are linked to property rights?
> 2. Which issues in your local area are linked to property rights?
> 3. Could your students benefit from exploring these issues?
> 4. How could an understanding of property rights support the move towards a more sustainable society?
> 5. Which careers are linked to understanding these kinds of issues?

> **DIGGING DEEPER – FOUR RESOURCES TO DELVE FURTHER**
> - Short video about basics of property rights and the environment: https://www.youtube.com/watch?v=Y8kDv4ZP5Os (Accessed: 28/05/2025)
> - Ostrom on property rights: https://www.youtube.com/watch?v=BDEAgmkINyE (Accessed: 28/05/2025)
> - Free market environmentalism: https://www.youtube.com/watch?v=ADowFfaeWoU (Accessed: 28/05/2025)
> - Carbon markets in practice: https://www.youtube.com/watch?v=m5ych9oDtk0 (Accessed: 28/05/2025)

REFERENCES

Alchain, A. & Demsetz, H. (March 1973) The property right paradigm. *Journal of Economic History*, 33(1), 16–27.

Becker L. (1977) *Property Rights: Philosophic foundations*. London: Routledge and Kegan Paul.

Bishop of Oxford. (1913) *Property, Its Duties and Rights: Historically, philosophically and religiously regarded*. California, USA: University of California Libraries.

Bollier, D. (2014) *Think Like A Commoner: A Short Introduction To The Commons*. Gabriola Island, Canada, New Society Publishers.

Bromley, D. (1993) *Environment and Economy: Property Rights and Public Policy*. Oxford: Blackwell.

Gillespie, C. & Thomas J. (1989) *Property Rights and the Environment*. London: IEA.

Dobson, A. (1990) *Green Political Thought*. London: HarperCollins.

Grubb, M. (1999) *The Kyoto Protocol: A Guide and Assessment*. The Royal Institute of International Affairs.

Jacobs, M. (1991) *The Green Economy*. Pluto Press.

Marsden, T., Murdoch, J., Lowe, P., Munton, R.C. & Flynn, A. (1993) *Constructuring The Countryside: An approach to rural development* (1st ed.). Routledge.

Mason P. (2015) *Postcapitalism: A guide to our future*. Allen Lane.

Newby H. (1980) *Green and Pleasant Land?: Social change in rural England*. Penguin.

Newby H. (1987) *Country Life: a social history of rural England*. Weidenfeld and Nicolson.

Ostrom, E. (2015) *Governing the Commons: The evolution of institutions for collective action*. Cambridge University Press.

Ostrom, E. (1990) *Governing the Commons: The evolution of institutions for collective action*. Cambridge University Press.

Peterson, D. J. (1993) *Troubled Lands: The Legacy of Soviet Environmental Destruction*. Westview Press.

Schumacher, E. F. (1993 edition) *Small Is Beautiful: A study of economics as if people mattered*. Vintage Books.

Scruton, R. (2012) *Green Philosophy: How to think seriously about the planet*. Atlantic Books.

Scruton, R. (2017) *Where We Are: The state of Britain now*. Bloomsbury.

Shaiko, R. (1987) Religion politics, and environmental concern: A powerful mix of passions. *Social Science Quarterly*, 68(2).

Taylor, A. (1998) Property rights and the differentiating countryside: A case study in South West England. Unpublished Ph.D. Thesis, Cardiff, University of Wales College of Cardiff.

Taylor, A. (2004a) Trading hot air: backwards, forwards or any which way you like. *European Environment* Nr. 14.

Taylor, A. (2004b) Government cackhandedness in water management policy is money – and water – down the drain. *Vivid*, 63.

Taylor, A. & Bronstone, A. (2019) *People, Place and Global Order: Foundations for a networked political economy.* Routledge.

Taylor, A. & Bronstone, A. (2019) *Reconstructing the Global Network Economy: Pathways to local resilience.* Routledge.

Varoufakis J. (2023) *Technofeudalism: What killed capitalism.* Penguin Random House.

40. HOPEFUL GEOGRAPHY
DAVID ALCOCK
@DAVIDALCOCK1 / @HOPEFULED / WWW.ALCOCK.BLOG

'We have a seldom-told, seldom-remembered history of victories and transformations that can give us confidence that yes, we can change the world because we have, many times before.' (Rebecca Solnit, 2016, pp. xxii-xxiv)

Geography education should be alive to the possibility of hope. After all, education is about preparing young people for the future, and the geography curriculum deals with many themes relating to personal, local, national and global futures.

It is therefore crucial for geography educators to pay attention to the issues involved in fostering a more hopeful geography. These issues include an awareness of the affective dimension of our teaching, an understanding of how our curriculum choices might allow for, or deny, the possibilities of hope, and the practical techniques that we might employ in the classroom.

THE AFFECTIVE DOMAIN AND HOPE
There is increasing realisation of the importance of the affective domain in geography teaching. However, the demands of a knowledge-heavy curriculum and the pressures of accountability have arguably entrenched the situation outlined by Hicks and Bord over twenty years ago, whereby 'many educators … may make things worse for students by teaching

about global issues as if this were solely a cognitive endeavour' (2001, p.424). Simultaneously, psychologists have noticed a rise in eco-anxiety among young people (Hickman et al, 2021).

Partly as a response to this, increasing attention has been paid to the role of 'hope' in geography education. David Hicks has been at the forefront of this; he urges educators to pay more attention to global environmental issues, asks us to consider the feelings of our students, and provides us with many tools to incorporate a 'futures dimension' into our teaching (see Hicks, 2014).

However, I argue that geography teachers shouldn't just teach with hope *in mind*. We should also offer our students *grounds* for hope, of which I argue there are plenty. This article draws inspiration from the psychologist Maria Ojala, who argues that fostering feelings of 'constructive hope' can enhance students' engagement with issues of sustainable development:

> 'This process can be started by discussing young people's view of the global future. If it is very pessimistic, are there different ways of looking at it? Are there any positive trends to focus on? How has humanity solved large and seemingly uncontrollable problems historically?' (Ojala, 2012, p.638)

THOUGHTS TOWARDS A MORE HOPEFUL GEOGRAPHY CURRICULUM

Many teachers have lived through times characterised by 'progressive' ideas and technological breakthroughs which have saved lives or made them longer, more bearable, or even more pleasant. Depending on your age, these might include the end of apartheid, the crumbling of the Iron Curtain and the associated surge in democracies, the increasing recognition of human rights, and the scientific breakthroughs which have led to, for example, more resilient species of crops and the elimination of smallpox. Indeed, it is sometimes instructive for us to refamiliarise ourselves with such breakthroughs in order to put the stories from the contemporary news cycle into context. These events will not have been so salient in our students' lives as they have been in ours.

While there have been scientific and other 'advances' in the course of our students' lives – genomic sequencing, the tipping point of vehicular electrification, the rapid deployment of vaccines, and the advent of artificial intelligence, for example – the young people in our class will not have lived through quite so many headline-grabbing breakthroughs as their teachers have. They will therefore not have experienced the sense of optimism which many people (in the West at least) associated with these and many other trends in recent decades. While the ability of geographers to critically evaluate the impact of such advances should be actively encouraged, we should be wary of how *decontextualised* criticism might lead to a loss of faith in humanity to prepare ourselves for current and future challenges.

Indeed, those same young people in our class may not yet have developed the critical abilities needed to evaluate and filter the information overload and the salience of predominantly bad news associated with early twentieth-century life. No wonder that when questioned, so many young people express fears about the future.

In order to give students space to consider what influences their world views and to offer them 'scope for hope', since 2020 I have held Grounds for Hope workshops in my school and in two others. I begin by surveying the students in order to gain an insight into some of their hopes and fears about the future of the world. Fears outweigh hopes in almost every global issue raised; in one workshop, 58% of 86 Year 10 students were either 'somewhat worried' or 'very worried' about 'the state of the world'. It also takes critical insight to register some societal changes: when we teach health geography, do we pay enough attention to what Steven Johnson calls 'non-events'?

> *'There's this strange property about progress in human health that is different from other, more traditional forms of technological progress: it's measured in non-events, things that didn't happen. With technological progress, the achievements are easy to see—the smartphone in your hand, the electric car in your garage—but the biggest breakthroughs in health appear in the form of events that didn't happen: the smallpox infection that didn't kill you at the age of two because we invented a vaccine; the glass of drinking water that didn't give you a fatal case of cholera when you were 15 because we chlorinated the drinking water.'* (Johnson, 2021)

It is also difficult for us and our students to register long-running and widespread trends (positive or negative), because they are drowned out in our media by unusual, short-term events: a study reported that for every person killed by a volcanic eruption, nearly 40,000 people have to die of a food shortage to get the same probability of coverage in US televised news (Eisensee and Stromberg, 2007); 'good news' trends such as the virtual elimination of tapeworms also struggle to get airtime.

Nevertheless, changes to the geography curriculum necessitated by a more hopeful orientation should not come at the expense of studying the immense, complex and often 'wicked' problems faced by humanity, and the urgent necessity to continue to educate our young people about the impact of humanity on the planet and its ecology. (See chapter 37.)

KEEPING OUR CURRICULUM UP TO DATE

It is incumbent upon us as global educators to frequently update our subject knowledge, and that includes understanding global trends easily achieved through widespread access to information. There are two key online resources which will help geography teachers with such understanding: Gapminder and Our World in Data – see 'Digging deeper'.

Both websites (as with all resources) should be viewed with a critical eye: Gapminder in particular has been critiqued for underplaying the persistence of economic inequalities through the usage of logarithmic axes when plotting income data: Jason Hickel has written critically about both sources (see 'Digging deeper').

If we accept the premise that one of the roles of education is to foster the development of politically aware critical thinkers and active citizens of the world, then it follows that students should be aware not only of up-to-date global trends but also of their capacity to shape the future. These aims chime with those of advocates of Geocapabilities – the belief that geography can play a powerful role in helping students to think beyond themselves and their everyday experiences (see https://www.geocapabilities.org; Mitchell, 2022).

It is a short step from this argument to formulate a threefold concept, not only of a more hopeful geography, but also of a wider concept of a more 'hopeful education'. Both involve advocating for a curriculum which enables our students to do three things:

1. **Evaluate progress** – interrogate the notion of 'progress', understand and explain positive trends, put these gains in the context of regress in other areas, and use this understanding to inform debates about future progress.
2. **Believe in humanity** – seek to reaffirm the potential of human nature to work collectively for the common good, whether that be in the classroom, inter-generationally, locally, nationally, or in the context of global governance and co-operation.
3. **Create a sustainable future** – champion and facilitate futures thinking and education for sustainable development.

So, what techniques might be employed to introduce more 'scope for hope' into the teaching of geography?

PEDAGOGIES TO ASSIST IN FACILITATING A MORE HOPEFUL GEOGRAPHY

HOPEFUL GEOGRAPHY - PRACTICAL IDEA 1 - FUTURE TIMELINE

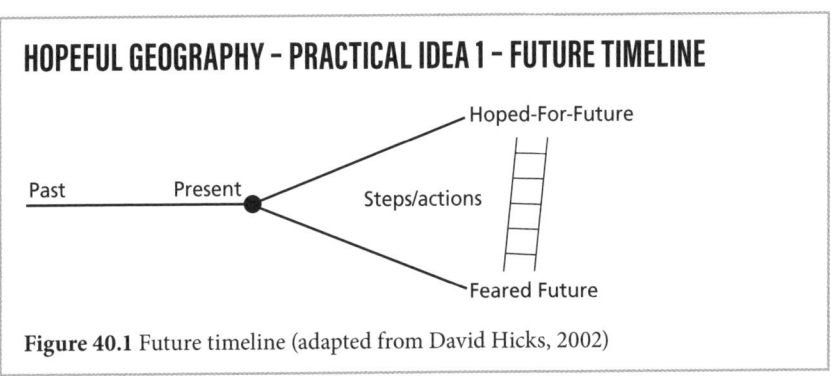

Figure 40.1 Future timeline (adapted from David Hicks, 2002)

This idea, adapted from techniques adopted by futures educators such as David Hicks, allows students to consider the world in a temporal context. Issue the timeline (Figure 40.1) on A3 or flipchart paper to groups of two or three. Tell them it's about the general state of the world, not their personal lives. Ask them to annotate each section, one part at a time: How do they feel about the past? What were the good and bad points about it? Repeat for the present, feared future and hoped-for future. Some groups may get onto the steps/actions that society could take in order to reach the future.

Students then return to the 'hoped-for' and 'steps/actions' part of their timelines following a discussion and/or other learning activities based on the theme of 'Hopeful Geography' – see Figure 40.2 for an example.

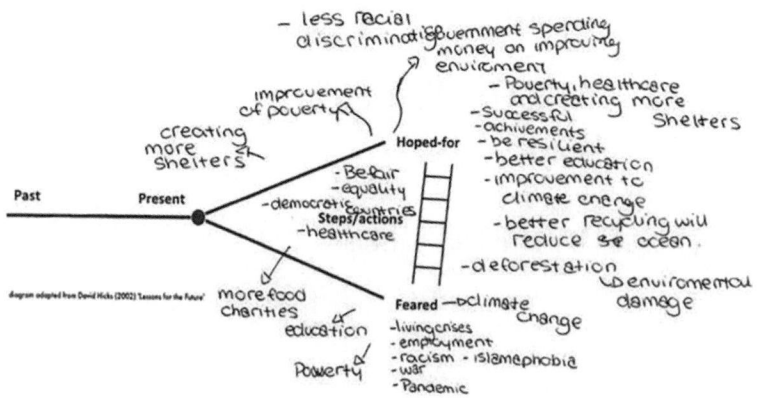

Figure 40.2 A completed future timeline from a student at Hathershaw College, Oldham

(Source: Lisa Lott)

HOPEFUL GEOGRAPHY – PRACTICAL IDEA 2 – THE NEWS OF THE CENTURY

Ask students: *What would be the headline story if a newspaper or a single web page was to be written about the last one or two hundred years? What would the other news stories on the front page be?* Showing 'The world as 100 people over the last two centuries' graphic may help (critically analyse it first): https://ourworldindata.org/a-history-of-global-living-conditions-in-5-charts.

HOPEFUL GEOGRAPHY – PRACTICAL IDEA 3 – FUTURE HISTORIES / POSTCARD FROM 2050

Showing students that apparently intractable, large-scale, historical challenges have been overcome will help them to believe that today's similarly daunting challenges can also be overcome.

A powerful way of doing this is to ask them to put themselves in the position of someone 125 years ago, when no women could vote. Ask them whether they would have foreseen a world in which women could vote. Many people said that it would never happen – and yet it did. But how? Repeat this for other social, technological and environmental breakthroughs.

Ask them to fast forward to the planet that they hope to see in 2050, considering what the society, economy, environment and politics might look like. They could even sketch such a society. Ask them to look back to the years leading up to 2050 and tell the story, via a postcard or vignette, of what happened to make their hoped-for future come true – just like the visionaries behind votes for women or the banning of CFCs did.

CHALLENGES TO HOPEFUL GEOGRAPHY

A valid criticism of hopeful geography is that it might slide into complacent optimism, despite the best intentions of the educator. This could be guarded against by repeatedly reminding students that the 'progress' seen in recent decades in so many areas of our lives has not come about automatically. Instead, individuals and collectives (like unions, co-operatives and scientific research groups) worked to make these changes happen.

Hopeful geography might seem divorced from global reality – one need only think about ongoing wars, prevarication over climate action, political polarisation and other retrograde steps. However, a successful hopeful geography seeks to take a big picture – one which will put these setbacks in the context of more numerous steps forward – without being seen to be insensitive.

A related concern is that hopeful geography might turn students' attention away from the most pressing needs of our contemporary world, including climate change, ecological crises, and income and other inequalities. It may even be accused of paying attention to the future at the expense of paying attention to the pressing needs of today. But applied judiciously, the converse may apply. Max Roser's framing is useful here:

> 'The world is awful. The world is much better. The world can be much better. All three statements are true at the same time.
>
> ... But to see that a better world is possible, we need to see that both are true at the same time, the world is awful, and the world is much better.' (Roser, 2022, n.p.)

He illustrates this with reference to infant mortality rates (see Figure 40.3).

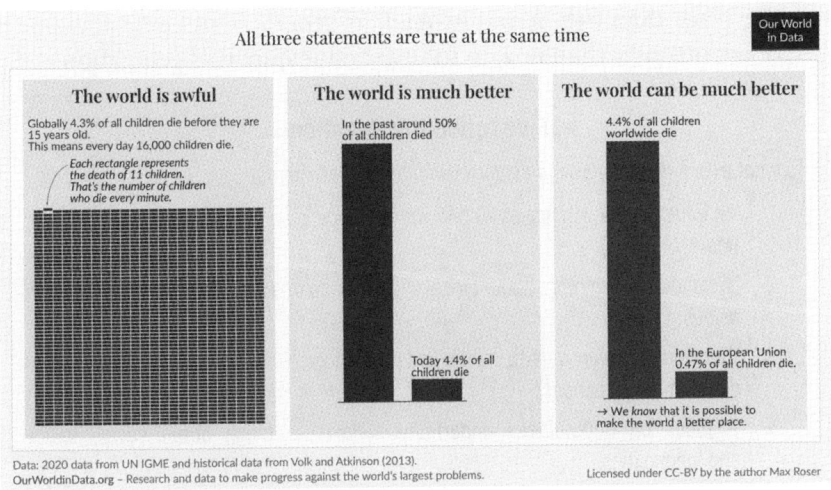

Figure 40.3 The world is awful, the world is much better, and the world can be much better
Source: https://ourworldindata.org/much-better-awful-can-be-better

There is a risk that hopeful geographies which concentrate too much on climate change adaptation may make us lose focus on the crucial need for mitigation. A related concern posed by radical geographers is that hopeful geography might overplay techno-solutions over the social movements that have made many transformations possible. Looking through a decolonial lens, while it could be argued that hopeful geography can break the stereotype of a stationary, polarised world, it needs to be approached with attention paid to different conceptions of the term 'progress', as well as the inclusion of voices from across a broad swath of society – there is certainly scope for reflection here (see chapter 31).

CONCLUSION

There is a need for conversations about what a more hopeful geography education might mean. The stakes for the planet, the discipline and the mental health of our students are high.

However, using the subtitle of Hannah Ritchie's book (2024), is it not more helpful to open the possibility to our students that they might be 'the first generation' to achieve sustainability and to leave the world in a better place than they found it? And, as geographers, could there be a better opportunity than now to inform and inspire this generation?

Five reflection questions

To what extent do you agree or disagree with these statements?

1. 'An education which leaves a child without hope is an education which has failed.' (Mary Warnock)
2. 'We should teach more about global social trends to avoid presenting the situation as being static.'
3. 'Alongside an overdramatic media and psychological biases, education can help to create misconceptions.'
4. 'A focus on hopefulness undermines efforts to teach about social justice and inequalities.'
5. 'Students turn away from geography because it tends to focus on problems rather than solutions.'

DIGGING DEEPER – THREE RESOURCES TO DELVE FURTHER

- Gapminder – www.gapminder.org – includes an updated version of 2017's 'Ignorance test' and links to talks from Hans and Ola Rosling. Annually updated bubble charts show how the relationships between different social, economic and environmental indicators change over time.
- Our World in Data – https://ourworldindata.org/ – accessible portal to reports, clear maps and other visualisations, summarising research and analysis. Helps to understand the world as well as "make progress against the world's largest problems". Includes a tool which tracks the United Nations Sustainable Development Goals: https://sdg-tracker.org/ (Accessed: 28/05/2025)
- For a more critical commentary on global trends – https://www.jasonhickel.org/ (Accessed: 28/05/2025)
- For more from the author – www.alcock.blog (Accessed: 28/05/2025)

REFERENCES

Eisensee, T. & Strömberg, D. (2007) News droughts, news floods, and US disaster relief. *The Quarterly Journal of Economics*, 122(2), 693–728.

Hickman, C., Marks, E., Pihkala, P., Clayton, S., Lewandowski, R. E., Mayall, E. E., Wray, B., Mellor, C. & Susteren, L. van (2021) Climate anxiety in children and young people and their beliefs about government responses to climate change: A global survey. *The Lancet Planetary Health*, 5(12), e863–e873. https://doi.org/10.1016/S2542-5196(21)00278-3.

Hicks, D. (2002) Lessons for the Future: the missing dimension in education. Abingdon: Routledge.

Hicks, D. (2014) *Educating for Hope in Troubled Times*. London: Trentham Books, IoE Press.

Hicks, D. & Bord, A. (2001) Learning about global issues: Why most educators only make things worse. *Environmental Education Research*, 7(4), pp.413–425.

Johnson, S. (2021) Extra life: A short history of living longer. https://nextbigideaclub.com/magazine/extra-life-short-history-living-longer-bookbite/28159/ (Accessed: 28/05/2025)

Mitchell, D. (2022) GeoCapabilities 3—knowledge and values in education for the Anthropocene. *International Research in Geographical and Environmental Education*, DOI: 10.1080/10382046.2022.2133353.

Ojala, M. (2012) Hope and climate change: The importance of hope for environmental engagement among young people. *Environmental Education Research*, 18(5), pp.625–642. https://doi.org/10.1080/13504622.2011.637157.

Ritchie, H. (2024) *Not the End of the World: How we can be the first generation to build a sustainable planet*. London: Chatto and Windus.

Roser, M. (2022) The world is awful. The world is much better. The world can be much better. https://ourworldindata.org/much-better-awful-can-be-better

Rosling, H., Rosling, O. & Ronnlund, A. R. (2018) *Factfulness: Ten reasons we're wrong about the world – and why things are better than you think*. London: Sceptre.

Solnit, R. (2016) *Hope in the Dark: Untold histories, wild possibilities*. Edinburgh: Canongate.